AF595582

# Sustainable High Volume Road and Rail Transport in Low Income Countries

# Sustainable High Volume Road and Rail Transport in Low Income Countries


Editors

**Michael Burrow**
**Gurmel Ghataora**
**Bruce Thompson**
**Bernard Obika**



MDPI • Basel • Beijing • Wuhan • Barcelona • Belgrade • Manchester • Tokyo • Cluj • Tianjin

*Editors*
Michael Burrow
University of Birmingham
UK

Gurmel Ghataora
University of Birmingham
UK

Bruce Thompson
IMC Worldwide
UK

Bernard Obika
IMC Worldwide
UK

*Editorial Office*
MDPI
St. Alban-Anlage 66
4052 Basel, Switzerland

This is a reprint of articles from the Special Issue published online in the open access journal *Sustainability* (ISSN 2071-1050) (available at: https://www.mdpi.com/journal/sustainability/special_issues/High_Road_Rail_Transport_Low_Income).

For citation purposes, cite each article independently as indicated on the article page online and as indicated below:

LastName, A.A.; LastName, B.B.; LastName, C.C. Article Title. *Journal Name* **Year**, *Article Number*, Page Range.

**ISBN 978-3-03943-088-8 (Hbk)**
**ISBN 978-3-03943-089-5 (PDF)**

Cover image courtesy of Ben Walker.

# Contents

# About the Editors

**Michael Burrow** is senior lecturer in the School of Engineering at the University of Birmingham, and specialises in pro-poor infrastructure asset management. He has 20 years' experience in multi-disciplinary applied road research and capacity building with partners in Africa and South Asia and has published more than 100 peer-reviewed papers and professional reports on this work. Projects are on-going in Ethiopia, The Gambia, Ghana, India, Kenya, Liberia, Sierra Leone and Uganda. As director of the MSc Road Management and Engineering programme, Michael lectures in road asset management, risk management and rural roads for development. He currently supervises some 40 PhD research students in these and related topics. His excellence in teaching and mentoring was recognised with the 2018 Teaching Innovation Award by the Higher Education Futures institute; and the award for Excellence in Doctoral Research Supervision (College of EPS; 2018-19).

**Gurmel Ghataora** has a PhD in geotechnical engineering and some 50 years' experience in academic and industrial research on the geotechnics of transport infrastructure. His international experience includes materials' testing (laboratory and field), ground improvement, use of out-of-specification materials in construction, improvement of roads and railways, and laboratory research. His international experience includes on-going projects in Ethiopia, The Gambia, Ghana, India, Liberia, Sierra Leone and Uganda. He lecturers in geotechnical engineering, rural roads and soil mechanics, and sustainable transport. He has published two books on materials and transportation geotechnics, several book chapters, and over 150 peer-reviewed journal and conference papers. Dr Ghataora has supervised over 60 research students.

**Bruce Thompson** is an independent transport policy adviser to international development clients. Currently, he is working for the DFID Applied Research Programme in High Volume Transport in Low-Income Countries in Africa and South Asia. While Head of Infrastructure Policy at DG Development at the European Commission in Brussels, he wrote the Commission's transport sector guidelines and EU policy communications on transport in EU development cooperation. He was a founding Board Member of the Africa Transport Policy Program (SSATP) and has also worked in close coordination with the African Development Bank, the African Union Commission, UN Economic Commission for Africa, and the World Bank. His in-depth knowledge in the transport sector reflects his work in the private and public sectors and insights gained from living in developing countries, mainly Africa. He is a Chartered Civil Engineer with over 30 years' experience in developing transport policy, strategy, sectoral reform programmes, and in managing investment programmes.

**Bernard Obika** is leading the DFID funded Applied Research Programme in High Volume Transport in Low-Income countries in Africa and South Asia. The research programme aims to deliver an evidence base for transport that is affordable, efficient, green and safe. Dr Obika has a PhD from the University of Birmingham and has published some 40 technical papers and research reports on infrastructure and institutional development issues. Topics range from local contractor capacity development, and labour-based works, to soluble salt damage to bituminous surfacing of roads and runways in tropical environments. Committed to international development, Dr Obika has more than 30 years' experience in transport engineering, working for multilateral organisations, national governments and the private sector. During this time, he has been responsible for over 200 major infrastructure and institutional development projects in Africa, Asia, South America, the Caribbean, Europe and United Kingdom. In 2017, Powerlist nominated Dr Obika as one of the top 100 most influential black persons in the UK. In 2015, he was a finalist in European CEO of the Year in recognition of his contribution to transport and infrastructure development for governments in Africa and Asia.

*Editorial*

# Advancing Applied Research in High Volume Transport in Low-Income Countries in Africa and South Asia

**Bruce Thompson * and Bernard Obika**

IMC Worldwide Ltd., 64–68 London Road, Redhill, Surrey RH1 1LG, UK; bernard.obika@imcworldwide.com

* Correspondence: hvtinfo@imcworldwide.com

Received: 7 May 2020; Accepted: 13 May 2020; Published: 16 May 2020

**Abstract:** The Department for International Development (DFID) is funding the High Volume Transport (HVT) Applied Research Programme. This programme is an integral component of the UK response to delivering transport and mobility that is accessible, efficient, safe, and green in the low-income countries (LICs) in Africa and South Asia. The first part of the HVT programme produced an up-to-date and comprehensive state of knowledge on high volume transport in these countries. This Special Issue presents a selection of papers to cover key research priorities identified in road and rail transport, low carbon transport, and gender and inclusive transport. The state of knowledge has produced a sound basis for setting priorities for applied research in the second part of the programme. Applied research is directed to delivering high volume transport that contributes to economic growth and social development, and that is more resilient to the impact of climate change in LICs in Africa and South Asia.

**Keywords:** accessible road and rail infrastructure; efficient transport services; transport corridors; green transport; low carbon transport; road safety; disability and mobility; gender disparity; vulnerable groups

---

## 1. Background

Research on transport in low-income countries (LICs) is a strategic priority for the UK Department for International Development (DFID) because accessible, efficient, safe, and green transport is imperative to economic growth, social development, and the environment. Since the transport sector is one of the largest recipients of aid in almost every LIC and DFID focus countries, DFID is committed to ensuring that spending on transport is effective. Every one percent gain on the cost effectiveness of transport infrastructure is worth potentially US $1 billion each year in Africa. Thus, there is an ongoing and urgent need for research that will contribute to meeting the emerging and changing conditions in many LICs, and to ensuring the transport sector works better in these countries to achieve the UN Sustainable Development Goals (SDGs).

## 2. Research Objectives

DFID development priorities are aligned with the UN SDGs and to this end, DFID is a leading partner in the Sustainable Mobility for All (Sum4All) Consortium. Sum4All partners are committed to ensuring transport plays its critical role in supporting LICs to reach their SDGs by seeking innovative ways to make transport accessible, efficient, safe, and green.

### 2.1. Universal Access

One billion people in LICs have no access to an all-weather road [1]. This means about 70% of Africa's rural population—some 450 million people—are isolated from economic activity, employment, and basic provisions of health and education [2]. Meanwhile, the urban population in developing countries, which is expected to rise by two billion people by 2045, has rapidly outstripped the capacity growth of public transport [2]. Thus, transport infrastructure and services urgently need adapting and extending to meet these pressing mobility and access needs of people in LICs in Africa and South Asia. Adequate access includes safe mobility for women, people with disabilities, and other vulnerable groups.

### 2.2. *Efficiency*

Missing links in strategic transport networks connecting LICs in Africa and South Asia make transport more costly, render trade uncompetitive, and stifle economic growth. LICs in Africa, particularly the landlocked countries, are the hardest hit, with intra-African trade only 15% compared to 61% in South Asia [3]. This means African consumers pay more for goods because 40% of the price of goods is transport costs. Although vehicle and mainly fuel efficiency improved in the period between 2005 and 2015, progress has stalled recently, reducing the potential benefits to goods and passenger transport particularly in urban areas. Thus, better transport connectivity and lower transport prices will benefit manufacturers, commerce, farmers, consumers, and passengers alike.

### 2.3. *Road Safety*

Low and middle-income countries account for 90% of global road fatalities, yet own only 50% of the world's road vehicles [1]. The fatality risk is 20 times higher for motorcyclists than for car occupants, and 7 and 9 times higher for cyclists and pedestrians, respectively. By far the highest number of global traffic fatalities (40% to 50%) occur in urban areas, and is rising faster in LICs [2]. These high fatality and injury rates cost 3% of GDP in most LICs and place a heavy demand on hospital and health budgets. They also damage the livelihoods of countless families in LICs in Africa and South Asia.

### 2.4. *Green Transport*

Between 2000 and 2016, greenhouse gas (GHG) emissions from all transport modes increased by 86% in Africa and 92% in Asia, primarily driven by increases in motorised passenger and freight transport; whereas transport emissions decreased by 2% in OECD countries [4]. Consequently, air and noise pollution is a health hazard in many LIC cities. Traffic congestion, low quality fuel, and poorly maintained trucks and motorcycles contribute to pollutant levels in excess of levels set by the World Health Organisation. Thus, LICs need to adapt transport policies, to adopt new technologies, and to boost urban transport efficiency in ways that will reduce GHG emissions and improve health in urban cities and urban areas.

## 3. Methodology

In November 2017, DFID launched the High Volume Transport (HVT) Applied Research Programme as part of the UK response to deliver transport and mobility that is accessible, efficient, safe, and green. HVT is a five-year research programme backed by UK Aid funding of £14 million. The first part of the programme assessed the State of Knowledge in four thematic areas: (1) long distance strategic road and rail transport; (2) urban transport; (3) low carbon transport; and (4) gender, inclusion, vulnerable groups, and road safety.

Following a competitive call, HVT commissioned research suppliers to review current research in order to establish future directions for research that will contribute to meeting the development goals for transport in LICs. The findings of the literature reviews were examined and tested with LIC stakeholders in Africa and South Asia using qualitative and quantitative questionnaires, as well as physical and virtual interviews. Participatory workshops were held in Bangladesh, Kenya, Tanzania, and Malaysia. This comprehensive testing and analysis resulted in the identification of priority research areas and key topics for applied research to be carried out in the second part of the HVT programme that started in January 2020.

## 4. Future Research Directions

The first part of the HVT programme revealed a wide range of research topics, as shown in the representative papers presented in this Special Issue. The selected papers cover key research priorities in road and rail transport, low carbon transport, and gender and inclusive transport with regard to universal access, efficiency, safe, and green transport in LICs in Africa and South Asia.

### 4.1. Road Infrastructure

A technical and contractual review of the use of recycled road pavement materials and inexpensive, non-conventional materials identified a range of road designs based on these materials in temperate and tropical climates [5]. However, little agreement was found on the use of non-conventional materials and on appropriate contractual arrangements to mainstream their use in road pavement construction. The contractual methods assessed were mainly public-private partnerships (PPPs) and "alliancing", a new form of PPP. However, to gain cost and environmental benefits of using recycling road pavement and non-conventional materials, two research areas were identified that would contribute to green transport. One is to gain more insight into the design and specifications of these materials for water-resilient roads, and the other is to establish guidelines for financial, institutional, and contractual arrangements for sustainable PPPs.

### 4.2. Railway Infrastructure

LICs in Africa and South Asia are investing, mainly with Chinese funding, in new railway lines to replace deteriorating infrastructure built in the 19th and 20th centuries [6]. However, many rail infrastructure projects are driven by international supplier preference to create continent-wide networks, rather than effective interoperable regional networks. A new technical strategy that focuses on national and regional networks would contribute to more efficient and greener alternatives to road transport. To this end, technology and standards need to be adapted to increase railway interoperability and thus to improve access to more competitive railway services. To support this strategy, the research priorities have been identified as comprehensive data collection on rail network conditions and performance; development of affordable technologies for LICs; and harmonisation of technical standards to facilitate cross-border operations.

### 4.3. Road and Rail Transport Services

A comprehensive review identified multiple barriers to the efficiency and safety of road and rail services in LICs in Africa and South Asia. These barriers range from unregulated driver hours, inadequate road safety standards, vehicle overloading, inappropriate railway concessioning; they also include low skills and capacity in the public sector, and poor cross-border road and rail services. Overcoming these barriers will improve efficiency, widen public access to road and rail services, and contribute to better road safety. Based on best practice worldwide, research priorities have been identified as systematic study in LIC transport services, development of new technology and e-border systems at border crossings, improvement of management of rail infrastructure and services in concessions, and assessment of railway regulatory frameworks and railway authorities [7].

### 4.4. Transport Corridors

The review of transport corridors in LICs in Africa and South Asia focused on corridor development in the light of trade investment, governance, and management of cross border and corridor operations [8]. The review also covered the impact of transport corridor development on the rural economy and inclusive employment, trafficking of women, and the spread of HIV/AIDS and other sexually transmitted diseases. Research in these priority areas would contribute to improving efficiency, increasing access, and providing safer mobility for women and vulnerable groups. As few studies have been conducted on the socio-economic impacts of transport corridors, the research priorities identified focus on the uneven corridor investment outcomes across geographical locations and population groups. The research priorities include modelling distributional impacts of regional transport investment, regulatory corridor governance and management, and gaining better insight into the negative social externalities of transport corridors.

### *4.5. Low Carbon Transport*

In an assessment of the current status, feasibility, and potential of low-carbon transport measures, ten "quick-wins" for sustainable development and green transport were identified [9]. These quick wins range from better fuel economy and pricing incentives as well as the introduction of electric two- and three-wheelers to making urban mobility more sustainable and the promotion of non-motorised transport. Directed to integrating national transport and climate change policies, the applied research priorities identified are to assess the design and implementation of cost effective quick wins. In addition, greater coordination between transport agencies, vehicle manufacturers, and energy suppliers would build a cohesive low carbon strategy in LICs.

### *4.6. Mobility of Vulnerable Groups*

#### 4.6.1. Gender Disparity

The review of gender inequality in mobility and transport focused on spatial and transport planning, and mobility in newly emerging smart cities [10]. Gender disparities in transport were found to have impacts on young girls' and women's access to education, employment, health services, and well-being and consequently affect income levels and livelihood outcomes. Transport disparities were found to be higher in cities where more women pedestrians were involved in road traffic accidents, and innovations in smart mobility tended to benefit men more than women. Furthermore, women were found to be under-represented in transport and urban planning, thus further compounding transport gender disparities. Thus, delivering accessible transport and safe mobility that would ensure better access to health and education services and jobs, creating better livelihoods for women, is important. The review identified 11 priority research areas to tackle this gender disparity in transport, highlighting the need to integrate quantitative and qualitative data in knowledge generation and decision making.

#### 4.6.2. Disability and Mobility

Equitable and inclusive transport for people with disabilities were identified as key issues in urban, rural, and long distance journeys, as well as affordability of transport services and availability of special transport services for people with mobility difficulties. The findings of the literature review showed knowledge gaps on the barriers to accessible and inclusive transport and an urgent need for extensive research on inclusive transport in LICs and on effective ways to monitor it [11]. The research priorities include ensuring the needs of adults and children with disabilities are better understood, as well as how to involve them in planning and setting standards for transport services. In addition, training in disability awareness needs to be developed for public and private sector transport providers, as well as adapting technology to better facilitate access and safe mobility for all people with disabilities.

#### 4.6.3. Older People

Little is known about the mobility needs of older people in LICs [12]. The literature review has shown how health, gender, and social well-being are influenced by their mobility and transport. Transport can be a barrier to access to health services and may involve multiple journeys to different health centres. Furthermore, socially isolated old people without inadequate support in urban and rural areas call for innovative transport services. Consequently, research priorities that would benefit the access and mobility of older people include collection of gender and age-disaggregated data, gaining insight into transport's role in affordable access to health services and job opportunities, new approaches to involve older people actively in research, and the use of new technology to explore spatial barriers to urban services.

#### 4.6.4. Children and Youth

Limited access to transport was found to restrict the mobility of children and youth from poorer households in LICs and thus impact on their livelihood potential and life chances [13]. The literature review covered journey patterns of schoolchildren in rural and urban areas, transport access and affordability for young workers, and the safe mobility of vulnerable children, cyclists, and motorcyclists. Research priorities for better access to safe and affordable mobility include engaging the young in research and the outcomes, assessing transport infrastructure and services interventions; developing approaches to transport subsidies; assessing job opportunities for young people, particularly the poor and women; and the use of mobile technology to change travel patterns.

### 4.7. Road Safety

The review of road safety focused on data collection and management, traffic engineering, and safety policy aspects in 10 key areas of safety, thus providing a holistic approach to road safety [14]. Key research focuses identified are to gain more insight into underreported road crashes; traffic injuries and resulting disabilities; the cost of road crashes; vehicle safety and risks in the different composition of vehicle fleets; social and behavioural approaches to road safety in LICs; and capacity building in all aspects of safety. Four initial steps for better road safety were identified: improving the quality of data collection and analysis; raising public awareness of road crash reporting, as well as examining data analysis techniques; and using proactive measures to prioritise investments.

## 5. Conclusions

The literature reviews have produced up-to-date and comprehensive state of knowledge on high volume transport in LICs in Africa and South Asia. The review findings focus on the DFID priorities of universal access to transport, efficient transport networks and services, and safe and green transport for all. Thus, the first part of the programme provides a sound basis for setting applied research priorities. These priorities are directed to delivering high volume transport that is more accessible, more efficient, more inclusive, and more resilient to the impact of climate change in LICs across Africa and South Asia.

**Author Contributions:** Writing, B.T.; review, B.O. All authors have read and agreed to the published version of the manuscript.

**Funding:** This research was funded by UK AID through the UK Department for International Development under the High Volume Transport Applied Research Programme, managed by IMC Worldwide.

**Conflicts of Interest:** The authors declare no conflict of interest.

## References

1. World Bank. Available online: https://www.worldbank.org/en/topic/transport/overview (accessed on 28 March 2020).
2. Sustainability Mobility for All (Sum4All). Available online: https://sum4all.org/priorities/universal-access (accessed on 28 March 2020).
3. Africa Transport Policy Program (SSATP). Available online: https://www.ssatp.org/topics/regional-integration (accessed on 28 March 2020).
4. SLoCaT (2018). Transport and Climate Change Global Status Report 2018. Available online: http://slocat.net/tcc-gsr (accessed on 11 May 2020).
5. Thom, N.; Dawson, A. Sustainable Road Design: Promoting Recycling and Non-Conventional Materials. *Sustainability* **2019**, *11*, 6106. [CrossRef]

6. Blumenfeld, M.; Wemakor, W.; Azzouz, L.; Roberts, C. Developing a New Technical Strategy for Rail Infrastructure in Low-Income Countries in Sub-Saharan Africa and South Asia. *Sustainability* **2019**, *11*, 4319. [CrossRef]
7. Wheat, P.; Stead, A.; Huang, Y.; Smith, A. Lowering Transport Costs and Prices by Competition: Regulatory and Institutional Reforms in Low-Income Countries. *Sustainability* **2019**, *11*, 5940. [CrossRef]
8. Quium, A. Transport Corridors for Wider Socio–Economic Development. *Sustainability* **2019**, *11*, 5248. [CrossRef]
9. Bakker, S.; Haq, G.; Peet, K.; Gota, S.; Medimorec, N.; Yiu, A.; Jennings, G.; Rogers, J. Low-Carbon Quick Wins: Integrating Short-Term Sustainable Transport Options in Climate Policy in Low-Income Countries. *Sustainability* **2019**, *11*, 4369. [CrossRef]
10. Uteng, T.; Turner, J. Addressing the Linkages between Gender and Transport in Low- and Middle-Income Countries. *Sustainability* **2019**, *11*, 4555. [CrossRef]
11. Kett, M.; Cole, E.; Turner, J. Disability, Mobility and Transport in Low- and Middle-Income Countries: A Thematic Review. *Sustainability* **2020**, *12*, 589. [CrossRef]
12. Gorman, N.; Jones, S.; Turner, J. Older People, Mobility and Transport in Low- and Middle-Income Countries: A Review of the Research. *Sustainability* **2019**, *11*, 6157. [CrossRef]
13. Porter, G.; Turner, J. Meeting Young People's Mobility and Transport Needs: Review and Prospect. *Sustainability* **2019**, *11*, 6193. [CrossRef]
14. Heydari, S.; Hickford, A.; McIlroy, R.; Turner, J.; Bachani, A. Road Safety in Low-Income Countries: State of Knowledge and Future Directions. *Sustainability* **2019**, *11*, 6249. [CrossRef]

*Review*

# Sustainable Road Design: Promoting Recycling and Non-Conventional Materials

**Nicholas Thom * and Andrew Dawson**

Department of Civil Engineering, University of Nottingham, Nottingham NG7 2RD, UK; andrew.dawson@nottingham.ac.uk
* Correspondence: nicholas.thom@nottingham.ac.uk; Tel.: +44-115-9513901

Received: 20 September 2019; Accepted: 28 October 2019; Published: 2 November 2019

**Abstract:** Many factors impact on the sustainability of road maintenance, including the organization of road authorities, contract forms used, financing structure and, unfortunately, political interference and corruption. However, this paper reviews the opportunities to increase sustainability by utilizing less environmentally damaging material sources, and also the associated challenges. It is a field that has seen advances in recent decades, for example in the effectiveness of cold-mix asphalt binders. Nevertheless, the opportunities are not being taken up in many countries, and this reflects uncertainty in predicting performance. This paper reviews the different design methods available, developed in both temperate and tropical climates, and highlights the lack of agreement with regard to non-conventional materials. The different sources of uncertainty and risk are then discussed, together with ways of limiting them. It is found that, while advances in performance prediction are highly desirable, the key to encouraging recycling and the use of inexpensive but non-conventional materials lies in development of the right contractual arrangements, specifically partnering and risk/reward sharing. The paper concludes with a discussion on approaches to partnering in the construction industry and the prerequisite climate of trust without which innovation is almost inevitably stifled.

**Keywords:** road; materials; recycling; non-conventional; risk; design; partnering

---

## 1. Introduction

The need to minimize use of scarce primary resources is becoming ever more urgent in most industries as humanity relentlessly exhausts this planet's ability to satisfy its demands and carelessly discards waste to the detriment of the environment. The roads industry is no exception to this. In the UK, for example, roads consume some 25% of all materials extracted from the ground [1] and, while most of these sources are not in immediate danger of becoming exhausted, the impact on the environment is substantial. Furthermore, even if exhaustion of resources has not yet occurred, the costs are real. For example, in relation to Malawi, Kamanga and Steyn [2] list material cost and/or shortage as a key reason for project delays. They suggest this is often because designs and specifications do not allow for the use of a possibly inferior but more readily available material, which would include recycled materials. And if, as reported by Oke et al. [3] in the case of Nigeria, failed asphalt is routinely discarded rather than being recycled, this in itself represents a direct and negative environmental impact.

Furthermore, the cost of road materials delivered to a construction site comprises two parts: The cost of the raw material at the quarry or gravel pit; and transport costs, both financial and environmental, which are frequently the higher of the two. Locally-available materials are obviously to be preferred. To this must be added the fact that if in situ recycling can be achieved, there are substantial time savings, beneficial for both the road authority and the user. Thus, the drivers are strong, both economic and environmental, in support of recycling and/or the use of locally-available materials.

The first objective of this paper is, therefore, to explore the reasons why many road authorities find it difficult to move from primary to secondary (e.g., industrial by-products), recycled or locally-available but possibly inferior sources. A further objective is to suggest the changes that have to be put in place if this move is to be made.

The paper identifies two areas where change is necessary. The first is technical and relates to the developments needed in design methodology in order to incorporate materials with non-standard properties. The second, and certainly more significant, is organizational and relates to the type of contractual arrangement that best enables such materials to be used. It will be suggested that management of the risk accompanying the use of non-conventional materials requires effective partnership between public and private sectors in a transparent and no-blame environment.

## 2. Barriers to Non-Conventional Technologies

It cannot be denied that road authorities are grappling with extremely difficult engineering issues. The passage of time and the seemingly ever-increasing numbers of commercial vehicles that use our road networks cause accumulated damage to road materials, damage which cannot be reversed. Road renewal/reconstruction will therefore continue to be required. One technology that is becoming increasingly conventional, and where the only barrier to uptake is investment in adapted plant, is so-called hot in-plant asphalt recycling [4]. Just as waste steel can be taken back to a foundry and incorporated into a new product, so waste asphalt can be taken back to an asphalt plant and incorporated into a new mixture. Researchers have established [5] that, so long as the percentage of recycled material is kept to a reasonable limit (30% in the case of the Federal Highway Administration in US) there should be no measurable loss of performance compared to a 100% virgin (i.e., primary aggregate) mix. However, the gain in terms of sustainability is modest. Resources are saved, and there is some reduction in energy demand but, depending on asphalt plant location, transport costs are still present.

Another conventional process is to hot-recycle an asphalt surface course in situ. Machines exist that have the capability to re-heat the surface to a depth of several centimeters to sufficient temperature for the asphalt to become pliant. This allows cracks to heal and the material to be re-compacted into a dense intact mat, usually with additional material added. This is proven technology and is used extensively. For example, Finlayson et al. [6] report on 25 years of successful experience in Canada. In terms of sustainability, this process is partially successful in that it cuts out transport costs. However, the energy cost is very high, and it has technical limitations in that it cannot treat problems deeper than a few centimeters.

The real 'game-changer' is cold in situ recycling. It has been done very successfully all over the world, e.g., in South Africa [7], India [8], Poland [9], with established design guidance available (e.g., [10]). Troeger and Widyatmoko [11] illustrated cost savings of 35–40% for different in-situ recycled solutions compared to conventional reconstruction, while a parallel estimate from Canada [12] was a 42% saving. One London borough [13] quoted a cost-saving ratio of over three between the two processes, and alongside financial savings reductions in environmental and disruption costs have also been documented [11].

But the problem with a game-changer is that new rules have to be formulated to govern the 'game'. Companies such as Wirtgen [14] have been producing the necessary plant for decades, and there is a long-associated history of production of high-quality in situ recycled pavements. However, the resulting materials, though often subjectively 'high-quality', are distinctly different in terms of their engineering properties from conventional new materials [15], which means they do not fit easily into traditional pavement design methods.

The same difficulties arise with other non-conventional materials, notably those using industrial by-products such as fly ash or blast-furnace slag as partial binder replacements. Different materials with different properties require different designs. The following sections will review the range of non-conventional material types and the technical issues that currently present barriers to use.

### 2.1. In Situ and Ex Situ Recycling

Cold in situ recycling [16] is conceptually simple. A large rotary milling device breaks up the pavement to a specified depth and at the same time mixes in a binding or stabilizing agent, which may be cementitious or bituminous. At the back of the machine, a paving screed gives a reasonably smooth finish to the newly-created recycled layer, and this is followed by conventional roller-compaction. Depths up to about 250 mm can be treated in this way.

The term 'ex situ' is commonly used to describe an alternative, better controlled but more costly process in which the milled products are transported to a mobile mixing plant located on, or adjacent to, the site. Conventional mixing then takes place and the mixed material is transported, placed back onto the road, and compacted.

The key problem that recycling introduces is variability [17]. With both in situ and ex situ recycling, variability in the source material, i.e., the existing road, is unavoidable. In the case of in situ, there is likely to be additional variability due to differences in binder application rate and mixing efficiency. Increased variability relative to a virgin mix means that the performance of a cold recycled material will never exactly match that of conventional materials, thus requiring adjustment to design standards.

### 2.2. Cementitious Binders

Cementitious binders include conventional Portland cement, and if this is used to stabilize a material, then the resulting layer is in effect a weak concrete. This is a well understood class of material and is covered by standard specifications and design methods (e.g., [18]). However, in the context of renewal or rehabilitation of a road it comes with restrictions, and these limit its use.

If the new cement bound layer is designed to remain substantially intact under traffic, then it needs to be handled carefully. Typically, it must be left for seven days before it is strong enough to allow paving of another layer on top, and another layer is certainly needed in order to achieve the required surface level tolerances for anything other than the lowest of speeds. Furthermore, shrinkage due to hydration reactions combined with diurnal thermal cycles will eventually cause the new layer to crack into discrete lengths [19], and there is then the likelihood that these cracks will 'reflect' through overlying asphalt and require maintenance. Thus, this solution, though used, risks the need for significant future maintenance expenditure.

An alternative is to opt for a weaker material and accept that it will crack under construction traffic. No delay is then necessary before paving an overlying surface, and the danger of reflective cracking is diminished since cracks in the cement stabilized base, though numerous, will be individually less severe. On the other hand, the value of the recycled material is also diminished, relegating its properties to little more than those of an excellent granular base. This means that an increased thickness of new material has to be imported to site for overlying layer construction.

A more radical and much less conventional solution is to opt for a slow-setting binder, often a blend of hydrated lime with industrial by-products such as fly ash or ground granulated blast furnace slag [20,21]. Again, there is no need to delay construction, but the advantage here is that immediately after construction, the material is still in the relatively early stages of strength gain, and so long as early traffic loading is not too severe, there is every prospect that it will achieve a good final strength. The obvious problem is that this is difficult to tie down in terms of a conventional specification. If a material is not expected to reach its potential until several weeks after construction, how is it possible to control quality? In the UK, Highways England [22] have opted to test in situ and to assume a future strength gain, but any such approach inevitably carries risk. Yet the possibilities offered by this technique are highly attractive both technically and in terms of reduced environmental footprint. This difficulty represents another key point where design standards need to differ from those of conventional solutions.

### 2.3. Bituminous Binders

Since cold in situ recycling is carried out at ambient temperature, there is no opportunity to heat and form conventional hot-mix asphalt (HMA). The bitumen has to be delivered cold, and two widely-used products have been developed to do this: Bitumen emulsion and foamed bitumen. The binder arrives into the mixing zone in the form of tiny droplets (emulsion) or fine flakes (foamed bitumen) carried by water. The material is still effectively unbound during compaction. The compaction process itself then compresses the droplets or flakes of bitumen between aggregate particles, forcing them to adhere and thereby beginning the process by which a cold-mix asphalt is formed [23]. The process is a gradual one, and it is only as the water evaporates that the bitumen becomes ever more effective at binding the particles together, especially fragments of old asphalt surfacing that already contain bitumen.

The problem here is similar to that with slow-setting cementitious binders. If the construction is carried out well, the final state of the recycled layer will be that of an intact material with reasonably high strength, but this is not possible to verify during construction. Furthermore 'reasonably high strength' is still unlikely to be truly equivalent to a conventional HMA; for instance, a specific problem is that it is likely to have a reduced resistance to water attack [24]. Once again, conventional standards cannot be applied.

## 3. The Design/Specification Challenge

Both cold recycling and stabilization of locally-available, often secondary, materials are processes that can produce a cheap new pavement base with minimized environmental impact. The problem is that the materials with the potential to deliver the greatest economic and sustainability benefits also present the greatest challenge to engineers.

In essence, there are four significant technical barriers to implementation:

1. Recycled materials are inherently more variable than virgin mixes.
2. Material behavior, e.g., crack resistance, differs from conventional mixtures.
3. Water-susceptibility is often higher for recycled or stabilized mixtures.
4. Full strength can take many weeks or months to develop.

Undeniably each of these issues contribute to there being an appreciably higher risk attached to many non-conventional materials than is usually considered acceptable, and this risk has to be taken into account in design. For example, the widely used AASHTO (1993) method [25] requires the use of a coefficient to quantify the effectiveness of each material, and it has been suggested [26–28] that cold recycled materials should be assigned coefficients somewhere between 0.2 and 0.36. This compares to around 0.44 for conventional HMA, and means they would have to be 1.2–2.2 times as thick. Similarly, one of the highway authorities in the UK [29] recommends a thickness 1.33 times that of HMA. Others have taken a still more cautious approach and consider cold recycled materials as high quality granular layers, with equivalence factors of 1.4 or 1.5 (e.g., [30]—relating to Californian practice) times a conventional granular base.

In a given country or region, with materials and climate specific to that region, such an experience-based approach may be satisfactory. But it is not automatically transferable elsewhere. A more flexible, but potentially riskier, approach to design is given by so-called analytical methods, in which materials are typically defined by a stiffness modulus and a fatigue cracking law.

Valentin et al. [31] have made a thorough review of the way these methods have been applied to in situ recycled materials, revealing a large variety of approaches. In the UK, the design advice most commonly followed was developed by the Transport Research Laboratory [10], and three grades of material are specified in terms of their differing characteristic stiffness moduli. However, thickness is then determined from a chart, which is itself largely based on experience. In France [32], different stiffness modulus values are suggested depending on the proportion of Recycled Asphalt Pavement (RAP) included in the cold recycled asphalt layer. However, the calculations that follow do not include fatigue cracking of the recycled layer; i.e., it is treated as an already-cracked material. Similarly, in New

Zealand [33], in situ recycled materials are treated as superior unbound bases rather than having any intrinsic fatigue strength. In contrast, in Australia they are generally treated as slightly inferior asphalt layers, and a fatigue life is computed [33].

Thus, it is fair to say that many road authorities have found ways around the four barriers listed above. However, there is a lack of consistency, partially explicable by different typical road structures being used in different parts of the world. It is also fair to say that over the years there have been advances in prediction of the performance of roads incorporating non-conventional materials such as cold-mix asphalt [34–37] or in situ recycled layers [38].

Nevertheless, the four barriers listed above still significantly inhibit the take-up of recycled and other non-conventional materials. The following subsections consider each barrier in more detail.

### 3.1. Material Variability

Material variability is undeniably a negative feature of any in situ recycled material. Traditionally material variability has been dealt with in the same way as any other source of uncertainty, i.e., it contributes to the overall reliability of a design, and a client has to select a certain level of reliability appropriate to each class of road. In some methods, e.g., AASHTO (1993) [25], the client has freedom to choose; in others, e.g., Highways England [18], a certain probability of achieving the design life is built in, 85% in that case. But these methods were all developed based on experience of variability in conventional materials.

However, variability in recycled materials is different. The nature of the mixing process combined with variations in material type/quality along the road mean that there can be very large differences between small elements of material. These differences then tend to even out over larger areas, e.g., the >1 m diameter area stressed by a heavy goods vehicle tire. In a South African context, Lynch and Jenkins [39] report an increased variability between closely spaced test points in in situ recycled material. This is supported by the authors' own experience of a comparative trial on a newly-reconstructed pavement, in situ recycled against plant-mixed. Table 1 shows the results obtained.

**Table 1.** Comparison of tests on in situ recycled and plant-mixed base pavements.

| Description | Test Method | Plant-Mixed | | In-Situ Recycled | |
|---|---|---|---|---|---|
| | | Mean | Coefficient of Variation | Mean | Coefficient of Variation |
| Modulus from tests on cores | Indirect tensile; BS-EN 12697-26 [40] | 4960 MPa | 17% | 3930 MPa | 58% |
| Modulus over a larger area | Falling weight deflectometer | 3890 MPa | 24% | 2460 MPa | 28% |

The implication is that defects that are the result of combined effects from a relatively wide area, e.g., rutting, will show similar variation to that expected with plant-mixed materials; on the other hand, localized defects, e.g., cracks, will be much more varied in terms of when they first appear. Thus, in the common case of an in situ recycled layer overlaid by a relatively thin asphalt surface, localized surface cracking may appear over localized weak spots in the recycled base. For in situ recycled material, this negative feature has to be understood and either a reduced reliability has to be accepted or else the design needs to be modified to give the same reliability as for plant-mixed materials. In either case, there will be increased uncertainty and this has to be managed, both technically and contractually.

### 3.2. Unconventional Material Behavior

Unconventional material behavior is another difficult problem and is the subject of ongoing research, e.g., [41]. In essence, the materials under discussion can be classed into one of three material types:

- Strong, cementitious binder;
- Strong, bituminous binder;

- Weak, either binder.

The first of these is effectively equivalent to a normal cement-bound base and so can be considered as conventional.

The second however, if cold-mix binder (emulsion or foamed bitumen) is used, is not equivalent to conventional asphalt. Cold-mix asphalts are a class of material that can be described as partially bound [28], which means they start life with ready-formed and well distributed micro-cracks. In one sense, this is a disadvantage since the initiation phase of fatigue cracking is effectively bypassed. But in another sense, it is an advantage because the cracking that eventually occurs tends to be well distributed. This avoids occurrence of discrete large cracks [37], and reduces the stresses and strains felt by an overlying surfacing. Unfortunately, this is a level of complexity which the art of pavement performance modelling is not yet capable of addressing confidently, and this introduces additional uncertainty in performance prediction.

The third type will display unbound material behavior, but with a higher strength and stiffness than conventional unbound layers. In many design approaches, e.g., AASHTO (1993) [25] or analytical methods, this presents no problem so long as a realistic long-term modulus can be assigned, although this is something that currently relies more on experience than pavement science. However, in more restrictive design methods that rely on non-numeric descriptors for materials, recycled materials often do not fit easily into any conventional category. In such a case, it has to be accepted that the method cannot be directly applied.

The problem of unconventional behavior is one which is still being researched, bringing inherent risk, additional to that already identified due to material variability. Whilst the fruits of research may reduce these risks in the future, there is nevertheless a clear need for effective risk management.

### 3.3. Water Susceptibility

The problem of water susceptibility applies chiefly to cold asphalt mixes, whether recycled or not [24]. Many asphalts have a degree of susceptibility to water since aggregates are often hydrophilic. This means that if water can reach the bond between aggregate and bitumen it will gradually destroy it [42], although adhesion promoters incorporated into the asphalt can be effective at combating this problem.

Water ingress is a serious design challenge for all pavement types. Trapped water when pressurized by traffic loads softens soil, reduces the stiffness and shear strength of unbound materials, and also leads to breakage of cement and bitumen-bound materials [43]. In a conventional hot-mix asphalt, each particle of aggregate is fully coated by bitumen and it is difficult for water to gain access; in a cold-mix, the particles are not fully coated and water has a ready route in. It is therefore particularly desirable to keep water out of cold-mix asphalts, and the potential benefits will not be fully realized if water is allowed to gain access in large quantities. Preventative measures could include:

- Increasing the camber on the road;
- Installing/repairing functioning sub-surface drainage prior to recycling/re-construction;
- Re-sealing the road surface as necessary.

These are all practical steps that should lead to the water content within the pavement being controlled. However, the key point is that yet another source of risk is introduced, one that is difficult to design out completely and which therefore has to be managed contractually. And in a world in which the climate is changing rapidly in many locations, the magnitude of this risk is only likely to increase.

### 3.4. Delayed Strength Gain

In several of the materials under discussion strength may continue to develop for upwards of a year [44–46]. The long-term gain in properties therefore needs to be estimated as part of pavement design [47]. This is probably the hardest problem of all to deal with and it brings the issue of risk into sharp focus. This sub-section will therefore begin to introduce non-engineering aspects of risk management. The options appear to be:

- Specify tests on accelerated-cured laboratory specimens in order to predict in-road properties. However, material curing in the road depends heavily on ambient temperature, moisture availability, exposure to air, and level of compaction achieved. This means that the laboratory value can only give an indication of what would be possible under ideal conditions [48].
- Ask for contractor guarantees. However, there are many factors outside the contractor's control, which means that the guarantee will be expensive to the client. Furthermore, there may often be arguments the contractor can make to cast doubt on his responsibility for any perceived lack of performance.
- Partnering and shared risk/reward. This approach [49] acknowledges the inherent unknowns involved and is designed to avoid the confrontations and disagreements that are almost inevitable with either of the first two options. It removes the risk of punitive claims or penalties and allows engineers to make relatively unimpeded judgments.
- All risk is taken by the client. If both design and construction are carried out in-house by the client, then in theory this gives even more flexibility since there are no externally imposed requirements to satisfy.

It would seem unavoidable that simple reliance on a specification based on tests on laboratory-cured specimens—even accepting that the engineering community knows which tests to apply—is a recipe for uncertainty, early failure, and contractual dispute. And while contractor guarantees are logical and workable if restricted to defects that become apparent within a year or so of completion, this does not easily apply to road base materials. Thus, in the opinion of the authors, this technical difficulty, on top of the others introduced previously, simply cannot be overcome without first setting in place a means of taking and managing risk.

Figure 1 summarizes the above discussion relating to the four barriers identified. The next section will discuss further the critical issue of risk management.

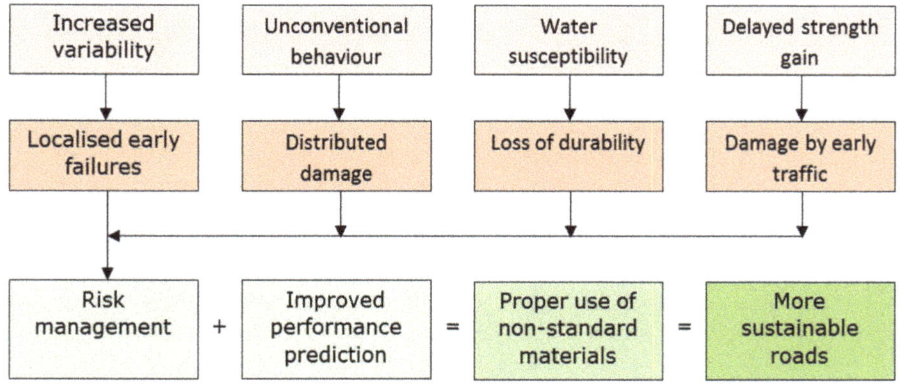


**Figure 1.** Summary of problems and solutions related to use of non-conventional materials.

## 4. Managing the Risk

Foregoing sections have made the point that improved engineering alone cannot overcome the risks associated with in situ recycling or the use of various types of non-conventional material. Thus, it is necessary to explore the means of providing an appropriate contractual climate to allow the risk-benefit balance to be managed properly.

Turning to the broad types of contractual arrangement possible, there is some evidence [50] that internal corruption is often less in the private sector than in the public sector, which suggests that keeping all activities within the public sector will not usually deliver an optimum result. It has also been noted that the use of sub-contractors often increases efficiency [51], which further suggests the benefits of private sector involvement. Furthermore, internal constraints from a risk-averse hierarchical client organizational structure may be hard to overcome and, it has to be conceded, many also lack technical expertise. Experience also suggests [52] that a performance-related incentive of some sort to an external organization (the contractor) leads to considerable benefit, and many would argue, e.g., [53] that partnering and the sharing of risk and reward is the optimum way to achieve this. Many studies have also come to the conclusion that public–private partnering in one form or another is the best way to stimulate innovation as well as handling project risk [54–56].

Public–private partnerships (PPPs) are widespread, but they vary enormously. In many cases, the motivation is finance, tying in banks as well as construction companies themselves, and these can be a means of procuring infrastructure developments that would have otherwise been unaffordable. For example, in an African context, Ajacaiye and Ncube [57] evaluate the potential of PPPs to contribute to development, and they are strongly supportive, based largely on the lack of available public finance in many African countries.

However, that has nothing to do with encouraging innovation; a bank may be even more risk-averse than a road administration. Leiringer [58] discusses the concept of 'design freedom' commonly promoted as a benefit of PPPs, and finds that in practice this freedom is easily stifled. There are pressures to control uncertainty at the bid stage; restrictions are often written into the contract, for example, to follow an established standard; and the more that risk is placed on the contractor's shoulders, the greater is the incentive to fall back on standard solutions. Compared to other aspects of PPP, design innovation is easily forgotten [59] despite it featuring prominently in perceived risk factors [60].

Thus, if innovation such as the use of non-conventional materials is to be encouraged, then the PPP has to be deliberately set up to achieve this. Issues such as speedy delivery and meeting environmental or safety targets can of course still feature, but it is also necessary to build in measures that reward design whose benefits can only be seen in the long term. To achieve this, the nature of the partnership has to be deep and long-lasting.

'Alliancing' is a form of PPP [61] that has risen to prominence over the last decade, leading for example to the UK Institution of Civil Engineers' NEC4 Alliance Contract, brought out in 2018. Highways England have declared their intention to use alliancing on all future 'smart motorway' projects. Alliancing brings in the concept of zero blame, zero claims, and a pre-agreed cash flow, as well as risk/reward sharing. The concept is that a partnership (the alliance) is set up very early, with an integrated team from all the main parties involved [62]. This requires considerable up-front work by the client organization and self-evidently can only function effectively in a very transparent culture in terms of bidding and contract award.

Love et al. [63], reporting on Australian experience, suggest that the normal way that alliancing contracts work is that contractors are rewarded in three ways: (a) All direct project costs are fully reimbursed, whatever the outcome of the project; (b) an agreed percentage overhead is also paid; (c) performance incentive payments are paid (or penalties levied) according to success against a number of Key Result Areas (KRAs). These can often be primarily concerned with speed of delivery [62], but this is also the area where long-term performance-related measures can be written in, potentially with reward or penalty being deferred for several years. In the context of recycling or using local materials, KRAs related to environmental damage/preservation would also appear to be appropriate.

However, Love et al. [63] also evaluated by means of interviews the actual factors that drove innovation by individual members of staff. Their conclusion was that the details of the alliancing contract itself were important only in allowing a collaborative and transparent culture to develop. The real drivers for innovation by individuals were accountability, credibility, pride, and reputation. The implication appears to be that engineers will come up with innovative solutions, but only if they are given the right no-blame environment to work in. This is the real challenge that procurement agencies face.

They also face the continuing battle against corruption if such an environment is to be created, a subject for which there is a large body of literature related to the roads sector, e.g., [64]. Links to political patronage have been documented [65] as has endemism within the procurement sector [66]. As an example, Ntayi et al. [67] note that an estimated $107 M is lost to procurement-related corruption each year in Uganda, and they provide a detailed and thoughtful study reflecting on the causes, for example poor public sector pay. Snaith and Khan [68] even found that effective unit rates for road works varied as a function of the source of funding and they developed a model to quantify the effects of this corruption on national wealth.

However, a wide body of literature suggests that transparency makes corruption more difficult and therefore almost inevitably increases cost-effectiveness. For example, e-procurement systems can be used to avoid the danger of deals being done in secret. Its introduction in India and Indonesia has been analyzed [69], leading to the conclusion that either quality goes up (India) or delays are reduced (Indonesia). Neupane et al. [70] also report positive experience in Nepal, particularly an increase in the level of trust in the procurement process. They concluded that e-procurement cuts down on the opportunities for secret meetings between bidders and public sector officials.

Thus, the types of partnering suggested as being the logical means of encouraging use of non-conventional materials depend greatly on there being an appropriate level of transparency. The very real benefits of PPPs, particularly the alliancing model, have to be offset against the dangers of collusion and cartels [71] in the letting of PPP contracts, implying that achieving the desired outcome demands a transparent environment and suitable public scrutiny, supported by meaningful penalties.

## 5. Conclusions

This paper has set out the key issues that currently hold back the use of recycled and other non-conventional materials in road construction, concentrating particularly on the economically and environmentally attractive in situ recycling options. Several of these options have been shown to have real long-term benefits and to be highly cost-effective. It is therefore essential that the industry finds ways of making these benefits a reality, even if this means taking a rather different approach to procurement and design than has historically been the case. Research should target the following areas if in situ recycling and the use of secondary and other non-conventional materials is to develop as it should:

- Research that delivers guidelines covering the financial, institutional, procurement, and contractual arrangements necessary to facilitate sustainable forms of public-private partnerships.
- Research that delivers design and specifications for using recycled and non-conventional materials for building and maintaining water-resilient road pavements.

Of these, the first is considered to be the real key. Once risk and reward sharing are established within working partnership arrangements, then engineers and researchers will find ways to improve their predictive capabilities and therefore their designs, leading to greatly improved sustainability in road construction and maintenance.

**Author Contributions:** Investigation, N.T. and A.D.; writing—original draft, N.T.; writing—review & editing, A.D.

**Funding:** This research was funded by UK AID through the UK Department for International Development under the High Volume Transport (HVT) Applied Research Programme, managed by IMC Worldwide.

**Acknowledgments:** The authors are particularly grateful for the advice of Bruce Thompson, the leader of Theme 1, Long Distance Strategic Road and Rail Transport, of the HVT Programme. The authors also acknowledge the assistance of representatives of Cardno.

**Conflicts of Interest:** The study upon which this paper is based followed the requirements of the funder and submission of this paper was also their requirement. However, the content of the paper and its conclusions are those of the authors. The authors themselves declare no personal conflict of interest.

## References


1. BGS. *Construction Aggregates. Mineral Planning Factsheet*; British Geological Survey: Nottingham, UK, 2011.
2. Kamanga, M.J.; Steyn, W.J.V.D.M. Causes of delay in road construction projects in Malawi. *J. S. Afr. Inst. Civ. Eng.* **2013**, *553*, 79–85.
3. Oke, O.L.; Aribisala, J.O.; Ogundipe, O.M.; Akinkurolere, O.O. Recycling of asphalt pavement for accelerated and sustainable road development in Nigeria. *Int. J. Sci. Technol. Res.* **2013**, *2*, 92–98.
4. Sivilevičius, H.; Bražiūnas, J.; Prentkovskis, O. Technologies and principles of hot recycling and investigation of preheated reclaimed asphalt pavement batching process in an asphalt mixing plant. *Appl. Sci.* **2017**, *7*, 1104. [CrossRef]
5. Stephanos, P.; Pagán-Ortiz, J.E. *Reclaimed Asphalt Pavement in Asphalt Mixtures: State of the Practice*; Report FHWA-HRT-11-021; Federal Highway Administration: Washington, DC, USA, 2011.
6. Finlayson, D.; Nicoletti, C.; Pilkington, I.; Sharma, V.; Teufele, B. British Columbia's success with hot-in-place recycling—A 25 year history. *Proc. Can. Tech. Asph. Assoc.* **2011**, *56*, 185–201.
7. Jenkins, K.J.; Collings, D.C. Mix design of bitumen-stabilised materials–South Africa and abroad. *Road Mater. Pavement Des.* **2017**, *182*, 331–349. [CrossRef]
8. Bhavsar, H.; Dubey, R.; Kelkar, V. Rehabilitation by in-situ cold recycling technique using reclaimed asphalt pavement material and foam bitumen at vadodara halol road project (SH87)—A case study. *Transp. Res. Procedia* **2016**, *17*, 359–368. [CrossRef]
9. Dolżicki, B. Polish experience with cold in-place recycling. *Mater. Sci. Eng.* **2017**, *236*, 012089.
10. Merrill, D.; Nunn, M.; Carswell, I. *A Guide to the Use and Specification of Cold Recycled Materials for the Maintenance of Road Pavements*; TRL611; Transport Research Laboratory: Crowthorne, Berks, UK, 2004.
11. Troeger, J.; Widyatmoko, I. Development in road recycling. In Proceedings of the 11th International Conference on Pavement Engineering and Infrastructure, Liverpool, UK, 15–16 February 2012.
12. Alkins, A.E.; Lane, B.; Kazmierowski, T. Sustainable pavements: Environmental, economic and social benefits of in situ pavement recycling. *Transp. Res. Rec.* **2008**, *2084*, 100–103. [CrossRef]
13. Construction News. Enfield Saves Big with Insitu Road Recycling. *Construction News*, June 2012.
14. Wirtgen. *Wirtgen Cold Recycling Technology*; Wirtgen GmbH: Windhagen, Germany, 2011.
15. Karlsson, R.; Isacsson, U. Material-related aspects of asphalt recycling—State-of-the-art. *J. Mater. Civ. Eng.* **2006**, *18*, 81–92. [CrossRef]
16. Milton, L.J.; Earland, M. *Design Guide and Specification for Structural Maintenance of Highway Pavements by Cold in-Situ Recycling*; Transport Research Laboratory Report TRL386: Crowthorne, Berks, UK, 1999.
17. Zaumanis, M.; Oga, J.; Haritonovs, V. How to reduce reclaimed asphalt variability: A full-scale study. *Constr. Build. Mater.* **2018**, *188*, 546–554. [CrossRef]
18. Highways England. *Pavement Design. HD26/06 Design Manual for Roads and Bridges 7(2) Part 3*; Highways England: London, UK, 2006.
19. Pittman, D.W.; McCullough, B.F. Development of a roller-compacted concrete pavement crack and joint spacing model. *Transp. Res. Rec.* **1997**, *1568*, 52–64. [CrossRef]
20. Juenger, M.C.G.; Winnefeld, F.; Provis, J.L.; Ideker, J.H. Advances in alternative cementitious binders. *Cem. Concr. Res.* **2011**, *41*, 1232–1243. [CrossRef]
21. Hicks, J.K.; Caldarone, M.A.; Bescher, E. Opportunities from alternative cementitious materials. *Concr. Int.* **2015**, *37*, 47–51.
22. Highways England. *Design Guidance for Road Pavement Foundations (Draft HD25)*; Interim Advice Note 73/06 revision 1; Highways England: London, UK, 2009.
23. Miljkovic, M.; Radenberg, M. Effect of compaction energy on physical and mechanical performance of bitumen emulsion mortar. *Mater. Struct.* **2016**, *49*, 193–205. [CrossRef]
24. Valentin, J.; Montschein, P.; Fiedler, J.; Mollenhauer, K.; Batista, F.; Freira, A.C. *Report on Durability of Cold-Recycled Mixes: Moisture Susceptibility*; Conference of European Directors of Roads (CEDR): Brussels, Belgium, 2015.
25. AASHTO. *Design of Pavement Structures*; American Association of State Highway and Transportation Officials: Washington, DC, USA, 1993.
26. Davis, K.; Timm, D. *Recalibration of the Asphalt Layer Coefficient*; Report 09-03; National Centre for Asphalt Technology: Auburn, AL, USA, 2013.

suspension systems, lowering the centre of gravity of the wagon, or adjusting the lateral forces on the tracks [19].

Figure 6 illustrates the greater connectivity found in the southern region of the continent, where countries connect to the more developed network in South Africa. A metre gauge line from Mombasa in Kenya to Dar es Salaam in Tanzania connects the East Africa region. Until recently, little has been done to improve the connectivity and interoperability of rail infrastructure in LICs in Sub-Saharan Africa. Early attempts with concessionary frameworks transferred the responsibility of infrastructure maintenance and renewal to operators. Following that, the lines were not extended and connectivity remained very limited [6].

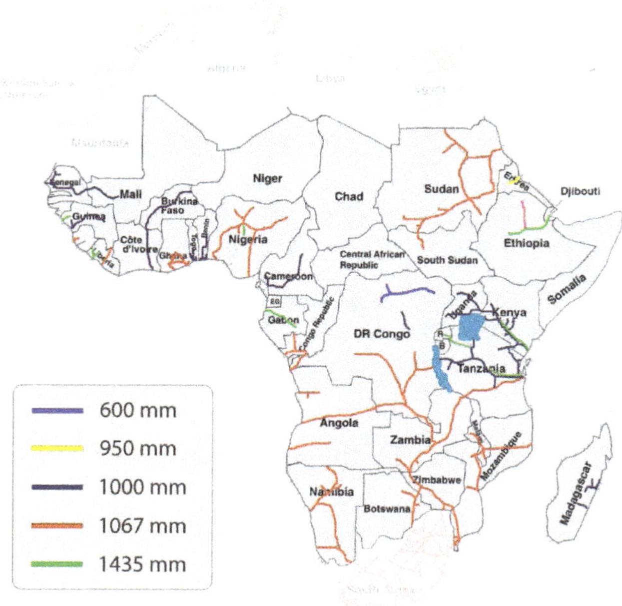


**Figure 6.** Track gauges operated in Sub-Saharan Africa (adapted from creative commons media Attribution-ShareAlike 4.0 International Public License).

Literature shows that not only track density but also track quality is below international standards in Sub-Saharan Africa. Many railway assets in LICs are more than 100 years old, and thus the quality of track materials is considerably outdated [6]. In much of Sub-Saharan Africa, the original rail tracks are ballasted but have not been appropriately or regularly maintained. The combination of asset age and inadequate track maintenance and tamping has severely affected the operational performance of railways in the region. Firstly, the maximum axle load that the railway track structure can withstand in Sub-Saharan Africa is approximately 15 tonnes [3]. In comparison, European standards for axle loads are greater than 25 tonnes [20]. The poor track quality affects the maximum speed achievable. In 2009, freight trains in Sub-Saharan Africa reached only an average speed of 18 km/h [3].

Almost all lines are single track. Data on double track sections or passing loops were not available for a more detailed investigation. Sidings and platforms are outdated and date back to the original lines first built. In many cases, signaling on networks still relies mostly on manual systems, whether with mechanical signals or manual train orders [3]. Manual systems are adequate for the very low traffic currently achieved on most lines, but human error often creates safety problems. For traffic to be increased in a safely manner, signaling systems may prove to be a barrier that needs to be overcome.

27. Labuz, J.; Tang, S.; Cao, Y. *Structural Evaluation of Asphalt Pavements with Full-Depth Reclaimed Base*; Report MN/RC 2012-36; University of Minnesota: Minneapolis, MN, USA, 2012.
28. Khosravifar, S.; Schwartz, C.H.; Goulias, D. Mechanistic structural properties of foamed asphalt stabilised base materials. *Int. J. Pavement Eng.* **2015**, *16*, 27–38.
29. Babtie. *Road Pavement Design Guide*; Kent County Council: Maidstone, Kent, UK, 2000.
30. Jones, D.; Fu, P.; Harvey, J.; Halles, F. *Full-Depth Pavement Reclamation with Foamed Asphalt: Final Report*; Research Report UCPRC-RR-2008-07; University of California: Berkeley, CA, USA, 2008.
31. Valentin, J.; Montschein, P.; Fiedler, J.; Mollenhauer, K.; Batista, F.; Freira, A.C. *Report on Incorporation of Cold-Recycled Pavement Layers in Empirical and Mechanistic Design Procedures*; Conference of European Directors of Roads (CEDR): Brussels, Belgium, 2014.
32. LCPC. *Retraitement en Place à froid des Anciennes Chausées*; Guide Technique; Laboratoire Centrale des Ponts et Chaussées: Paris, France, 2003.
33. Browne, A. Foamed bitumen stabilization in New Zealand—A performance review and comparison with Australian and South African design philosophy. In Proceedings of the 25th ARRB Conference, Perth, Australia, 28–30 August 2012.
34. Liebenberg, J.J.E.; Visser, A.T. Towards a mechanistic structural design procedure for emulsion-treated base layers. *J. S. Afr. Inst. Civ. Eng.* **2004**, *463*, 2–8.
35. Kuna, K.; Thom, N.; Airey, G. Structural design of pavements incorporating foamed bitumen mixtures. *Proc. Inst. Civ. Eng. Constr. Mater.* **2018**, *171*, 22–35. [CrossRef]
36. Perez, I.; Medina, L.; Del Val, M.A. Mechanical properties and behaviour of in situ materials which are stabilised with bitumen emulsion. *Road Mater. Pavement Des.* **2013**, *14*, 221–238. [CrossRef]
37. Lacalle-Jiménez, H.I.; Edwards, J.P.; Thom, N.H. Analysis of stiffness and fatigue resistance of cold recycled asphalt mixtures manufactured with foamed bitumen for their application to airfield pavement design. *Mater. De Construcción* **2017**, *67*, 127. [CrossRef]
38. Paige-Green, P.; Ware, C. Some material and construction aspects regarding in situ recycling of road pavements in South Africa. *Road Mater. Pavement Des.* **2006**, *73*, 273–287. [CrossRef]
39. Lynch, A.; Jenkins, K. Materials recycled using foamed bitumen stabilisation: What is their long-term load-spreading capacity? In Proceedings of the 15th AAPA International Flexible Pavements Conference, Brisbane, Australia, 22–25 September 2013.
40. British Standards Institution. *Bituminous Mixtures-Test Methods for Hot Mix Asphalt*; BS EN 12697: Part 26; British Standards Institution: London, UK, 2012.
41. Nasser, A.I.; Mohammed, M.K.; Thom, N.; Parry, T. Characterisation of high-performance cold bitumen emulsion mixtures for surface courses. *Int. J. Pavement Eng.* **2018**, *19*, 509–518. [CrossRef]
42. Canestrari, F.; Cardone, F.; Graziani, A.; Santagata, F.A.; Bahia, H.U. Adhesive and cohesive properties of asphalt-aggregate systems subjected to moisture damage. *Road Mater. Pavement Design* **2010**, *11*, 11–32. [CrossRef]
43. Wang, W.; Wang, L.; Xiong, H.; Luo, R. A review and perspective for research on moisture damage in asphalt pavement induced by dynamic pore water pressure. *Constr. Build. Mater.* **2019**, *204*, 631–642. [CrossRef]
44. Avirneni, D.; Peddinti, P.R.; Saride, S. Durability and long term performance of geopolymer stabilized reclaimed asphalt pavement base courses. *Constr. Build. Mater.* **2016**, *121*, 198–209. [CrossRef]
45. Beeghly, J.H. Recent experiences with lime-fly ash stabilization of pavement subgrade soils, base and recycled asphalt. In Proceedings of the International Ash Utilization Symposium, University of Kentucky, Lexingston, KY, USA, 19 October 2003; pp. 20–22.
46. Kazmierowski, T.; Marks, P.; Lee, S. Ten-year performance review of in situ hot-mix recycling in Ontario. *Transp. Res. Rec.* **1999**, *1684*, 194–202. [CrossRef]
47. Tabakovic, A.; McNally, C.; Fallon, E. Specification development for cold in-situ recycling of asphalt. *Constr. Build. Mater.* **2016**, *102*, 318–328. [CrossRef]
48. Serfass, J.-P.; Poirier, J.-E.; Henrat, J.-P.; Carbonneau, X. Influence of curing on cold mix mechanical performance. *Mater. Struct.* **2004**, *37*, 365–368. [CrossRef]
49. Dewulf, G.; Kadefors, A. Collaboration in public construction—Contractual incentives, partnering schemes and trust. *Eng. Proj. Organ. J.* **2012**, *2*, 240–250. [CrossRef]
50. Kenny, C. Transport construction, corruption and developing countries. *Transp. Rev.* **2009**, *29*, 21–41. [CrossRef]

51. Wales, J.; Wild, L. *The Political Economy of Roads: An Overview and Analysis of Existing Literature*; Overseas Development Institute: London, UK, 2012.
52. ADB. *Guide to Performance-Based Road Maintenance Contracts*; Asian Development Bank: Manila, Philippines, 2018.
53. Osipova, E.; Eriksson, P.E. How procurement options influence risk management in construction projects. *Constr. Manag. Econ.* **2011**, *29*, 1149–1158. [CrossRef]
54. Lu, W.; Liu, A.M.M.; Wang, H.; Wu, Z. Procurement innovation for public construction projects: A study of agent-construction system and public-private partnership in China. *Eng. Constr. Archit. Manag.* **2013**, *20*, 543–562.
55. Nawaz, A.; Waqar, A.; Shah, S.A.R.; Sajid, M.; Khalid, M.I. An innovative framework for risk management in construction projects in developing countries: Evidence from Pakistan. *Risks* **2019**, *7*, 24. [CrossRef]
56. Tang, W.; Duffield, C.F.; Young, D.M. Partnering mechanism in construction: An empirical study on the Chinese construction industry. *J. Constr. Eng. Manag.* **2006**, *132*, 217–229. [CrossRef]
57. Ajakaiye, O.; Ncube, M. Infrastructure and economic development in Africa: An overview. *J. Afr. Econ.* **2010**, *19*, 3–12. [CrossRef]
58. Leiringer, R. Technological innovation in the context of PPPs: Incentives, opportunities and actions. *Constr. Manag. Econ.* **2006**, *24*, 301–308. [CrossRef]
59. Raisbeck, P. Considering design and PPP innovation: A review of design factors in PPP research. In Proceedings of the 25th Annual ARCOM Conference, Nottingham, UK, 7–9 September 2009; pp. 239–247.
60. Chan, D.W.M.; Chan, A.P.C.; Lam, P.T.I.; Yeung, J.F.Y.; Chan, J.H.L. Risk ranking and analysis in target cost contracts: Empirical evidence from the construction industry. *Int. J. Constr. Manag.* **2011**, *29*, 751–763. [CrossRef]
61. Clifton, C.; Duffield, C.F. Improved PFI/PPP service outcomes through the integration of alliance principles. *Int. J. Proj. Manag.* **2006**, *24*, 573–586. [CrossRef]
62. Ibrahim, C.K.I.C.; Costello, S.B.; Wilkinson, S.; Walker, D. Innovation in alliancing for improved delivery of road infrastructure projects. *Int. J. Manag. Proj. Bus.* **2017**, *10*, 700–720. [CrossRef]
63. Love, P.E.D.; Davis, P.R.; Chevis, R.; Edwards, D.J. Risk/reward compensation model for civil engineering infrastructure alliance projects. *J. Constr. Eng. Manag.* **2011**, *137*, 127–136. [CrossRef]
64. Zedillo, E.; Cattaneo, O.; Wheeler, H. *Africa at A Fork in the Road: Taking off or Disappointment Once Again*; Yale Center for the Study of Globalization: New Haven, CT, USA, 2013.
65. Lehne, J.; Shapiro, J.N.; Eynde, O.V. Building connections: Political corruption and road construction in India. *J. Dev. Econ.* **2018**, *131*, 62–78. [CrossRef]
66. Mahmoud, S.A.I. Public procurement and corruption in Bangladesh: Confronting the challenges and opportunities. *J. Public Adm. Policy Res.* **2010**, *26*, 103–111.
67. Ntayi, J.M.; Ngoboka, P.; Kakooza, C.S. Moral schemas and corruption in Ugandan public procurement. *J. Bus. Ethics* **2013**, *112*, 417–436. [CrossRef]
68. Snaith, M.S.; Khan, M.U. Deleterious effects of corruption in the roads sector. *Proc. Inst. Civ. Eng. Transp.* **2008**, *161*, 231–235. [CrossRef]
69. Lewis-Faupel, S.; Neggers, Y.; Olken, B.K.; Pande, R. Can electronic procurement improve infrastructure provision? Evidence from public works in India and Indonesia. *Am. Econ. J. Econ. Policy* **2016**, *83*, 258–283. [CrossRef]
70. Neupane, A.; Soar, J.; Vaidya, K. Evaluating the anti-corruption capabilities of public e-procurement in a developing country. *Electron. J. Inf. Syst. Dev. Ctries* **2012**, *552*, 1–17. [CrossRef]
71. World Bank. *Curbing Fraud, Corruption and Collusion in the Roads Sector*; World Bank: Washington, DC, USA, 2011.

*Review*

# Developing a New Technical Strategy for Rail Infrastructure in Low-Income Countries in Sub-Saharan Africa and South Asia

**Marcelo Blumenfeld *, Wendy Wemakor, Labib Azzouz and Clive Roberts**

Birmingham Centre for Railway Research and Education, University of Birmingham, Birmingham B15 2TT, UK
* Correspondence: m.blumenfeld@bham.ac.uk

Received: 5 July 2019; Accepted: 6 August 2019; Published: 9 August 2019

**Abstract:** Low-income countries (LICs) in Sub-Saharan Africa and South Asia are investing in new railway lines to replace deteriorated infrastructure from the 19th and 20th century. These actions, despite financial and economic constraints, have been justified in common visions of continent-wide efficient networks to cope with the demands of growing populations. However, most of the recent rail infrastructure projects are driven by international suppliers' preferences and financing rather than creating railways that match the requirements of interoperable regional networks. This paper therefore explores the current status of rail infrastructure in these LICs and the operational performance achieved to understand specific capability gaps in each regional network. Drawing from the experience of European countries in transforming regional future visions into applied research, a technical strategy for rail infrastructure in LICs is proposed. The strategy captures the key capabilities to be addressed in order to achieve future performance goals, while emphasizing the need for emerging technologies to be used in fit-for-purpose solutions. It is envisioned that the strategy will provide the basis for the development of continental technical strategy programs with specific technology roadmaps towards a common goal.

**Keywords:** capability plan; low-income countries; railways; railway technical strategy; South Asia; Sub-Saharan Africa; sustainability

---

## 1. Introduction

While road transport has been the dominant mode for the movement of people and goods since the mid-1900s, railways are now staging a global comeback, illustrated by growth in the total length of rail tracks and overall rail traffic units per year (passenger-km and tonne-km) [1]. Among the various reasons for their renaissance are the concerns over the significant social, economic, and environmental externalities created by the transport sector. Global greenhouse gas emissions from the transport sector have more than doubled since 1971, and over three quarters of this increase has come from road vehicles [2]. Rail transport is potentially more environmentally friendly than its road counterpart in terms of energy consumption and emissions per traffic unit. Railways are costly to build and maintain but can produce significantly lower external costs than other modes of transport, particularly when carrying freight.

Either powered by diesel or electricity, the railways can reduce external costs of transport by at least 47.5% per passenger-km, and 75.4% per tonne-km when compared to road modes (see Figure 1) [3]. The biggest savings are found in environmental impacts, such as air pollution and climate change. Rail freight produces between 75% and 85% less greenhouse gas emissions per transport unit when compared to articulated trucks, monetizing its benefit at around 0.1–0.4 cents per net tonne-km [4].

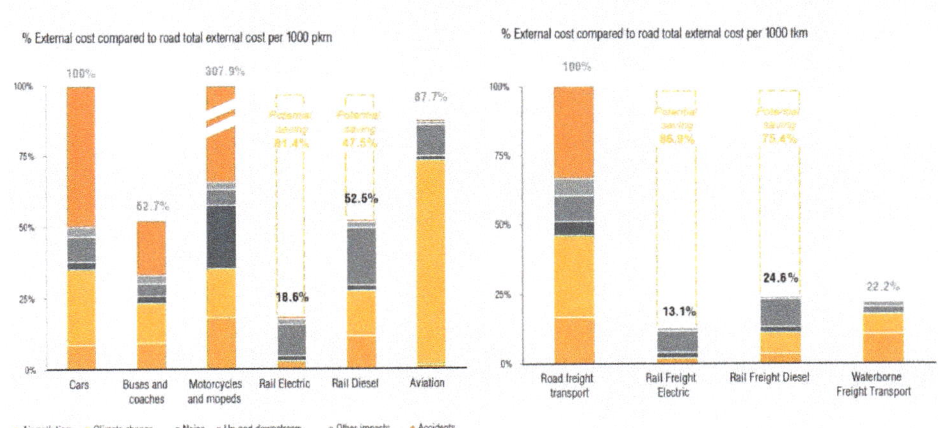


**Figure 1.** External costs of railways compared to other modes of transport [4].

The compound annual growth of rail freight transport has been positive around the world in the last three decades [4]. For passenger services, the compound annual growth rate (CAGR) has varied across different continents, as shown in Figure 2. However, the inferior performance of most low-income countries in both passenger and freight transport is evident. As a whole, with few exceptions, low-income countries have experienced low growth in freight transport and a drop in the passenger railway market. The graph for the South Asia region does not illustrate the situation in its low-income countries because the growth in traffic units has happened mostly in India (which is not considered a low-income country by the World Bank).

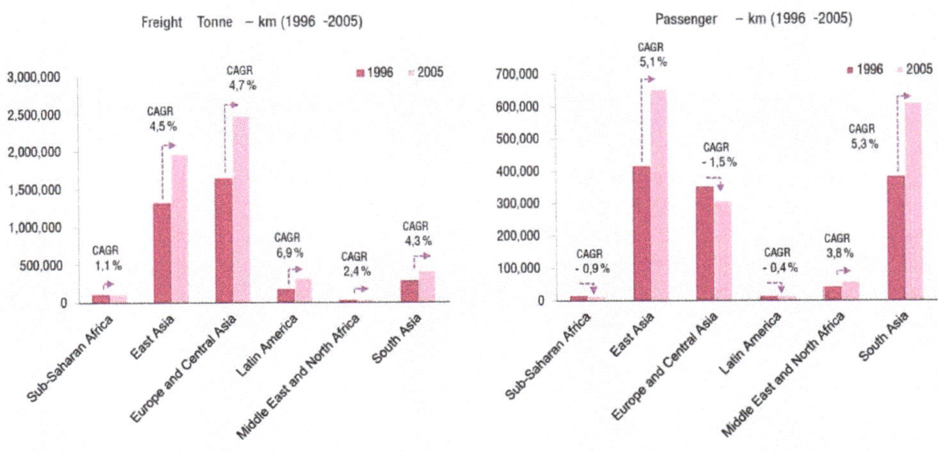


**Figure 2.** The evolution of the world rail market between 1996 and 2005 [4].

The increase of road transport in low-income countries has been driven by the higher costs of rail operations, maintenance, and renewals, and a lack of strategic planning resulting from political instability and conflicts [3]. In Africa, road transport is responsible for 90% of passenger traffic and 80% of the movement of goods [5]. Railways require dedicated corporations to operate passenger and freight services. The centralised costs for maintaining and upgrading capital assets create the impression that the railways are too costly to compete with road transport, but rail is considerably cheaper when economies of scale are possible [4].

Despite the financial constraints, many low-income countries in Sub-Saharan Africa and South Asia have followed through and started prioritizing railway infrastructure in their plans for the upcoming decades. The current moment seems appropriate, as most low-income countries are experiencing higher economic growth than the global average, leading to promising prospects [6]. When this growth is combined with an emerging middle class and untapped resources, it is clear that the potential for railway development is timely [6]. In line with this, the African Union has published a vision of a continent-wide rail network to facilitate inter- and intra-regional trade and meet the travel needs of its growing population [7]. In South Asian countries, future goals for railway development are of a similar scale.

Therefore, there seems to exist important gaps between the current status of rail infrastructure, the common vision of regional development, and short-term projects in implementation. While investment in rail infrastructure is seen as an essential feature for the achievement of Sustainable Development Goals, projects need to be chosen carefully, especially in financially constrained countries [5].

## 2. Aims, Objectives, Scope, and Limitations

### 2.1. Aims and Objectives

This paper builds upon the research conducted for the U.K. Department for International Development (DFID), as part of the High-Volume Transport (HVT) project. The project was established to identify key areas for capacity building in order to improve access to more affordable, safer, and lower carbon transport services in low income countries. More specifically, the objectives of the rail sub-theme were to determine the current state of the infrastructure in low-income countries to identify pathways towards regional interoperability goals. Subsequently, the research targeted the gaps in the continental visions with present-day capabilities in order to develop key guidelines for technical development in each region.

### 2.2. Scope

As a starting point, the HVT project focused on low-income countries in Sub-Saharan Africa and South Asia. According to the World Bank [8], low-income countries (LICs) are those where the Gross National Income (GNI) per capita is less than US$1005. Lower middle-income economies are those with a GNI per capita between $1005 and $3895. Besides these countries, the project also included DFID priority projects [9]. Some of these countries are in fact in the lower middle-income bracket, but present similar levels of Human Development Indices (HDI) to their low-income counterparts.

Not all countries that passed the criteria contain rail infrastructure or operate railway services. With these criteria, the initial list of 41 countries was then reduced to the 27 countries, which are listed in Table 1. The rail infrastructure sub-theme then adopted these as the scope of the research project.

**Table 1.** List of countries contemplated in the project.

| | | | |
|---|---|---|---|
| Afghanistan | Ethiopia | Mali | Sudan |
| Bangladesh | Ghana | Mozambique | Tanzania |
| Benin | Guinea | Myanmar | Togo |
| Burkina Faso | Kenya | Nigeria | Uganda |
| Cote d'Ivoire | Liberia | Pakistan | Zambia |
| Democratic Republic of the Congo (DRC) | Madagascar | Senegal | Zimbabwe |
| Eritrea | Malawi | Sierra Leone | |

Of the 27 countries contemplated, 23 are in Africa and 4 are in South Asia, as illustrated in Figure 3.

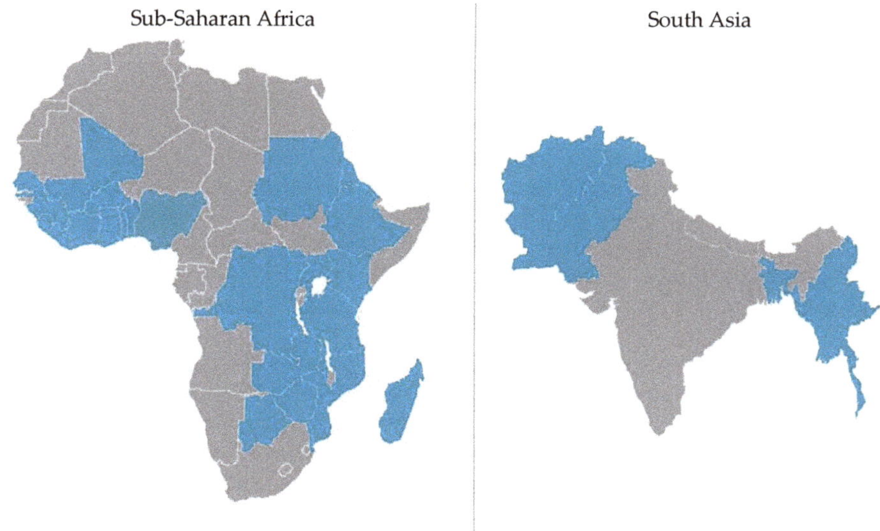


**Figure 3.** Low-income countries in Sub-Saharan Africa and South Asia contemplated in the project.

## 3. Methodology

This research reviewed the literature on the current state of rail infrastructure to identify development areas that have the greatest potential for increasing future affordability and sustainability of railways in low-income countries (LICs) in Sub-Saharan Africa and South Asia. Findings were also derived from a workshop conducted in Nairobi, Kenya, in November 2018, with key stakeholders from Sub-Saharan African countries.

The review looked at data available from primary and secondary sources. Extensive databases held by organisations such as the World Bank and the International Union of Railways (UIC) were used as the main sources. Secondary sources were reports that provided supporting information [10,11]. The availability, age, and robustness of databases found posed a challenge to the systematic process. In South Asia, more up-to-date information was available through reports from governments and international agencies [11–15].

Due to the significant differences in use of rail infrastructure between Sub-Saharan Africa and South Asia, the process of data collection adopted two separate fronts. Breadth of scope was prioritised in this review in order to provide a comprehensive understanding of key areas needing development. Under each front, themes were defined where data was available for both regions. Subsequently, each sub-theme and the list indicators for infrastructure condition and operational performance were defined according to the information available. The process is illustrated in Figure 4.

Information on rail infrastructure and performance in low-income countries was found to be generally fractured and outdated, in accordance with previous works [3,4]. Large discrepancies were found among sources. It was found that the most comprehensive database where indicators are available for most countries analysed stopped being updated between 2005 and 2008 [1,10]. More recent information, where available, was used to calibrate the reliability of sources [11]. Some countries listed infrastructure indicators until 2011, with one listing indicators until 2017 (Democratic Republic of the Congo). In the twelve years separating the data for DRC, no significant changes in indicators were found.

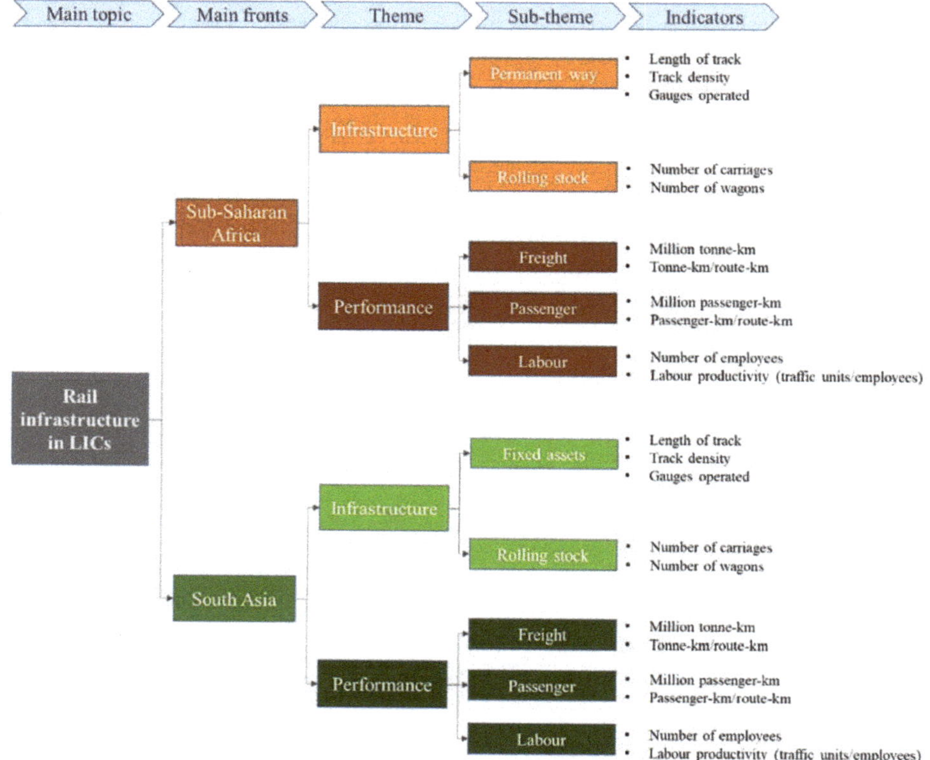


**Figure 4.** Review process and list of selected indicators.

## 4. Current State of Rail Infrastructure in LICs in Sub-Saharan Africa and South Asia

Please refer to Appendix A for a table comprising the indicators for each country

### 4.1. Overview

While the railways have the potential to be more cost effective where economies of scale permit, low-income countries have struggled to reap the benefits. The general image of railways in low-income countries reflects dilapidated infrastructure, outdated technologies, and low-quality operations in terms of performance and safety. After decades of poor maintenance and little investment, many countries now have major railway sections that are not in operation and require renewal. In general, rail infrastructure in Sub-Saharan African low-income countries seems to be in poorer condition than in South Asia. For instance, 23% of railways in Benin and 91% in Uganda are not operational [3]. In other countries, many parts of the rail network are idle, such as 60% of the network in Ghana [3].

The limited length and quality of railway routes has particularly impacted the regional connectivity of landlocked countries and imposes higher freight costs for two reasons. Firstly, the limited size and quality of the network reduces the freight volumes that the line can transport, negatively affecting labour productivity and efficiency. Secondly, the limited reach of rail routes means that freight has to be transferred to road transport to reach most destinations, which increases the costs compared to road-only transport. Thirdly, the variety of different gauges encountered within the regions prevents a steady traffic flow on the railway networks. It is not surprising that transport costs in LICs in Sub-Saharan Africa are the highest worldwide, with freight charges 20% higher than in low-income countries in other regions [16].

For many of these countries, the length of railway infrastructure has an impact on the potential to exploit their substantial deposits of natural resources. Furthermore, rail infrastructure in many LICs dates back to colonial times and much of the network has not been upgraded since. Each nation post-independence retained the track gauges that had been selected by their colonisers [6,10].

Track densities in LICs are generally lower than the global average. The exceptions are small countries, such as Bangladesh, and countries with very small networks, such as Afghanistan. In LICs in Sub-Saharan Africa (SSA), average track density is even lower at 2.76 km per 1000 km$^2$, while LICs in South Asia (SA) are served by a relatively higher density of 7.8 km of track per 1000 km$^2$ [10]. There are expected variations where either area or population is significantly high, yet LICs in both regions fare poorly when compared to countries with robust rail traffic. For instance, Germany has 121 km of track per 1000 km$^2$, which possibly explains why its railways carry 43% of the country's freight [17].

### 4.2. Sub-Saharan Africa

#### 4.2.1. Infrastructure

In Sub-Saharan Africa (SSA), rail infrastructure is predominantly used to transport freight (Figure 5). This can be partly explained by the vast reserves of minerals and bulk commodities in many countries in the region. Most railways connect major ports to large cities and mining areas because 90% of African imports and exports are transported by sea [4,18]. However, the movement of people and goods through the continent is limited because there is little connectivity between railway networks. Only a few lines cross borders, such as the Tanzania–Zambia Railway Authority (TAZARA) connecting Tanzania and Zambia, and Sitarail between Burkina Faso and Cote d'Ivoire. Interoperability is also constrained. Several countries operate different track gauges that range from 600 mm to standard (1435 mm). Cape gauge lines of 1067 mm predominate, especially in the south, followed by metre gauge (1000 mm) lines and standard gauge (1435 mm), which are more recently built [10]. Efforts in building new standard gauge tracks seem to overshadow interest in existing infrastructure.

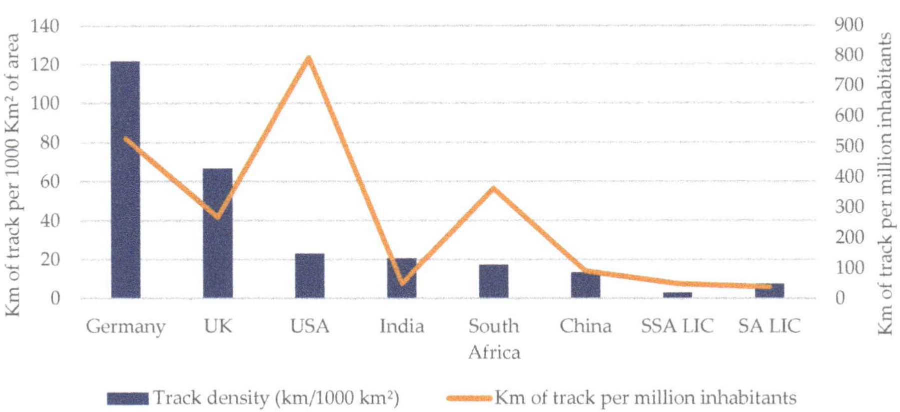


**Figure 5.** Track density in Sub-Saharan African (SSA) and South Asian (SA) low-income countries (LIC) compared to middle and high income countries [1,8].

Although the commonly used narrow gauge tracks are said to marginally limit the loading gauge of trains, greater loads are carried on South African Cape gauge tracks than on standard gauges in the United Kingdom [10]. However, smaller gauges tend to be associated with smaller curve radii, so rolling stock design and performance are crucial. The main concern with narrow gauges is the lateral oscillation (hunting) because of short-wheelbase bogies. The greater lateral oscillation of narrow gauges impacts passenger comfort, yet there are measures to address this issue, such as improving

None of the original lines are electrified [6]. Electrified lines are only available on a few routes built in recent projects, but these present challenges. The electrical grid in many low-income countries in Sub-Saharan Africa is not robust enough to accommodate an electrified railway network. In Ethiopia, for example, a separate grid was built to cater for the large amount of power required to run electric trains [21]. Where electrically powered signaling is installed, often it cannot be used because of short circuits, no electrical power, and degraded cable networks that are susceptible to cable theft. Telephone exchanges in many railways are obsolete, with limited capacity and with spare parts being virtually impossible to obtain [3].

#### 4.2.2. Operational Performance

Railway traffic consists mainly of freight services in low-income countries in Sub-Saharan Africa. The movement of goods accounts for almost 90% of the traffic in the region. This can be explained by the fact that it is more difficult to recover the costs of passenger operations without subsidies, and with limited track availability, more cost-effective choices have to be made.

The combination of low maximum axle loads and speed limits within limited routes has a significant impact on traffic volumes achievable in LICs in Sub-Saharan Africa. Traffic volumes are an important measure of the sustainability and affordability of rail operations because they highlight infrastructure usage. Higher volumes generate economies of scale in the costs of infrastructure maintenance and improvements, which in turn increase the competitiveness of rail transport against other modes.

With the exception of Guinea, where private mining companies extensively use their own railway lines, all LICs show traffic densities below or around 1 million traffic units (tonne-km plus passenger-km) per route-km (Figure 7). Data for Eritrea, Liberia, and Sierra Leone were not available. In contrast, railway networks in the United Kingdom, France, and Germany operate around 4.5 million units per route-km. Global leaders in traffic volumes such as Russia, the United States, and China reach densities up to 40 million units per route-km (Figure 8) [1,17,22].

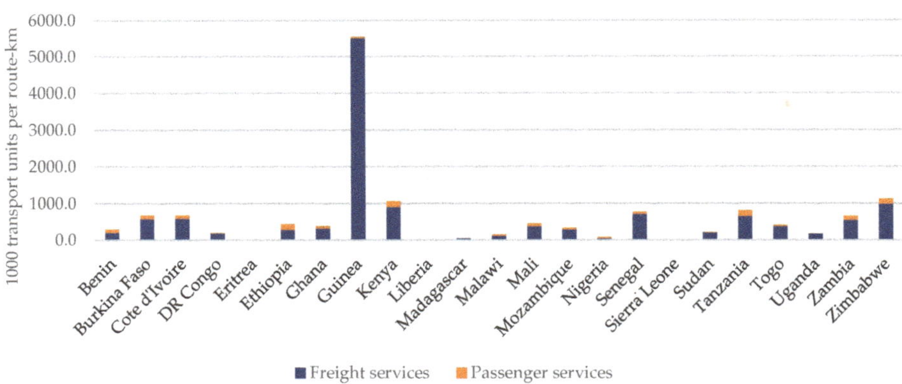


**Figure 7.** Rail traffic density in SSA LICs [10].

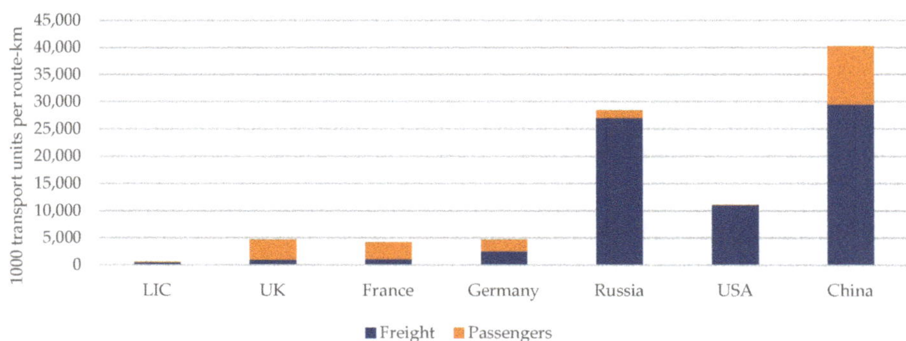


**Figure 8.** Rail traffic density in SSA LICs compared with high income countries and global leaders [1,10].

Due to the low volumes achieved, labour productivity is significantly lower in most LICs in Sub-Saharan Africa than in developed nations. On average, employees in the United Kingdom produce four-times more traffic units per employee than SSA LICs. In a more extreme comparison, while an average employee in LICs in Sub-Saharan Africa produces 150,000 traffic units per year, a U.S. employee produces 14.2 million (Figure 9) [8,22]. Under these circumstances, fixed costs are spread over fewer traffic units, which in turn become expensive in comparison to other transport modes. In exchange, that leads to a vicious cycle where higher unit costs lead to lower demand, which affects the unit costs and return. The exceptions in the region are Burkina Faso, Senegal, and Zambia, which reach approximately 500,000 units per employee and are close to the productivity of German railways [10]. No data was found for Guinea.

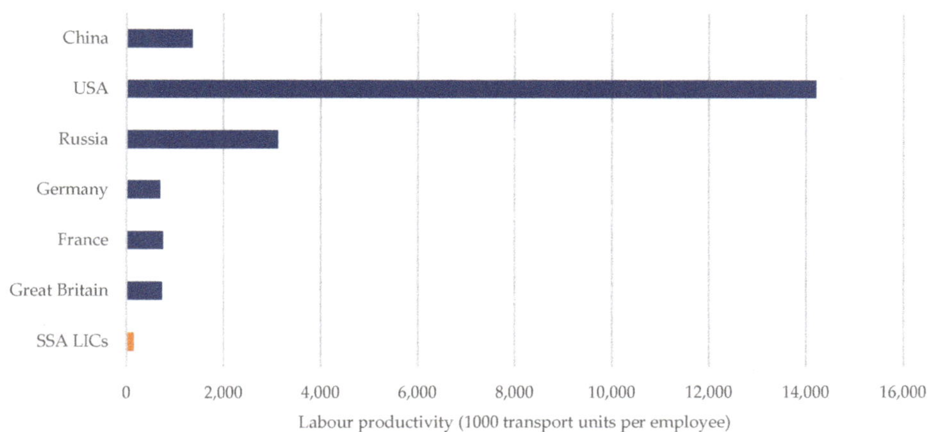


**Figure 9.** Comparison of labour productivity between SSA LICs and selected countries [10,22].

Safety on railways in low-income countries has been a matter of significant concern. There is very little data on safety records, as well as unclear standards and compliance, and the issue has been raised in the literature [3–5]. In addition, safety was the most mentioned aspect during the workshop with key stakeholders in Nairobi. Respondents listed trespassing, derailments, and theft of infrastructure as the main problems faced in the region. From the data available, safety records in LICs in Sub-Saharan Africa are significantly worse when compared to other countries [11].

Sample safety records were found for DR Congo (2015), Cote d'Ivoire (2002), Kenya (2002), and Nigeria (2003), and used to infer a comparison against more developed economies [11]. With

0.545 accidents for every million traffic units, DR Congo had an accident rate 545-times higher than Germany or Great Britain (0.001 accidents per million traffic units), and 32-times the average in the developing world [11,13]. Other LICs performed better but still experienced over 100 more accidents than the two European counterparts, as shown in Table 2.

**Table 2.** Safety record of sample SSA LICs compared to Germany and the United Kingdom [11,22].

| Country | Traffic Units (million) | Accidents | Accidents per Million Traffic Units |
|---|---|---|---|
| DR Congo (2015) | 697 | 380 | 0.545 |
| Cote d'Ivoire (2002) | 436 | 91 | 0.209 |
| Nigeria (2003) | 268 | 108 | 0.403 |
| Kenya (2002) | 2191 | 249 | 0.114 |
| Germany | 208,262 | 346 | 0.0016 |
| Great Britain | 68,912 | 71 | 0.0010 |

### 4.3. South Asia

#### 4.3.1. Infrastructure

There are significant differences between railway usage in South Asian LICs and those in Sub-Saharan Africa. Railway systems in South Asia's LICs are mostly used for passenger transport. Since the railways are owned and operated by the government in all countries analysed, services are heavily subsidised for social reasons. Railway lines connect major cities in commuting networks across regions and sometimes between countries.

Possibly due to closer geographical proximity and a more common colonial influence, track gauges in South Asia are more homogenous than in Sub-Saharan Africa (see Figure 10). Countries bordering India have adopted mainly the same broad gauge (1676 mm), which assists with connectivity and interoperability. Nepal has broad gauge track infrastructure but has not operated railway services since 2008. East of India, trains in Myanmar and east Bangladesh run on meter gauge tracks (1000 mm). In Pakistan and west Bangladesh, broad gauge tracks are the standards [23]. Further west, a short line in Afghanistan runs on Soviet gauge (1520 mm). Bangladesh also has 365 km of dual gauge track (metre gauge and broad gauge) connecting the broad gauge network in the west to the meter gauge network in the east [12,24]. Pakistan runs a short 312 km metre gauge line [15].

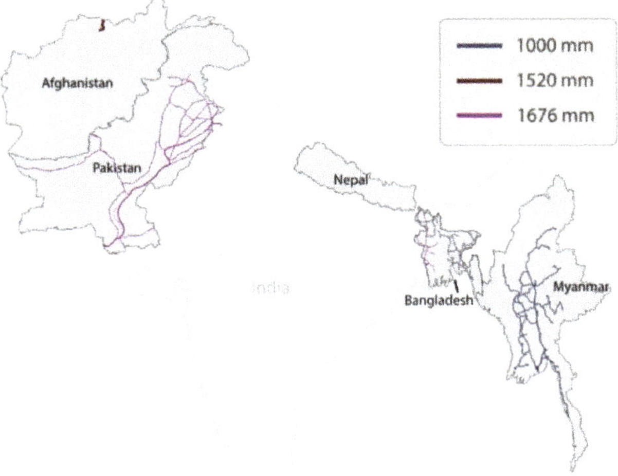


**Figure 10.** Illustration of track gauges in use in South Asia (based on previous work [10–13]).

With a total of 7.8 km per 1000 km$^2$, LICs in South Asia are almost three times more densely crossed by railway lines than low-income countries in Sub-Saharan Africa [10,22]. Bangladesh has more railway tracks per area than Pakistan and Myanmar but a lower track density per head of population. Despite the higher track density, age of assets and infrastructure quality in South Asian LICs are a similar issue to those found in Sub-Saharan Africa. In all four countries studied, most lines are single track. In Pakistan, only approximately 10% of the track is double track.

Track quality varies according to country, and has an important impact on operational performance. Poor condition of the track infrastructure, which includes aspects such as track alignment, ballast condition, and the presence of cracks on the rail surface, not only limits the maximum load and speed that can be safely achieved but also increases the risk of derailment. For instance, railway infrastructure in Myanmar has an axle load limit of only 12.5 tonnes [14]. Moreover, the infrastructure condition in the country limits the maximum speeds of freight trains to 48 km/h and passenger trains to 60 km/h. Pakistan and Bangladesh have upgraded rail tracks to support 22.5 tonnes per axle in certain sections. In other sections, axle load is limited to 17.87 tonnes [23]. The variance in track quality is problematic in two ways. The lower threshold limits rolling stock loads and wastes the investments in the sections with better track quality, while there is a risk that rolling stock loads match the improved sections and present a risk of overloading the track. Available traffic speeds have not improved. As an example, the average speed of freight trains in Bangladesh is only 18.5 km/h [12].

In the region, none of the lines are electrified and signaling is still mainly manual. Ahmed et al. [24] have listed the outdated systems as a threat to safe and efficient train operations in Bangladesh. In Myanmar, contracts to upgrade signaling to electronic systems have been signed recently but are still not available [25].

#### 4.3.2. Operational Performance

Traffic densities in South Asia are considerably higher than in LICs in Sub-Saharan Africa. Crowded passenger volumes make traffic densities (1000 traffic units per route-km) in Pakistan and Bangladesh comparable to high-income European countries, such as Great Britain, France, and Germany, but still far from world leaders, such as Russia and China (Figure 11). According to recent data, Bangladesh and Pakistan operated approximately 3.0 and 3.6 million traffic units per route-km, respectively [12,15]. Myanmar has a traffic density below 1 million units per route-km, and therefore is considered a low volume network similar to most low-income countries in Sub-Saharan Africa [14]. Contrary to the African continent, only a small proportion of total rail traffic in South Asian LICs is dedicated to freight. Passenger services account for 80% of the total traffic in the region [10]. No traffic data was available for the Afghan line, which operates only freight services to Uzbekistan.

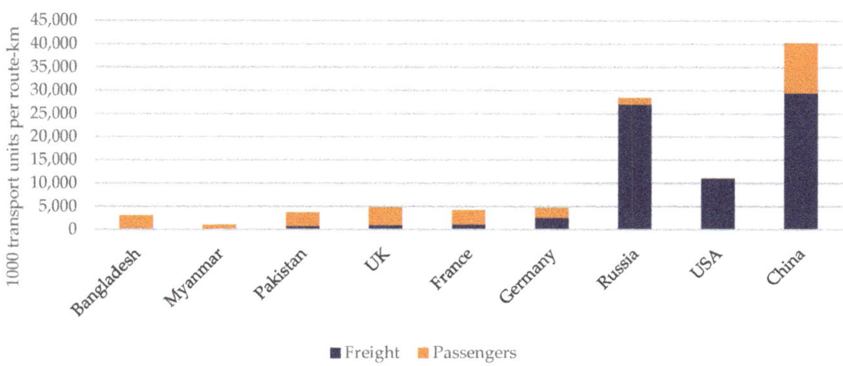


**Figure 11.** Traffic density of South Asian LDCs compared with high income countries and world leaders [9–13,15].

While there are clear differences in traffic densities between Myanmar and the other two countries (Bangladesh and Pakistan), the gap in labour productivity seems to be less wide. The three countries are more productive than the majority of Sub-Saharan African LICs, but the average output per employee is still half of the German counterpart (Figure 12). Bangladesh and Pakistan show similar output levels at 343,000 and 389,000 traffic units per employee [12,14]. Myanmar achieves lower productivity levels at 213,000 traffic units per employee, which can possibly be linked to the lower speeds and loads available on its network [13]. In railways where passenger services are dominant, as in South Asian LICs, operations are usually subsidised because they are unlikely to break even. Low productivity adds to the issue by increasing the operational expenditure per transport unit [3].

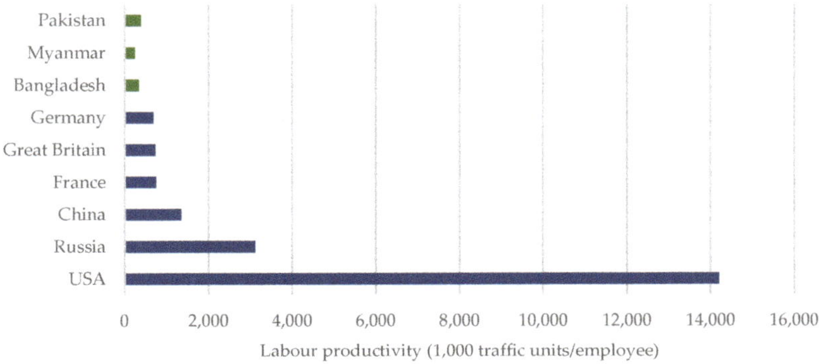


**Figure 12.** Comparison of labour productivity between SA LICs and selected countries [11–15,17].

Safety records in South Asian LICs are better than those found in low-income countries in Sub-Saharan Africa (Table 3). Bangladesh reported 707 accidents in 2005, which translates to 0.142 accidents per million traffic units in that year [11]. Pakistan has reported better records recently, with 139 accidents over a much greater volume, resulting in an accident rate of 0.004 incidents per million traffic units [15]. This performance positions Pakistan much closer to the rates found in other LICs. Safety records in Bangladesh still reflect similar values to LICs [12].

**Table 3.** Safety record of SA LICs compared to Germany and the United Kingdom [12–15,17].

| Country | Traffic Units (million) | Accidents | Accidents per Million Traffic Units |
|---|---|---|---|
| Bangladesh (2005) | 4981 | 707 | 0.142 |
| Myanmar (2013) | 4261 | 500 | 0.117 |
| Pakistan (2017) | 28,549 | 139 | 0.004 |
| Germany | 208,262 | 346 | 0.0016 |
| Great Britain | 68,912 | 71 | 0.0010 |

Secondary data from the Asian Development Bank (ADB) estimate that Myanmar Railways experienced around 500 accidents per year, which results in a rate of 0.117 accidents per million traffic units [13]. The same document states the accident rate as 0.852 per million traffic units, which highlights the fragmented nature of datasets. In the country, most accidents are derailments caused by poor track quality [13].

## 5. Recent Sino-African and Sino-Asian Projects

The review of rail infrastructure and operations in low-income countries in Sub-Saharan Africa and South Asia has highlighted the disparity between the current conditions and national and regional visions for the future. With financial challenges in working towards these goals, various low-income countries have recently signed resource-for-infrastructure agreements, mostly with China, to replace dilapidated infrastructure with new lines using modern technologies [16]. From the literature, Sino-African projects exceed those found in South Asia. This may be explained by the large investments from the ADB in the region to develop a Trans-Asian Railway network [26]. Conversely, projects in Sub-Saharan Africa seem to be conducted on an ad hoc basis.

Since its beginning in 2006, Sino-African trade volumes grew rapidly to reach more than US$200 billion in 2013 [16]. These resource-for-infrastructure investments include several railway projects across the continent, with some examples listed in Table 4. Most projects consist of single track, non-electrified, standard gauge lines [18,21,27–29]. Lines are built with freight and passenger services in mind and permit an average maximum speed between 100 km/h and 120 km/h [18,21,27–29]. There are exceptions to these standards: (1) the Addis Ababa–Djibouti line, which is electrified and includes 151 km of double tracks; (2) the Mali–Senegal line, which is a renovation of the existing infrastructure; and (3) the Abuja–Kaduna section of the Lagos–Kano line, which is double-track and will permit speeds of up to 250 km/h [30].

**Table 4.** Sino-African railway development projects [18,21,27–30].

| Country | Project | Length | Cost (in US$) |
|---|---|---|---|
| Chad | Chad Railways [27] | 1364 km | 5.6 billion |
| Ethiopia-Djibouti | Addis Ababa–Djibouti line [21] | 751 km | 4 billion |
| Mali | Mali–Guinea Railway line [28] | 900 km | 11 billion |
| | Mali–Senegal Railway [18] | 1286 km | |
| Nigeria | Lagos–Calabar line [29] | 1400 km | 11.1 billion |
| | Lagos–Kano line [30] | 1124 km | 8.3 billion |
| Kenya | Mombasa–Nairobi line [18] | 485 km | 4 billion |

In South Asia, such projects are not as numerous. In Bangladesh, work has started on the Padma Bridge Rail Link, a 225 km project connecting regions to the port of Payra, at a total cost of the project expected to be US$3.14 billion with a loan of 80% of the amount [31]. Pakistan signed with China a US$8.2 billion overall investment in railways in Pakistan, including the renewal of broad-gauge tracks and the acquisition of rolling stock. It has since, however, been reduced by US$2 billion due to concerns over the costs of the loans [32].

At a first glimpse, these projects are reshaping the capacity of existing routes. For instance, the line connecting Addis Ababa to Djibouti has reduced the journey time between Djibouti and the dry port of Mojo in Ethiopia from 84 h to 10–15 h [21]. Similarly, the Standard Gauge Railway has reduced journey times between the port of Mombasa and the capital Nairobi to less than 5 h. In Nigeria, these projects are expected to provide a long-awaited expansion to reduce congestion on the country's damaged roads [33].

However, there is significant concern over the sustainability of these projects. From an economic perspective, concern is increasing that these large loans will tie low-income countries to a long-term dependency on China rather than promoting internal development [16]. These concerns have some support in the case of Sri Lanka, which handed over control of one of its deep-sea ports to ease its debt with China [34]. Moreover, the cost-effectiveness of such large-scale projects has been questioned. For instance, the Mombasa-Nairobi line in Kenya is reported to have cost close to three times the international standard and four times the original estimate [35].

From a technical standpoint, systems have been developed only within the national context, with little attention to compatibility and standardisation [6,36]. Rahmatullah [36] adds that regional rail use is likely to be constrained by differences in track gauge, track structure, signaling systems, and incompatible rolling stock. Delelegn [21] highlighted that lines within Ethiopia run on different signaling systems because they were built by different companies. Stakeholders at the Nairobi workshop shared similar concerns about the link between new rail infrastructure and wider development regional plans, where projects are not standardised with regard to maximum axle loads and speeds, and control and communication systems to be used in different sections. It was also highlighted that national masterplans were not available and very few documents were digitalised for common access.

Moreover, it seems that recent investments are not entirely aligned with the current gaps in capability. Technical decisions on new projects are being driven by external forces, including international suppliers and foreign governments, rather than following regional plans with a holistic perspective. This has resulted in investments in solutions that do not match the requirements of the specific context and development trajectory of low-income countries. For example, electrification is still not possible in many countries, as the national electricity grids are insufficient to support railway operations. When electrified lines were built in Ethiopia, a new grid had to be added to the project [21]. Furthermore, expensive and highly sophisticated complex systems have been specified where they are not required, as in the example of the deployment of advanced ERTMS Level 3 (European Rail Traffic Management System Level 3) technologies in Zambia, where traffic volumes are less than 1.5 million traffic units per year [37].

There are clear discrepancies in future goals and current actions. Since there is no established technology roadmap to lead national development towards a common goal, countries are buying off-the-shelf technologies that are too expensive for the near future traffic projected. In doing so, their pathway to development is likely to only follow the steps of developed countries decades ago and remain outdated in the long-term. Little attention has been given to the specific context of the regions and their needs for fit-for-purpose solutions that can leapfrog previous development curves and create fit-for-purpose solutions.

If these problems are left unaddressed, there is significant potential for scarcely available money to be wasted, and for new railway systems to underperform and lose money. Such an outcome is likely to slow the rate of development of railways in low-income countries and lead to less efficient and more environmentally damaging solutions being deployed in the future.

Therefore, it is imperative that scarcely available investments are made consistently and coherently in order to follow a pathway towards continental future goals. Experience in developed regions shows that technical strategies are important studies that can identify common capabilities and produce context-specific and fit-for-purpose technology roadmaps that connect future visions to current levels of development.

## 6. Technical Strategies in Europe

In 1995, the European Union had a similar vision to restructure the rail transport market and strengthen the position of the rail industry in relation to other transport modes [38]. Infrastructure at that point was in a much better condition than currently found in LICs in Sub-Saharan Africa and South Asia, yet there was a need for harmonised development to fulfil the goals of a continent-wide network. This implied opening-up of the rail transport market to competition, improving interoperability and safety of national networks, as well as developing better rail infrastructure [36]. These three main areas—crucial to the development of a strong and competitive rail transport industry—are no different from what is required to enable low-income countries to provide an efficient, reliable, and safe alternative mode of transport for its citizens.

To achieve these goals, the European Union started developing technical strategies to connect future visions to tangible development programs. The first edition was published in 2008 with the purpose of "foreseeing the kind of railway that the rail industry is capable of supplying in response to European and national needs and affordability criteria, to assess whether this railway can be delivered through 'natural' incremental change mechanisms, or whether some planned strategic changes are required" [39].

Technical strategies are roadmaps to connect the present state to the future goals of systems because the gaps between both are usually of considerable magnitude. In this, the complex context and the uncertainty that surrounds development require coordinated actions to achieve a common goal. Moreover, technical strategies are incremental, as new versions build on achievements of the previous plans.

For instance, the technical strategy in 2008 focused on interoperability and efficient and border-free acceptance [39]. Ten years later, the 2019 version of the European Technical Strategy envisioned further advancements, such as demand-responsive services, low-carbon solutions, and greater safety than any other transport mode [40].

The jump in capabilities illustrates the active role that technical strategies can have in linking outputs and outcomes. In 2014, the European Commission established a platform for coordinating research activities in the railway sector under the name of Shift2Rail [41]. Since then, efforts to support a Single European Railway Area have been carried out in the form of joint undertakings, where public–private partnerships conduct research and development projects. The research programs are defined by the Shift2Rail commission, in the face of the overarching regional goals of the European Commission, as well as the recent technological developments in the various transport sectors. This combination of top-down and bottom-up perspective ensures that local capability is improved in accordance with regional visions.

In the United Kingdom, similar work has been carried out with the Railway Technical Strategy (RTS). The RTS aims to support strategic planning processes of railways while guiding stakeholders on the deployment of technologies to steer the future of the railway industry's technical direction. The strategy's time frame is 30 years, balancing between the lifespan of rolling stock, which ranges from between 25 and 40 years, and stations' and assets' lifespans of 100 years or more. The RTS has six main themes, namely control, command and communication; energy; infrastructure; rolling stock; information; and customer experience. These themes reflect on strategies that address technical and operational domains within the rail industry.

To connect the vision to achievable milestones, the RTS makes use of Capability Delivery Plans (CDP). These are developed for the delivery of a holistic set of key capabilities that can achieve strategic goals while ensuring a sustainable market. The first edition in 2012 identified the areas of development, and later iterations of the plan can build on them as the systems evolve. Visions are broken down into key capabilities to be developed, which are then transformed into key programmes to achieve them (See Figure 13).

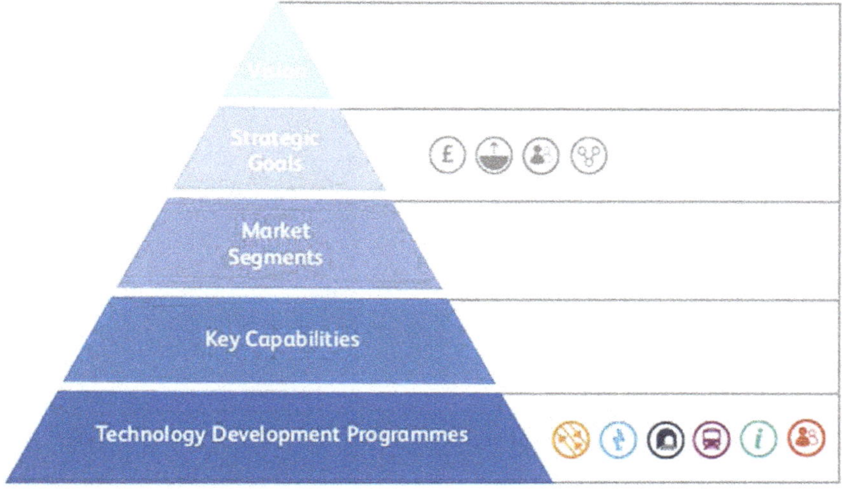


**Figure 13.** The U.K. Capability Delivery Plan [42].

## 7. Developing a Technical Strategy for Low-Income Countries

### *7.1. Overview*

As previously mentioned, there are visions of a continent-wide railway network, specifically high-speed rail, in Sub-Saharan Africa and South Asia to facilitate trade as well as to sustainably meet the travel needs of the growing populations. A more specific vision for the African railway sector in 2040 was published in 2013 [5]. Its main focus is on the regeneration of railway networks, highlighting the need to consider transport networks at regional and continental levels.

This paper suggests a framework for a technical strategy to be used in the development of rail infrastructure in LICs in Sub-Saharan Africa and South Asia (Figure 14). Assuming a future goal of interoperable and efficient rail networks, eight main capabilities were identified considering the current state of rail infrastructure found in the literature. These are infrastructure, signaling, interoperability, planning, standards, costs, data, and safety.

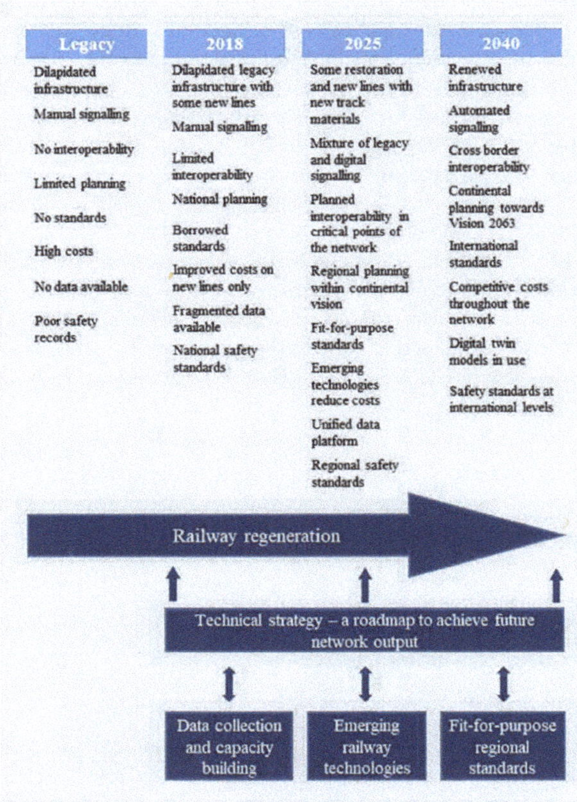


**Figure 14.** A technical strategy for developing rail infrastructure in low-income countries.

Currently, rail infrastructure in these LICs is mostly what is left from the legacy of the original networks of the 19th and 20th century, with little planning and almost no interoperability between countries. Most countries struggle with severely dilapidated infrastructure from years of negligence and poor maintenance. Traffic volumes are affected and so are the costs. Various track gauges have been found in bordering countries, challenging cross-border operations and limiting the competitiveness of railway transport with road-based modes. Some countries even operate different gauges within their own borders, creating operational difficulties that consume precious time and result in greater costs.

More important in the diagram are the intermediate milestones, which provide a systematic bridge between the current state and future goals of railway networks. It is crucial to acknowledge the incremental level of sophistication of the systems where appropriate cost-effective technologies can be implemented. Large scale projects such as railways take several years to be completed, meaning that the evolution of capabilities must be taken as a long-term process. Otherwise, localised initiatives such as those seen in Sino-African projects may become the norm, adopting unsustainable projects and unaffordable solutions for quick wins.

Based on regional documents outlining the vision for interoperable railway networks, we have established a maturity pathway for each of the capabilities. Similarly to processes conducted in Europe, they focus on outcomes and are kept "solution agnostic" in order to maintain the pool of potential technologies available. The framework also acknowledges previous and current projects of modernisation that have adopted international standards, but not in a cohesive manner.

There is a considerable gap between the desired state of rail infrastructure in 2040 and the current situation where countries are beginning to move away from legacy systems. However, there is also an advantage in the greater technological prowess available nowadays, which can help leapfrog the development path taken by high income countries. As an example, digitalisation and recent advances in traction and materials can reduce the costs of renovation, as well as achieve levels of interoperability and standardisation that have taken other countries much longer in the past.

The key outcomes of the strategy are found at the bottom of Figure 14 in the form of applied research programmes. Similarly to the European case, the role of a technical strategy for LICs is to bridge the desired capability development with applied and appropriate research programmes. The three main drivers identified in this research are: (i) data collection and capacity building; (ii) emerging railway technologies; and (iii) fit-for-purpose regional standards.

### 7.2. Data Collection and Capacity Building

Data was found to be a crucial capability, not only for the development of more efficient networks but also for the technical strategy itself. Many indicators of the current state of infrastructure and operational performance in low-income countries are not available. When they are, they are at least 10 years old, which renders them unreliable, especially in the context of accelerated economic growth seen in these countries. Little technical information has been found on the new projects from Sino-African and Sino-Asian agreements.

In this, data collection and capacity building become short-term priorities for the pathways of the technical strategy. However, the process must be systematic in identifying key performance indicators that can assess whether capability developments align with the future vision. Therefore, all countries in the region must agree on a list of indicators to be measured, the metrics to be used, and standard collection processes to ensure reliability in the results.

Building such capacity in data collection and benchmarking, LIC regions will become able to identify the different levels of each country and act accordingly. In addition, indicators can indicate gaps between current performance and each milestone, highlighting priority areas for research and development.

The advantage of using a technical strategy to guide data collection activities is that the type and amount of data is decided not on the current availability of the system but on the future vision of the networks. With this, when capabilities start to evolve, measures of effectiveness of past programs can be assessed, and new iterations of research and development (R&D) initiatives decided accordingly. On the other hand, this paradigm requires a high level of transparency between all countries that take part in the technical strategy. This means that efforts need to be leveled and a baseline amount of information shared.

### *7.3. Emerging Railway Technologies*

Improved databases can then be used as a basis for the technical strategy to achieve the other two medium- and long-term outcomes. The technical strategy should use performance indicators to identify differentiated technologies that can provide technological shortcuts to achieve similar outcomes without having to follow the same path of high-income countries. It is important that the capabilities defined are solution agnostic because one of the key elements of technical strategies is the differentiation approach to rail infrastructure. This perspective suggests that a more cost-effective plan for the development of railway networks is based on differentiation at an infrastructure level as part of a wider multi-purpose strategy, where different types of operations can share some routes, while others will be largely used by a single sector.

Track renovation and renewal seem to be an early milestone to restore traffic efficiency in areas where the network has suffered most. Specific geometries, track density, and gauge must follow careful consideration of demand forecasts to differentiate international corridors from local lines. By doing so, improvements towards interoperability can start with more cost-effective routes and use appropriate technologies for each section. In the meantime, research programs can be started on solutions for moving assets and control and communication.

For instance, emerging digital signaling technologies can offer a more cost-effective transition between the current state of the infrastructure (2018) and the next milestone (2025), leapfrogging track-side equipment. One important advantage of digital technologies is the easier and more cost-effective upgrading when traffic increases in the network.

Similarly, a deeper and broader knowledge of track conditions and geometry can give way to bespoke wagon design and materials to achieve a certain performance at appropriate costs. The role of emerging railway technologies can prove crucial to accelerate and integrate the regional development of rail infrastructure in LICs in Sub-Saharan Africa and South Asia. Nonetheless, the situation encountered in those countries is one of fragmented capabilities and distinct standards. It logically follows that off-the-shelf solutions may not suffice, and that research initiatives should also look at solutions at the technology level.

### *7.4. Fit-for-Purpose Regional Standards*

In the medium to long term, the adoption of emerging technologies in differentiated solutions can lead to agreements on regional standards. Fit-for-purpose standards can be elicited to cater for the specific operational paradigms encountered in low-development countries so that differentiated technologies can be used in a safe and reliable manner. Differentiated technologies comprise disruptive solutions that have the potential to save money (both in terms of initial cost and whole life cycle costs), while also providing greater capability, improved environmental impacts, and a better solution for customers. Adopting fit-for-purpose regional standards has the potential to create leading expertise regionally, and companies that are able to export products and expertise internationally.

In addition, having a coordinated approach to technology development creates markets for new fit-for-purpose technological solutions that consider the differentiated operating parameters in each region. Within this scenario, technical strategies such as the ones practiced in Europe help small and medium enterprises (SMEs) to access those specific regional markets.

There are challenges to be overcome. Regional standards that can foster local economies depend on collective and cohesive action between countries, and facilitated movement of products and people between borders. In addition, technical strategies usually channel funding streams centrally, requiring coordinated decisions from a central commission. These mechanisms are essential interfaces between the building blocks of the technical strategy that ensure its efficacy.

## 8. Conclusions and Discussion

Railways are staging a comeback around the world as governments are changing policies and strategies to address sustainable development goals. Railway transport has the potential to be more sustainable and more affordable than road transport when economies of scale are made possible. They can reduce emissions and energy consumption per traffic unit, both in passenger-km and tonne-km. However, low-income countries have been struggling to achieve this because of the current condition of their rail infrastructure. In most cases, tracks have been poorly maintained for decades, and little investment has been made to improve signaling systems and rolling stock technologies. As a result, most low-income countries analysed in this research produce low traffic densities as a measure of traffic units per kilometer of rail lines. With that, the operational costs per traffic unit becomes higher than in other parts of the world and reduces the competitiveness of rail against road transport.

This paper conducted a review on the current condition of infrastructure and operational performance of railways in low-income countries in Sub-Saharan Africa and South Asia. A total of 23 countries in Sub-Saharan Africa and 4 countries in South Asia were included in the study. Reliable datasets were found to be mostly outdated, and recent available data is fragmented. The review of the current state of the infrastructure confirms the general perception of the regions—assets dating back to colonial periods with little improvement since, and single-track routes with manual signaling systems limit the capacity of lines.

Despite the financial constraints, low-income countries in Sub-Saharan Africa and South Asia are following through and agreeing to various resource-for-infrastructure loans to build new lines to replace severely dilapidated existing networks. These plans seem to go in an opposite direction to a shared vision of continent-wide efficient rail networks that can provide a competitive alternative to road transport. Firstly, most of the lines planned have been driven by international suppliers' preferences and the financing available rather than as part of a coherent and coordinated development plan. As a result, some of the new lines suffer from the same interoperability issues as the original infrastructure from the beginning of the last century. Secondly, there have been concerns over the financial sustainability of these projects. They have been deployed with little consideration of specific requirements of the lines, resulting in technologies used in high-capacity networks being applied to the low traffic volumes currently found in the regions.

Further research is necessary to put such a technical strategy into practice:

1. As a starting point, data is a crucial aspect for a successful development of a technical strategy. There is very little data concerning the current state of railway infrastructure and the operational performance achieved. It is important that the current condition of assets and current performance levels are known so that specific technology roadmaps can be traced.
2. Specific research on emerging technologies that can provide cost-effective and more sustainable solutions for the specific operational context of low-income countries. Off-the-shelf technologies in use in countries with high volumes of traffic can be too costly and unsustainable in the long run. These include new traction solutions (e.g., hydrogen) that can bypass the need for expensive electrified networks, cost-effective materials for tracks and rolling stock, digitalised train-based signaling systems, etc.
3. A final part of the technical strategy relates to standards. In order to achieve efficient, interoperable, continent-wide rail networks, low-income regions need to ensure that the technologies adopted are standardised. If not, that they are at least compatible with cross-border operations. With the research on the use of emerging technologies for context-specific solutions, research should also develop fit-for-purpose regional standards in accordance with future visions. In addition, the creation of standards that are lacking in many LICs can promote improved operational performance, especially concerning safety.

Therefore, this paper suggests that low-income countries in Sub-Saharan Africa and South Asia need to develop a technical strategy to coordinate localised development towards a common goal in order to achieve their visions. Technical strategies used in high-income countries are used as a baseline to define key processes in developing a counterpart for low-development countries so that they can find technological shortcuts rather than just follow the previous development paths of high-income regions. A framework for a technical strategy for LICs in Sub-Saharan Africa and South Asia was presented, in which eight capabilities were identified for priority. In addition, the framework proposes a structured mechanism to transform the three main areas of further research into practice and achieve the milestones elicited.

Future work should initially scan the current condition and performance of rail infrastructure in LICs in Sub-Saharan Africa and South Asia to update and equalise databases. Subsequently, once the local and regional gaps are identified, research programs can be devised in order to address the building blocks suggested. Using emerging technologies and adopting fit-for-purpose standards will facilitate these regions to achieve their capability milestones and achieve their future visions more efficiently.

**Author Contributions:** Data collection, M.B., W.W., and L.A.; workshop organisation, M.B. and C.R.; authorship, M.B., W.W., L.A., and C.R.

**Funding:** This project is part of the High Volume Transport (HVT) applied research programme funded by the UK Department for International Development (DFID)

**Acknowledgments:** This research was funded by UK AID through the UK Department for International Development under the High Volume Transport (HVT) Applied Research Programme, managed by IMC Worldwide. The authors are particularly grateful for the advice of Bruce Thompson the leader of Theme 1, Long Distance Strategic Road and Rail Transport, of the HVT Programme.

**Conflicts of Interest:** The authors declare no conflict of interest.

## Appendix A

| | Total Track (km) | Track Density (km/1000 km$^2$) | Km of Track per Million Inhabitants | Number of Carriages | Number of Wagons | Labour Productivity (1000 Traffic Units per Employee) | Freight Traffic Volume (Million Net Tonne-km) | Passenger Traffic Volume (Million Passenger-km) | Traffic Density (1000 Transport Units per Route-km) |
|---|---|---|---|---|---|---|---|---|---|
| Benin | 579 | 5.0 | 54 | 20 | 326 | 40.4 | 117 | 52 | 291.9 |
| Burkina Faso | 622 | 2.3 | 34 | 29 | 657 | 481.0 | 360 | 58 | 672.0 |
| Cote d'Ivoire | 639 | 2.0 | 28 | - | - | - | 376 | 60 | 682.3 |
| DR Congo | 3641 | 1.6 | 47 | 233 | 3199 | 27.8 | 681 | 16 | 191.4 |
| Eritrea | 117 | 1.2 | 17 | - | - | - | 0 | 0 | 0.0 |
| Ethiopia | 681 | 0.7 | 7 | 27 | 493 | 70.9 | 189 | 111 | 440.5 |
| Ghana | 947 | 4.2 | 35 | 154 | 489 | 84.1 | 298 | 69 | 387.5 |
| Guinea | 1045 | 4.3 | 96 | - | - | - | 5760 | 38 | 5548.3 |
| Kenya | 2065 | 3.6 | 45 | 328 | 7140 | 203.5 | 1858 | 333 | 1061.0 |
| Liberia | 428 | 4.4 | 123 | - | - | - | 0 | 0 | 0.0 |
| Madagascar | 886 | 1.5 | 39 | - | 187 | - | 37 | 0 | 41.8 |
| Malawi | 797 | 8.5 | 47 | 17 | 478 | 131.0 | 91 | 27 | 148.1 |
| Mali | 641 | 0.5 | 44 | 44 | 509 | 339.0 | 240 | 54 | 458.7 |
| Mozambique | 3128 | 4.0 | 112 | 28 | 834 | 143.8 | 913 | 127 | 332.5 |
| Nigeria | 3557 | 3.9 | 21 | 236 | 1299 | 37.2 | 160 | 108 | 75.3 |
| Senegal | 1053 | 5.5 | 73 | 108 | 755 | 479.2 | 754 | 59 | 772.1 |
| Sierra Leone | 84 | 1.2 | 13 | - | - | - | 0 | 0.0 | 0.0 |
| Sudan | 4680 | 2.5 | 116 | 167 | 3949 | 76.7 | 946 | 58 | 214.5 |
| Tanzania | 3691 | 4.2 | 72 | 216 | 3600 | 305.5 | 2389 | 586 | 806.0 |
| Togo | 522 | 9.6 | 84 | - | - | - | 195 | 20 | 411.9 |
| Uganda | 1244 | 6.3 | 36 | 21 | 1289 | 155.9 | 211 | 0 | 169.6 |
| Zambia | 2164 | 2.9 | 140 | 21 | 1150 | 502.0 | 1176 | 248 | 658.0 |
| Zimbabwe | 3227 | 8.3 | 247 | 314 | 10713 | 389.9 | 3170 | 457 | 1124.0 |
| Afghanistan | 277 | 0.4 | 8 | - | - | - | - | - | - |
| Bangladesh | 2874 | 22.1 | 17 | 1476 | - | 343.6 | 677 | 8135 | 3066.1 |
| Myanmar | 6207 | 9.5 | 116 | 1375 | 3384 | 241.5 | 887 | 5307 | 997.9 |
| Pakistan | 7791 | 8.8 | 40 | 1636 | 16085 | 389.6 | 6073 | 22476 | 3664.4 |

## References


1. World Bank. Data Bank: Railways. Available online: https://data.worldbank.org (accessed on 18 April 2019).
2. Borin, E.; Guivarch, C. Transport infrastructure costs in low-carbon pathways. *Transp. Res. Part D* **2017**, *55*, 389–403. [CrossRef]
3. Bullock, R. *Off-Track: SSA Railways*; The World Bank: Washington, DC, USA, 2009.
4. AfDB. *Rail Infrastructure in Africa: Financing Policy Options*; African Development Bank: Abidjan, Cote d'Ivoire, 2015.
5. UIC. *Railway Revitalization in Africa—Destination*; International Union of Railways: Paris, France, 2013.
6. Olievschi, V.N. *Railway Transport: Framework for Improving Railway Sector Performance in SSA*; The World Bank: Washington, DC, USA, 2013.
7. African Union Commission. *Agenda 2063: The Africa We Want*; African Union Commission: Addis Ababa, Ethiopia, 2014.
8. The World Bank. World Bank Country and Lending Groups. Available online: https://datahelpdesk.worldbank.org/knowledgebase/articles/906519-world-bank-country-and-lending-groups (accessed on 2 April 2019).
9. DFID. Where We Work. Available online: https://www.gov.uk/guidance/where-we-work (accessed on 2 April 2019).
10. The World Bank. African Infrastructure: Railways. Available online: https://databank.worldbank.org/data/source/africa-infrastructure:-railways (accessed on 23 March 2019).
11. UIC. RAILISA Statistics. Available online: https://uic-stats.uic.org (accessed on 29 April 2019).
12. Bangladesh Railway. *Rail Master Plan*; Government of Bangladesh: Dhaka, Bangladesh, 2013.
13. Asian Development Bank. *Myanmar Transport Sector Policy Note: Railways*; Asian Development Bank: Mandaluyong, Philippines, 2016.
14. Ministry of Rail Transportation. *Developing a Myanmar's Rail Network that Meet Demands*; Government of Myanmar: Naympyitaw, Myanmar, 2015.
15. Pakistan Railways. *Yearbook 2016*; Ministry of Railways of the Government of Pakistan: Lahore, Pakistan, 2017.
16. Habiyareme, A. Is Sino-African trade exacerbating resource dependence in Africa? *Struct. Chang. Econ. Dyn.* **2016**, *37*, 1–12. [CrossRef]
17. Eurostat. Railway Transport Database. Available online: https://ec.europa.eu/eurostat/data/database (accessed on 15 April 2019).
18. Morlin-Yron, S. All Aboard! The Chinese-Funded Railways Linking East Africa. Available online: https://edition.cnn.com/2016/11/21/africa/chinese-funded-railways-in-africa/index.html (accessed on 15 April 2019).
19. Gopalakrishnan, S. In Search of a Smoother Ride on Narrow Gauge Track. Available online: https://www.railwaygazette.com/news/single-view/view/in-search-of-a-smoother-ride-on-narrow-gauge-track.html (accessed on 15 April 2019).
20. European Union. *Commission Decision 2004/446/EC of 29 April*; European Union: Brussels, Belgium, 2004.
21. Delelegn, M. The hottest projects in Ethiopia and what is being done to open for investments. In Proceedings of the East Africa Rail Conference, Nairobi, Kenya, 21–22 November 2018.
22. UIC. *Railway Statistics: 2015 Synopsis*; International Union of Railways: Paris, France, 2015.
23. JICA. *Pakistan Transport Plan Study in the Islamic Republic of Pakistan*; Japan International Cooperation Agency: Tokyo, Japan, 2008.
24. Ahmed, B.N.; van der Windt, N.; van Rijsbergen, B. *Railway Signalling and Interlocking in Bangladesh*; Erasmus University of Rotterdam: Rotterdam, The Netherlands, 2015.
25. Railway Gazette. Myanmar Signalling Contract Signed. Available online: https://www.railwaygazette.com/news/news/asia/single-view/view/myanmar-signalling-contract-signed.html (accessed on 10 December 2018).
26. Asian Development Bank. $1.5 Billion ADB Rail Project will Develop Southeast Bangladesh. Available online: https://www.adb.org/news/15-billion-adb-rail-project-will-develop-southeast-bangladesh (accessed on 29 April 2019).
27. Railway Gazette. Work to Begin on Chad Rail Network. Available online: https://www.railwaygazette.com/news/infrastructure/single-view/view/work-to-begin-on-chad-rail-network.html (accessed on 29 April 2019).

28. Samseer, M. Mali Signs $11bn Agreements with China for New Rail Projects. Available online: https://www.railway-technology.com/uncategorised/newsmali-signs-agreements-with-china-for-new-rail-projects-4375202/ (accessed on 29 April 2019).
29. Odditah, C. FG, CCECC Sign $11.117 Billion Lagos-Calabar Rail Contract. Available online: https://guardian.ng/news/fg-ccecc-sign-11-117-billion-lagos-calabar-rail-contract (accessed on 29 April 2019).
30. Railway Technology. Abuja-Kaduna Rail Line. Available online: https://www.railway-technology.com/projects/abuja-kaduna-rail-line (accessed on 29 April 2019).
31. Railway Gazette. Padma Bridge Rail Link Contract Awarded. Available online: https://www.railwaygazette.com/news/infrastructure/single-view/view/padma-bridge-rail-link-contract-awarded.html (accessed on 10 March 2019).
32. Bukhari, M. Pakistan Cuts Chinese 'Silk Road' Rail Project by $2 Billion due to Debt Concerns. Available online: https://www.reuters.com/article/us-pakistan-silkroad-railways/pakistan-cuts-chinese-silk-road-rail-project-by-2-billion-due-to-debt-concerns-idUSKCN1MB2V8 (accessed on 29 April 2019).
33. Ndukwe, L. Getting Nigeria's Railways Back on Track with China's Help. Available online: https://www.bbc.co.uk/news/world-africa-42172955 (accessed on 29 April 2019).
34. Mourdoukoutas, P. China is Doing the Same Things to Sri Lanka that Great Britain did to China after the Opium Wars. Available online: https://www.forbes.com/sites/panosmourdoukoutas/2018/06/28/china-is-doing-the-same-things-to-sri-lanka-great-britain-did-to-china-after-the-opium-wars/#298adeba7446 (accessed on 7 April 2019).
35. Kacungira, N. Will Kenya Get Value for Money from Its New Railway? Available online: https://www.bbc.co.uk/news/world-africa-40171095 (accessed on 7 April 2019).
36. Rahmatullah, M. Transport issues and integration in South Asia. In *Promoting Economic Cooperation in South Asia: Beyond SAFTA*; Ahmed, S., Kelegama, S., Ghani, E., Eds.; The World Bank: Washington, DC, USA, 2010; pp. 174–194.
37. Railway Gazette. ERTMS Regional Contract in Zambia. Available online: https://www.railwaygazette.com/news/infrastructure/single-view/view/ertms-regional-contract-in-zambia.html (accessed on 29 April 2019).
38. European Commission. Mobility and Transport. Available online: https://ec.europa.eu/transport/modes/rail_en (accessed on 3 April 2019).
39. EIM. *European Railway Technical Strategy: Technical Vision to Guide the Development of TSIs*; European Rail Infrastructure Managers: Brussels, Belgium, 2008.
40. UIC. *Railway Technical Strategy Europe*; International Union of Railways: Paris, France, 2019.
41. European Commission. Rail Research and Shift2Rail. Available online: https://ec.europa.eu/transport/modes/rail/shift2rail_en (accessed on 25 July 2019).
42. RSSB. *Rail Technical Strategy: Capability Delivery Plan*; Rail Safety and Standards Board: London, UK, 2017.

Review

# Lowering Transport Costs and Prices by Competition: Regulatory and Institutional Reforms in Low Income Countries

**Phill Wheat *, Alexander D. Stead, Yue Huang and Andrew Smith**

Institute for Transport Studies, University of Leeds, Leeds LS2 9JT, UK; a.d.stead@leeds.ac.uk (A.D.S.); y.huang1@leeds.ac.uk (Y.H.); a.s.j.smith@its.leeds.ac.uk (A.S.)
* Correspondence: p.e.wheat@its.leeds.ac.uk

Received: 26 July 2019; Accepted: 7 October 2019; Published: 25 October 2019

**Abstract:** High passenger and freight transport costs are a barrier to economic growth and social mobility, particularly in Low Income Countries (LICs). This paper considers the current state of knowledge regarding the barriers to achieving lower generalised transport costs. It considers both the road and railway modes across passenger and freight transport. These issues include a reform on the regulations for driver hours (preventing the road infrastructure from overloading), structuring rail concessions, increasing competition, and tackling corruption. Such reforms aim to deliver efficiency gains and service quality improvements at lower costs for users. This paper identifies the knowledge gap in previous research and concludes by setting out a research agenda that builds the evidence base for how the best practices from around the world can best be applied to the specific circumstances in Low Income Countries, with a particular focus on Sub-Saharan Africa and South Asia.

**Keywords:** transport costs; passenger and freight; road and railway; Low Income Countries

---

## 1. Introduction

Transport is an important enabler of economic growth and society's development [1–3]. Strategic transport along road and rail corridors presents several challenges from a cost, subsidy, and pricing perspective. This is particularly important in low-income countries (LICs), where limited studies have yet to explore economic and mobility needs. Efficient transport services require appropriate funding to meet user needs. These services must be offered at prices that are affordable to facilitate the free flow of goods and services and to increase the mobility of people. Furthermore, a strong oversight of road traffic transit regulations and rail concessions by public authorities is vital for regulatory compliance, including road safety.

This paper addresses a broad set of mode-specific issues, such as road versus rail, and issues common to both road and rail in LICs, specifically aimed at LICs in Sub-Sharan Africa and South Asia. The aim of the paper is to identify both outstanding issues not fully addressed by existing research and such issues that future research, in turn, can help address. Thus, our paper focuses on identifying opportunities for new research. This paper covers issues related to lowering transport costs in terms of the generalised cost of transport. This includes not only financial costs but also the monetary equivalent values of intangible aspects of transport services. Examples include the travel time for shipments and passenger journeys, the associated reliability of that transit time, and service quality aspects, such as the perceived comfort of passenger travel and wider safety considerations impacting users and the broader society (such as other road users in the case of road safety issues).

The role of road transport and rail transport differs both between LICs and high-income countries (HICs) and across different LICs. In HICs, both modes cover freight and passenger movements. Overall, road transport is dominant, but rail still has a non-trivial mode share in both the passenger and freight market. In LICs, experience varies. In Sub-Sharan Africa, the lack of railway infrastructure means that, for passenger flows outside of the long distance segment (which is small in any case), the road mode is dominant. As such, the rail transport that exists is biased towards freight. In South Asia, however, LICs can have substantive railway transport focused on passenger flows (for example in Bangladesh and Myanmar).

More broadly, road and rail services are different because roads are open to private users whilst rails have much stronger access restrictions. Studies on the regulation, concession design, and institutional skills requirements for planning, designing, and monitoring transport services have been reviewed. Given the access restrictions in rail, these issues are more applicable to the rail sector. The situation for roads is different, where the state usually provides the infrastructure either directly or indirectly via letting tenders, but private users are also free to use the infrastructure. The issues with respect roads are more focused on regulations on driver hours, road safety, and preventing the over loading of vehicles and tackling corruption. Such reforms aim to deliver efficiency gains and service quality improvements at lower costs for users.

Common to both road and rail are the long and costly clearance times at border crossings. Arvis [4] shows that the cost in the bottom three quintile LICs is three times higher, and its paperwork twice as high as that for the top two quintiles LICs. The literature also indicates that transit time and the reliability of transit impact freight choices and that cross-border delays significantly increase freight tariffs. Existing evidence on alternative technologies that can help improve cross border crossing is also considered. In many LICs, it is clear that the opportunities presented by strategic roads and rails are not being fully realised for a variety of reasons.

Specific experiences in LICs are considered (both in Sub-Sharan Africa and South Asian but also outside of this area), with conclusions identifying opportunities to enhance transport service delivery, costs, and pricing. Our literature search also reflected the experiences and lessons learned from HICs and MICs that can be applied to LICs, especially where the evidence from LICs is sparse on solutions to specific issues (an example being the regulation of driver hours in Section 3.1). The paper concludes by setting out a research agenda that builds the evidence base for how best practices from around the world can best be applied to specific circumstances in LICs. This research agenda will contribute to realising economic growth and social benefits by improving transport services and transit flows.

The structure of this paper is as follows. Section 2 provides an overview of our approach to reviewing the literature. Section 3 reviews road specific issues. Section 4 reviews rail specific issues. Section 5 reviews common issues across the road and rail sectors. Finally, Section 6 synthesises the state of knowledge review into a set of research issues.

## 2. Review Methodology

This paper is a synthesis of the literature. As part of this research project, we undertook a time limited review of literature pertaining to key issues in transport services in LICs and the best practices in responding to these issues across high income countries (HICs), medium income countries (MICs), and LICs. We supplemented our review of databases (described below) with literature on transport services by the project team. Social issues, such as the gender barriers associated with access to transport, were not reviewed as part of this project.

The literature on some of these topics is limited, so the criteria for inclusion were widened in several different ways, so that the review would encompass less specific, but still relevant, studies. Rather than look for literature pertaining specifically to individual countries, or to low-income countries, we searched for any literature on a given topic that was likely to be relevant. For example, several headings relate to rail franchising and regulation. In order to capture a wide range of studies, we entered a few relatively general search terms, such as "railway franchising", and selected publications that seemed relevant to one or more of the headings. These were studies involving low-income countries or studies of high-income countries that were felt to be particularly relevant.

During the search, we noticed that there is something of a trade-off (though not in all cases) between the relevance of publications and their quality. For example, there are a large number of relatively high-quality publications on the topic of franchising in high-income countries, but not all of these studies are of direct interest here. On the other hand, there tend to be a much smaller number of publications related to low-income countries, and some of these tend to be of lower quality or are not peer-reviewed (e.g., World Bank reports and articles in trade journals). A degree of judgement was made to determine which articles should be included in the review. Indeed, this selection is one of the conclusions of this review, namely the need for high quality, peer reviewed studies on LICs for many of these issues. A second issue was that much evidence from LIC countries was centred around specific case studies. The generalisability of the findings across LICs was unclear, and this result is, again, a conclusion of this review.

We used four databases to search the literature: The Transport Research Information Database (TRID), Web of Science, EconLit, and Google Scholar. TRID integrates the database owned by TRB's Transportation Research Information Services (TRIS) and the OECD's International Transport Research Documentation (ITRD) database. Web of Science is a general, multidisciplinary, citation indexing service, and should yield a comprehensive list of results for a given search term. EconLit is an economics-specific database which we expect will yield sets of results that are more limited, but also more specific and with less irrelevant material. Google Scholar is a freely accessible search engine that covers scholarly literature, including peer-reviewed papers, books, and technical reports across an array of disciplines.

In Table 1 below, we give examples of our approach to the literature search for two of the areas covered in this review: the value of time and the value of reliability for freight transport, and railway concessioning, franchising, and regulation, for two of the databases used (Web of Science and Google Scholar). In addition to deciding on the search terms used, it was necessary to decide which of the results to include in the review. In the case of Web of Science, which returned fewer but more relevant results, we used all relevant results, while for Google Scholar, which typically turns up a much greater number of results, including some which are less relevant, we decided on appropriate cut-off points based on the numbers of relevant results as we moved down the list of results.

**Table 1.** Example search terms.

| Topic | Web of Science Search | Google Scholar Search |
|---|---|---|
| **Value of Time and Reliability for Freight** | TS = ("value of time" freight transport) All relevant results | "value of time" freight transport Relevant material from first 15 pages of search results, ordered by relevance |
| | TS = ("value of travel time savings" freight transport) All relevant results | "value of travel time savings" freight transport Relevant material from first 15 pages of search results, ordered by relevance |
| | TS = ("value of reliability" freight transport) All relevant results | "value of reliability" freight transport Relevant material from first 15 pages of search results, ordered by relevance |
| **Railway Franchising, Concessioning, and Regulation** | TS = (railway concessions) All relevant results | railway concessions Relevant material from first 10 pages of search results, ordered by relevance |
| | TS = (railway franchising) All relevant results | railway franchising Relevant material from first 10 pages of search results, ordered by relevance |
| | TS = (railway concession failure) All relevant results | railway concession failure Relevant material from first 10 pages of search results, ordered by relevance |
| | TS = (railway infrastructure price regulation) All relevant results | railway infrastructure price regulation Relevant material from first 10 pages of search results, ordered by relevance |

In addition, the following sources were found to contain useful reports:

- World Bank and United Nations sponsored project reports.

- Articles in speciality magazines and media, such as engineering news and toll infrastructure Services.
- Reports by transport consultancies, which are made publicly available.
- Web articles, such as news articles. These are not peer reviewed but may contain insider views, case studies, and recent data.

The following are typical reasons for literature being excluded from the review:

- Only general information pertinent to the subject, e.g., overloading, was found.
- Reference to the data was not provided.
- The full text was not available (only the abstract).
- Information was only found second-hand; we were unable to retrieve the original file.
- A substantive amount of the research that we initially found was done in a European context and there were few implications for elsewhere.

The culmination of our review was to identify a set of issues, which are grouped in Table 2. We will, in turn, discuss in the remainder of this paper the issues that are specific to road transport, rail transport, and issues that are common across the two modes. Within this, approximately 100 papers were found to provide relevant unique contributions to this review.

**Table 2.** Summary of the issues in road and rail freight transport with references.

| | Issues | References | Issues | References | Issues | References | Issues | References |
|---|---|---|---|---|---|---|---|---|
| **Road Specific Issues (Section 3)** | 3.1 Driver hours | [5–10] | 3.2 Safety regulation | [11–17] | 3.3 Corruption | [11,12,18–20] | 3.4 Overloading | [21–30] |
| **Rail Specific Issues (Section 4)** | 4.1. Railway concessioning | [31–61] | 4.2. Economic regulation | [37,39,62–67] | | | | |
| **Road and Rail Common Issues (Section 5)** | 5.1. Public authority and skills capacity | [58,68–74] | 5.2. Cross border road and rail freight—The impact of delays | [4,75–96] | 5.3 Cross border road and rail freight solutions - e-border technology and standardised documentation | [97–103] | | |

While we consider our review to be extensive, it is not a systematic review. Instead, it was a time limited review with the scope and approach discussed above. As such we may have missed some relevant literature. However, we note that our review has been subject to review by academic experts selected by the Department for International Development in the United Kingdom. They have helped supplement the material in this review to be evaluated by the academic author team, thereby providing a secondary check, alongside our primary review methodology, to ensure we have not missed key literature.

## 3. Road Specific Transport Services Issues

In the road sector, the primary issues relate to regulation, governance, and the enforcement of regulations. This includes driver hour regulations, road safety regulations, efforts to combat corruption, and vehicle overloading.

### *3.1. Regulation of Driver Hours*

The link between working hours and rest time for drivers and the risk of collisions is well established. This suggests the need for regulating drivers' hours of service. Enforcing these hours of service and similar regulations rely upon authorities' ability to inspect company and vehicle records. Technology has progressed from log books, to analogue tachographs, to digital tachographs, which have become progressively more difficult to tamper with and easier to process. Increasingly, as in the UK, it is a requirement that new vehicles be fitted with digital tachographs. The use of digital tachographs suggests the possibility that the authorities may eventually be able to collect data and, therefore, detect infractions in real time.

McDonald [5] argued that regulating drivers' hours reduced the risk of collisions and highlighted the need for the enforcement of, and adherence to, these regulations. Baas [6] surveyed truck drivers in New Zealand and found high levels of fatigue and sleepiness and also that many drivers exceeded the allowed driving hours. In a review of the literature, Amundsen [7] emphasised the high proportion of professional drivers reporting incidents of fatigue and falling asleep while driving, as well as the link between the length and quality of sleep and accident risk. The authors attributed the low levels of compliance to the complexity of regulations and suggested that harmonisation across countries may improve compliance and safety. Based on a survey of UK truck drivers, Poulter [8] found that perceived behavioural control had the largest effect on compliance with regulations than other 'soft' factors.

Hall [9], using data from the Fatality Analysis Reporting System of the US National Highways Traffic Safety Administration, estimated that a 3–5% reduction in crashes could be achieved with perfect enforcement of hours of service regulations. Jones [10] compared the working hour regulations in several countries and suggested a 'hybrid' approach to regulation, combining prescriptive measures (e.g., hours of service) and a less prescriptive 'outcomes based' approach. However, overall, there is a lack of evidence related to LICs in this area, which could be studied in future research. The experiences and lessons learned from previous practices in developed economies can be applied to an LIC based on its prevailing conditions.

### 3.2. *Road Safety Regulation*

Road safety is a particular problem in developing countries, and there has been much discussion of the issue with regard to Africa specifically, where accident and fatality rates have increased along with vehicle ownership [11,12].

Driver licensing and vehicle licensing are generally the responsibility of government agencies. The enforcement of traffic laws and regulations is generally the responsibility of traffic police and courts, while some government agencies have a role in the roadside inspection of freight vehicles.

Assum [13] appraised the road safety initiatives in five African countries: Benin, Côte d'Ivoire, Kenya, Tanzania, and Zimbabwe. The author concluded that each country has the legal framework, organisation, technology, and institutions required for road safety and is aware of effective road safety measures, as well as the scope of the problem. However, a number of weaknesses were identified, such as their lack of political concern and funding, their lower value placed on human life, the weak political positions of their road safety boards, their overreliance on education and information approaches, and their corruption.

Khayesi [14] discussed the growth in road traffic injuries in Africa and the need for governments to improve data collection and analysis, share between agencies, and tackle the problem of under-reporting. Chen [15] noted an ongoing upward trend in road traffic injuries, and identified the lack of leading agencies with regulatory powers and public support as one of the key obstacles to reducing injury rates.

Sumaila [16] discussed the limitations of the Federal Road Safety Corps in Nigeria and recommended restructuring the agency and strengthening the strategic ties with other government departments and agencies to increase their effectiveness. The need for driver education in road safety was also emphasised.

Abegaz [17] analysed the effectiveness of new road safety regulations in Ethiopia. These regulations include bans on the use of mobile phones while driving, driving without a seatbelt, and riding a motorcycle without a helmet, as well as strengthening existing laws on drunk driving, speeding, and unsafe loading. Using data from the period of 2002–2011, the authors found considerable reductions in fatalities and accident rates following the first year after the introduction of these new regulations.

### 3.3. *Corruption and the Effectiveness of Road Safety Regulation*

Nantulya [11] discussed road traffic injury rates and the reasons for the differing rates and trends across countries, citing corruption as one of the main factors behind high injury rates in developing countries. Kopits [12] found a non-monotonic relationship between GDP per capita and traffic fatality rates, that traffic fatality rates increase initially with income, but begin to fall after a threshold income level. The authors noted that, since vehicle ownership rates increase monotonically with income, this turning point reflects reductions in fatalities per vehicle. This inverted-U shaped relationship was also found by Anbarci [18]. Importantly, Anbarci [18] also found that public sector corruption is associated with increased traffic fatalities. Various ways that corruption can affect traffic fatality rates were outlined: corrupt license examiners may allow drivers to bypass training and testing for a fee; corrupt vehicle inspectors may sell safety certificates for unsafe vehicles; and corrupt traffic police may accept bribes to overlook infringements and undermine faith in traffic regulations by extorting money from innocent drivers.

Tackling corruption by public officials is, therefore, important for road safety. Measures could include increased monitoring and detection and the punishment of offenders via some combination of dismissal, fines, and imprisonment. An alternative, or complementary, concept suggested by Becker [19] is to pay efficiency wages above the market-clearing rate, with mark-ups reflecting the potential benefits of corruption and the risk of being caught. This increases the costs of engaging in corruption and may be less costly than increased monitoring. Empirical support for an inverse relationship between wages and corruption was provided by Van Rijckeghem [20].

### 3.4. Road Freight Overloading

Vehicle overloading, where vehicle axle loads exceed pavement design limits, is a specific issue on African roads. This occurs because freight operators overload cargo to maximise their payload and operational efficiency and reduce fuel costs (per tonne-km output). However, overloading significantly affects the life of the highway pavement's asset. Rys [21] studied the effect and concluded that if 20% of trucks are overloaded, the life of the road pavement reduces by about 50%.

The reason that vehicle overloading is important for examining barriers to lower transport costs is three fold. Firstly, there is a cost impact on the whole system when overloading the road network because the infrastructure deteriorates faster. This ultimately results in a poor use of resources if sufficient resources do exist. At some point, the operators or society will have to pay for this increase in costs. Secondly, in an LIC environment, funds for maintenance and renewal of infrastructures are scarce. Thus, the impact of overloading tends to be that infrastructure quickly degrades into a state that leads to poor quality, resulting in the slowing down and unreliability of road transport. Thirdly, a lack of compliance with overloading regulations can result in the lack of a level playing field between transport providers, which skews competition and enhances corruption in the sector.

To provide insight into the extent of the problem, we highlight three case studies. The World Bank (2005) found that 30% to 40% of trucks in India are overloaded by 25% to 50%. This report contained data on truck operations in India, such as vehicle operating costs, toll rates, and journey times, that can be used in the economic appraisal of overloading and enforcement measures. A case study by Kolo [22] found that 53% of trucks using a rural road in Nigeria were overloaded. They took a similar approach to Bagui [23] and analysed the percentage and magnitude of overloading separately based on the number of axles and positions (e.g., front and rear) of the axles. Chan [24] showed that the weight limit was exceeded by as much as 100%, and some 50–70% of heavy trucks were overloaded in a central province of China, compared to 0.5–2% overloaded in the USA.

Turning to road safety, an overloaded vehicle is difficult to control because it is operated outside the vehicle's design parameters for steering and braking and thus becomes a serious threat to road safety. Overloading leads to increased congestion caused by damaged roads and accidents that increase logistics costs in the region. According to Toll Infrastructure Services [25], transport operators largely welcome overloading control as they value a level playing field that fosters fair competition between modes and operators.

For these reasons, overloading control is a priority, and static weighbridges are used to control overloading. Toll Infrastructure Services [25] have estimated there to be 260 weighbridges in Africa, with plans for another 40. This quantity of bridges has undoubtedly had a strong beneficial effect.

Pinard [26,27] has documented the steady reduction in overloading due to weighbridge use in South Africa, Namibia, Zimbabwe, and Zambia. In addition to the technical solution of weighbridges, the decriminalisation of related offences has also produced benefits, according to Pinard [27].

However, static weighbridges have their limitations. They are inflexible and control a load at a specified point but do not cover the whole network. An alternative that has been available for implementation for several years is portable weigh-in-motion (WIM) devices. These devices allow a road authority to set up temporary checks. Though less accurate than static installations, these portable devices are recommended by, for example, Bagui [23]. According to Cottineau [28], overload control using WIM was tried in Taiwan, with a 30% tolerance margin due to potential measurement errors. This tolerance margin is higher than is desirable. Current research in France [28,29] is aiming to achieve a 10% tolerance. This, then, is an area where progress can be made, potentially leading to better overloading control.

Alternative technologies are available to monitor and enforce overloading. For example, a framework and field studies have recently been developed for truck overloading and novel control and monitoring methods in China and South Africa. The measures proposed by Hu [30] include weight sensors installed in vehicles, wireless device sending data to a GPS installed in the driver's cab, and a remote control terminal to receive and process the information sent by the GPS.

Overall, vehicle overloading can be combated by monitoring technology, and there has been instances of success, such as in South Africa, Namibia, Zimbabwe, and Zambia. As we will discuss in Section 5.1, there is also the need for the public authority to have sufficient skills and processes to support technology.

## 4. Railway Transport Service Specific Issues

Unlike access to roads, access to railways must be heavily regulated to ensure the safe and equitable scheduling of services. Railways feature a high fixed cost and low marginal cost and are suited to high volume passenger and freight flows. These high flows present a set of challenges to ensuring viable services. These include concessioning passenger services to achieve better financial value, reducing state subsidy requirements, and stimulating passenger demand. This also raises issues surrounding the extent to which it is optimal to separate transport services from the management of the infrastructure (track and signals) and the regulatory needs around such a separation.

### *4.1. Railway Concessioning*

Rail concessioning (also known as rail franchising) has been adopted in Great Britain (GB), Germany, Sweden, and (to a lesser extent) in the Netherlands and Norway. The EU's fourth Railway Package requires competitive tendering of public service railway contracts. A major study was published by the Centre on Regulation in Europe (CERRE) in 2016, drawing on the lessons from GB, Germany, Sweden, and offering suggestions for countries, such as France, who are about to start the tendering process. For the summary report, see Nash [31]. Individual country reports for GB, Sweden, Germany, and France are also available (see Smith [32], Nilsson [33], Crozet [34], and Link [35]).

Typically, rail franchising in Europe has produced cost reductions on the order of 20–30%. However, Britain has instead seen sharp cost increases (see Smith [32]). In Great Britain, an incoming operator takes over an existing company, rather than bringing his or her own rolling stock and staff. Combined with short franchises that are also net cost contracts, this incentivises a focus on revenue growth rather than cost reduction (see Smith [32]). In addition, British rail franchises may be too large, such that they face diseconomies of scale (see Wheat [36]). Costs and subsidies have risen substantially in France, where there has been no tendering Nash [31]. Affuso [38] used UK data to investigate the relationship between contract length and investment and found that short franchise lengths reduce incentives to invest in some assets.

Generally, demand has risen strongly in countries that have implemented tendering. However, part of this growth may be related to the "regionalisation" of responsibility for tendering, because France has also seen growth. Nash [31] lists key considerations to be made in future tendering:

- Who should be responsible for franchising (a regional or national body)?
- The size and length of franchises. Evidence suggests that longer franchises may reduce costs [35] and also that large franchises may increase unit costs [35,36].
- Risk Sharing: Evidence from Germany and Sweden suggests that gross cost contracts perform best, although some incentive mechanisms are needed to ensure service quality.

A recurring problem in Great Britain has been franchise failure. This has been less of a problem in other European countries, partly because of their greater use of gross-cost contracts, and possibly because their franchises are much smaller (see Nash [31]). Smith [39] discussed policy responses to franchise failures and their impacts on the performance of Train Operating Companies (TOCs) in GB. These have included placing TOCs on management contracts (these are essentially cost-plus regimes with extra subsidies and, hence, weak incentive properties), short term franchise renegotiations, and placing TOCs temporarily under public ownership. The authors found that the efficiency of TOCs on management contracts deteriorated markedly, but this was not the case with franchise renegotiation. Outside Europe, franchise failure has also been a problem (for example, in Melbourne during the early years of their reforms and in Latin America [40]).

Cruz [41] analysed renegotiations of road and rail concessions in Portugal, where renegotiations have been common, particularly in the very early years of the contract. The authors noted that renegotiation itself is a very costly process, and should be avoided where possible. The authors indicated that the main reasons for contract renegotiation were the inadequate monitoring and enforcement of contract terms and over-optimistic bids.

Outside of Europe, Carbajo [42] pointed to lower public subsidies and increased passenger numbers following rail concessioning in Argentina. They also defined early problems, such as difficulties in enforcing investment plans and imposing fines. Crampes [43] discussed issues in concession contract design, with reference to the Argentine experience, and recommended a form of menu regulation. Under menu regulation, companies choose from a menu of contracts offering different trade-offs between the rate of return, risk, and profit sharing. In choosing their optimal bundle, a company is forced to reveal information about itself to the regulator, which helps to mitigate information asymmetry (the fact that the company knows more about its activities than the regulator). This could potentially lessen the likelihood of regular and costly renegotiations or failures.

Campos [44] found that performance indicators generally improved rail privatisation in Mexico and that concessions may have contributed to reversing railway declines in developing countries. The author concluded that concessions can be a viable approach in developing countries, although care should be taken in concession design.

The analysis of around 1000 concessions in different industries (including the road, rail, energy, water, and telecommunications industries) in Latin America and the Caribbean concluded that concessions can work well [45]. However, the author found that the implementation of the contracts was flawed in many cases, and opportunistic renegotiations should have been dissuaded. Further, he found that most implementation mistakes could have been avoided by improved design and a greater intention to develop incentives.

Guasch [46] emphasised the importance of regulators in reducing the probability of renegotiations. Likewise, Guasch [47] found that regulators and arbitration processes both reduced the probability of government-led renegotiation, while price capping, longer contracts, elections, and growth shocks all increased the probability of renegotiation. Using data on road and rail concessions in Latin America, Estache [48] found that multi-criteria auctions (considering social goals, such as employment) increased the risk of renegotiation but that this problem could be mitigated by high-quality regulations and anti-corruption policies.

Rodríguez [49] concluded that rail (and road) concessions are more likely to fail if large-scale capital investments are involved, if they are in developing countries, if they involve urban transport, or if they demand uncertainty. Stern [50] emphasised the complementarity of contracting and regulation and the role of an independent regulatory agency in building trust between stakeholders, resolving conflicts, and enabling ordered renegotiations.

Tam [51] discussed three successful and three failed transport infrastructure BOT (build–operate–transfer) concessions in Asia. These were road concessions but are included here, as the findings can transfer to the design of rail concessions, and this study features specific comparisons of LICs and HICs. The authors identified reasonable rates of return, sound mechanisms for the adjustment of terms, strong and technically competent franchisees, and equitable and experienced governments and legal systems as the factors needed for success. Frequent changes in government, corruption, and political interference are identified as factors explaining failure. All three of Tam's (1999) success stories are from an HIC, such as the Hong Kong Special Administrative Region of the People's Republic of China, while all three failures are taken from Thailand, an LIC.

In Kazakhstan, rail reforms from 2005 separated train operation and infrastructure maintenance and encouraged competition in the passenger and freight markets via franchising and open access arrangements, respectively. Sharipov [52] finds that usage and safety both improved, while subsidy levels increased.

Gwilliam [53] stated that 14 of the 30 African countries with state-owned railways use concession arrangements, that four had begun the concession progress, and that only one was operating under a management contract. The rest were subject to political influence. Roy [54] noted that infrastructure investments have tended to remain the preserve of governments and aid agencies.

The introduction of railway concessions in Africa has been accompanied by significant investments in assets, including rolling stocks [54,55], and improvements in productivity [53–55] and allocative efficiency [55]. Budin [56] noted an impressive increase in freight traffic and a substantial improvement in quality of service during the first year of the Abidjan-Ouagadougou railway concession. Roy [54] concluded that there was no indication of an increase in freight rates or that travel for the poor has been made more expensive. Pozzo [57], reviewing railway concessions in Sub-Saharan Africa, found that, due to competition from road transport and clauses against excessive pricing in concession contracts, there was no clear evidence for an abuse of market power. Ndonye [58] evaluated the impact of different public–private partnership (PPP) strategies on the performance of concessions, focusing on the Rift Valley Railways concession in Kenya. A survey of managers indicated a perception that the most important strategic considerations for the performance of concessions are having a strong private consortium as a concession-holder, a sound financial strategy, having a sustainable risk allocation strategy, and investment in technology.

Railway concessions in Africa face significant risks, however. Roy [54] noted the disruptions caused by the civil war in the Ivory Coast and the collapse of a bridge in Malawi. Jones [59] cited political unrest in Kenya in 2008 as affecting the Rift Valley Railways concession, along with political pressure from governments. Pozzo [57] notes the generally weak financial performance of Sub-Saharan African railway concessions and the private sector reluctance to invest in transport in the region or to take on any risk.

Related to the design of rail transport services concessions is its ability to integrate into the management of rail infrastructures (the extent of vertical integration). In Europe, EU legislation has required some form of vertical separation. Countries such as Britain and Sweden have followed full legal separation, whereas others, such as Germany, have used a holding company structure. Mizutani [60] and van de Velde [61] reviewed the literature and provided both quantitative and qualitative assessments of different approaches to separation. A key finding is that, in congested networks, capacity constraints may magnify the misalignment of incentives, such that vertical separation increases costs. This suggests that a holding company or even full vertical integration are preferable. Partly as a result of these studies, European legislation permits both approaches.

Fair infrastructure access is important, especially when the incumbent operator also manages the infrastructure. This necessitates the independent oversight of track, station, and terminal access charges and capacity allocation in order to ensure that new operators can freely access the infrastructure while adequately compensating the infrastructure manager. This is crucial to enabling the emergence of new operators while ensuring the financial stability of the infrastructure manager.

### *4.2. Economic Regulation of Infrastructure and Services*

Transport services tend to operate as (i) directly provided state services, (ii) competition in-the-market (open access), or (iii) competition-for-the market (tendered services). These delivery mechanisms need to be investigated as to the level of protection they provide to final (passenger and freight) users from exploitation.

The 'natural monopoly' property of infrastructure management, i.e., the falling average up to a minimum efficient scale, implies that:

- For a given geographical area, a single firm can achieve lower unit costs than multiple firms.
- Average costs exceed marginal costs, so revenues under marginal usage cost pricing costs are not sufficient for full cost recovery.
- Average cost pricing, however, would make socially desirable services unviable.
- The key trade-off is between the financial viability of the infrastructure manager and the pricing of access for train operators.

A key study in the EU funded project, NETIRAIL-INFRA, examined how cross-industry innovation and cost reduction can be stimulated through a range of regulatory, contractual, and structural approaches (see Nash [37]). Benedetto [62] surveyed regulatory reforms across European railways based on a review of the literature on ideal regulatory characteristics. They found that while European rail regulators generally exhibit many of the features of ideal regulation, there is room for a more proactive approach to shaping the role of track access charges and in regulating the efficiency and quality of infrastructure managers.

Britain has had the most extensive powers of any economic regulator in Europe for many years. This regulator has conducted or commissioned many studies using econometric techniques to determine cost efficiency savings [39,63,64]. These have highlighted inefficiencies in Network Rail (the railway infrastructure manager) and have been used for setting cost efficiency targets. This form of benchmarking and price cap regulation (which refers to the setting of limits on the prices or revenue that a firm will receive) is considered to be successful for utilities but is arguably less successful for rail [65,66]. Smith [67] showed how the increased powers of economic regulators have contributed to efficiency savings and how these powers interact with other reforms.

## 5. Road and Rail Transport Service Issues

In addition to the issues specific to road and rail freight transport described in the previous two sections, there are issues common to both modes, which are described in this section.

### *5.1. Public Authority Capacity and Skills*

There are many instances where transport infrastructure in developing countries has been of low quality, and investment is biased towards 'prestige projects' that may not reflect the most efficient approach for investment [68]. Lee [69] noticed a tendency to focus on investment rather than institutional and regulatory setups and to identify poor financial and environmental sustainability but no comprehensive approaches to poverty, safety, and environmental issues.

Sohail [70] undertook case studies for transport regulation in Sri Lanka, Pakistan, and Tanzania, and highlighted the need for appropriate regulatory frameworks and more effective enforcement.

Mexico provides a useful case study for how public agency skills develop over time. The Agency for Rail Transport (ARTF) was recently established in response to complaints by freight shippers about tariffs and service levels. The International Transport Forum [71] compared the capabilities and functions of the ARTF to those of its US and Canadian counterparts, discussed issues in Mexican rail regulation, and outlined priorities for the new regulator. This report highlighted the considerable time needed for ARTF to develop an information base and capabilities comparable to more established regulators and report the recommended discussions with Mexican concession holders and the US and Canadian regulators to determine the financial and operating data that ARTF will need.

Until ARTF's capabilities are fully developed, the ITF recommends a low cost, less prescriptive approach to regulation based on negotiation and arbitration and the continuity of existing safety regulations in the short run. The development of safety regulations based on performance criteria, rather than prescribing certain behaviour and design requirements, was identified as a long-run priority. ITF identified the need for a "critical mass of human skills and management resources spread across law, economics, accounting, and engineering" comparable to that of US and Canadian regulators.

In discussing the regulation of railway concessions in Africa, Gwilliam [53] identified inadequate provisions for regulators as an impediment to effective regulation. The author stated that "many railway concessions in Africa lack formal regulatory structures with real power and are thus susceptible to abuse". In addition, the author listed factors impacting concession performance, including conflicts and delays with the governments over compensation, concession fees, time frames, staffing, administratively imposed salary increases, restrictions on access to container facilities, and unfunded public service requirements.

In the road transport sector, overloading of trucks provides an example of the challenges facing public authorities in LICs. Different institutions are responsible for overload control in different countries, and there are differences in institutional capacities and responsibilities for overload control. Strathman [72] highlights that authorities can influence compliance through changes in both enforcement intensity and the severity of penalties and finds that the marginal effects of these elements in deterring overloading are similar.

The role out of enforcement measures also creates challenges for public authority resources. In particular, alternative routes without weighbridges and the shortfalls in weighbridges necessitate adequate policing of the road network. Overload control strategies should focus on regions and not one stretch of road. Further, Chen [73] suggests that mobile weighbridges could be used to identify severely overloaded trucks before being escorted to weighbridges. However, this "first best" solution would create massive demands for a skilled work force.

Therefore, checking truck weight compliance places high demands on resources, and its efficiency is generally low. In 2014, 20% of the two million vehicles weighed in South Africa exceeded the weight limit. Only 2.6% of the 20% were prosecuted, according to Council for Scientific and Industrial Research reported in Engineering News [74]. Officers accepted bribes and operators paid their way through the weighbridges. Corruption is a major hindrance to compliance. Low fines are imposed, and the practice is seen as 'unimportant' by the Department of Justice [74].

Overall, there is no one solution to combat overloading. However, any solution requires both an investment in technology (weighbridges, etc.) and a strong institutional structure with clear and accountable roles of public authorities and the individuals working within them. This translates to other public authority enforcement activities (such as the road safety enforcement discussed in Section 3.2).

### *5.2. Cross Border Road and Rail Freight—The Impact of Delays*

The time to clear goods through customs rises sharply if the goods are physically inspected. Physical inspection is far more prevalent in lower-performing countries, as defined by Arvis [4], because of the quality of their logistics services, and the same shipment may be subject to repeated inspections by multiple agencies. Clearance times for the bottom three quintile (20%) performing countries are three times as much, and the paperwork twice as much, as those for the top two quintiles [4].

The key business and trade implication of such barriers is that export lead times are nearly four times as long for LICs than for HICs. Thus, the benefits of reducing the time and time uncertainty at borders include increased international trade and investment, new commercial opportunities, access to new materials, reduced input costs, and improved competitiveness.

Using data on landlocked Sub-Saharan African countries, Christ [75] conducted an analysis on the relative size of time costs versus money costs. Christ concluded that time costs (barriers to trade due to time taken to clear customs, for example) generally exceed the monetary costs of trucking. The study suggested that this helps explain why such countries have a comparative advantage in primary commodities, such as metals and agricultural products, while hindering the export of time-sensitive goods. Thus, border and transit delays reduce both the overall volume of trade and the impact of the mix of products traded, which may hinder development.

The extent to which benefits can be realised depends largely on the response of freight operators, and thus the extent to which such improvements reduce freight costs. It is, therefore, important to know the value of time (VOT) and value of reliability (VOR) from the perspective of freight operators.

While there is a large quantity of literature on the estimation of road users' VOT/VOR, much less research has been undertaken regarding the VOT/VOR for freight, particularly in the context of low-income countries. An overview of the methods by de Jong [76] distinguished between the factor cost methods that sum the monetary costs of the factors used (e.g., labour and fuel costs) and modelling approaches, including the revealed preference (RP) (based on an analysis of the impact of various characteristics on modal choices) or the stated preference (SP), which uses survey methods.

Summaries of VOT and VOR findings for rail and road, and overviews of the methods used, are provided by Feo-Valero [77], de Jong [76], and Zamparini [78–80]. A recurring narrative from this literature is that the reliability of journey times is very important to freight shippers [81] and is possibly more important than the journey time itself [82–85].

Shinghal [86] analysed a survey of firms sending freight via the Delhi–Bombay (now known as Mumbai) corridor and found that the reliability of transit times was particularly important for exporters. Larranaga [87] found that reliability and cost were the most important attributes in explaining freight mode choice in Rio Grande do Sul, Brazil. Norojono [88] concluded that safety and reliability were more important factors than travel time in explaining the choice between rail and road freight by Indonesian freight shippers. Similar findings are presented by Arunotayanun [89], who used the same data. These findings complement the HIC studies in suggesting that reliability is paramount.

However, Zamparini [90] found, from an SP survey of the logisitics managers of 24 Tanzanian firms, that reliability was one of the least important service attributes and that transit time and the monetary value of damage and losses is more important. The authors pointed out that this may reflect the challenges facing freight transport in Tanzania, where transport infrastructure is of low-quality compared to other Sub-Saharan African countries and LICs, leading to slow freight and high rates of damage and losses.

Ogwude [91] found that VOT and VOR vary significantly between the shippers of consumer goods and shippers of capital goods in Nigeria. Cundill [92] derived RP VOT estimates for road, rail, and pipeline freight traffic in Kenya and found that modal choice was sensitive to the cost and speed of delivery. In Malaysia, Thomas [93] found that journey time savings due to road improvements resulted in relatively small increases in vehicle utilisation because of other constraints.

Border issues with freight transport are of particular concern in landlocked developing countries (LLDCs), for whom the direct maritime transportation of international freight is not an option. This has been a major impediment to development for LLDCs, which face bottlenecks not only at the seaport but also at each transit country [94].

Zhang [95] analysed VOT in the context of rail freight from Tianjin Port in China to Ulan Bator, the capital of Mongolia. Compared to an average passenger journey time of 32 hours, the average rail freight journey time was 13.1 days: 4.3 days waiting at Tianjin Port, 3.8 days travelling from Tianjin to the Mongolian border, 1.2 days waiting at the border, and 3.8 days from the border to Ulan Bator. The authors described the route as being of great importance to landlocked Mongolia, as it is the shortest and most commonly used route to Mongolia's nearest seaport. Border issues derive from the need for the transhipment of freight from road to rail and from the standard gauge railways in China to the broad-gauge railways in Mongolia.

Banomyong [96] compared the various routes available to garment exporters from Laos, another landlocked country. The authors found that their preferred route (i.e., on the grounds of the estimated cost and time) via Malaysia was not the same as that of the most frequently used route via Thailand.

### 5.3. *Cross Border Road and Rail Freight Solutions: E-Border Technology and Standardised Documentation*

Given the discussion in Section 5.2, measures to improve the flow of goods across borders and through corridors are essential to enable low transport prices and stimulate trade. Customs procedures are becoming increasingly similar worldwide. However, customs is not the only agency in border management. Streamlining border procedures to facilitate trade and simplifying documentation for imports and exports have long been high on the trade development agenda, prompting initiatives to bring border agencies together and create a single window for trade. We have collated findings from research done for countries in Asia and Africa with regard to their barriers and recommended measures.

In Asia, a roadmap for the integration of logistics services was endorsed in 2007 by ASEAN (Association of Southeast Asian Nations) members. The promotion of IT both within and among the parties in the logistics chain are considered conductive to the processing of documents and other trade-related matters. The ultimate goal is to establish an ASEAN single window, adopt a 24 × 7 customs operation, and offer provisions in the World Trade Organization (WTO) agreement for customs valuation. Inefficient customs procedures and inspections are considered by De Souza [97] to be the biggest barriers to logistics services in ASEAN. Gupta [98] found that unwieldy customs procedures and inspections, a lack of coordination, and arbitrary rulings are barriers to freer cross-border trade within ASEAN. Tongzon [99] found that the implementation of measures to improve the competitiveness of ASEAN logistics industries has been limited and characterised by a significant perception gap between logistics firms and governments.

In Africa, Adaba [100] examined the effect of an e-government initiative to modernise customs procedures and facilitate trade in Ghana. Hinson [101] found that a 1% increase in internet use by a country's population is associated with a 2.2% increase in exports through a reduction in the entry and search costs associated with exporting from Africa. Hoffman [102] suggested that cross-border delays could be reduced by more than 80% through the interchange of information to facilitate the dynamic scheduling of customs processing capacities and operational changes.

Arvis [103] highlighted issues associated with border crossing in landlocked developing countries (LLDCs). These issues apply to both the Asian and African continent. In response to these challenges, several technologies/systems have been developed in LLDCs to facilitate trade via e-border technology [103]. These include:

- Automated Customs documentation. For instance, UN Conference on Trade and Development (UNCTAD) have developed transit add-ons to Automated System for Customs Data (ASYCUDA).
- The regional integration of national customs, such as the NCTS (New Computerised Transit System) in Europe, allowing the seamless exchange of information, thereby managing both documentation and guarantees.
- The carnet (passport for goods) barcode in e-TIR helps to validate the carnet at border crossings, using a central database accessible by each participating country. RFID technology may also facilitate the tracing of cargo on a corridor and speed up the controls at entry and exit checkpoints.
- Several groups of countries (such as those in Africa) have been experimenting with transit data interfaces facilitating the passing of advance information from country to country, yielding immediate benefits in terms of transit facilitation.

- The GPS tracking of merchandise. Benefits include helping to build confidence between customs and transit operators and leading to the disuse of inefficient control solutions, such as convoys.

Specifically, on rail transit, the new Annex 9 of the UNECE (United Nations Economic Commission for Europe) International Convention on the Harmonization of Frontier Controls of Goods (the UNECE Harmonization Convention), is expected to streamline railway border crossing procedures through:

- Introducing minimum requirements for border (interchange) stations;
- Co-operating at these stations;
- Moving the controls from borders to the departure or destination stations;
- Reducing the time required for controls;
- Eliminating paper documents;
- Using the CIM (Uniform Rules concerning the Contract of International Carriage of Goods by Rail) / SMGS (Agreement on International Freight Traffic by Rail) common consignment note as a customs document.

It can be seen from the above review that the measures recommended by UNECE will apply to, and improve on, current practices in Asia and Africa.

## 6. Discussion and Conclusions: Opportunities for Future Research

This paper has reviewed the broad state of knowledge for road and rail transport services in LICs. The aim of the paper was to identify outstanding issues not fully addressed by existing research that future research can help address. We have focused on the issues most prevalent to LICs in Sub-Sharan Africa and South Asia. This review has covered a wide range of topics, from franchising and concessioning, to various aspects of safety and economic regulation, to waiting times for border transit. This discussion makes it clear that LICs face considerable challenges in these areas and that resolving these issues will be important to their development. Based on this review, we have identified the following three areas for further academic research. This is not an exhaustive list but a prioritisation of what specific issues academic research can address.

Firstly, we note that this literature review has attempted to draw on the best practices that might be applicable to LICs. In doing so, evidence from LICs (both specifically in Sub-Sharan Africa and South Asia, as well as elsewhere in the world), MICs, and HICs has been considered. However, it is clear, overall, that there is a limited set of systematic evidence specifically for LICs. There are case study examples for LICs (with many examples focusing on the failures rather than successes of their applications, e.g., in concessioning). However, a general finding from the reviewed literature is that in the area of Transport Services for LICs, there is limited research that synthesizes experience across a range of countries. This type of systematic analysis is essential to yield best practice and generalizable findings for LICs. Thus, a key recommendation is the need for a greater quantity of, and more systematic, research in the area of Transport Services in LICs.

Secondly, transporting goods in Africa often takes three times longer and is three to five times more expensive than in Europe, Asia, and Latin America. This literature review has indicated that transit time and reliability affect freight transport choices. This is the case for both LICs and HICs, with evidence of border waiting times for cross-border freight having negative economic impacts on LICs.

Cross-border freight is a major concern for LICs and is stifling trade and economic development, particularly in landlocked countries. Research is needed to establish how new technologies and border process alignments can speed up and improve reliability at border crossings. The economic and social benefits arising from improved transit times would be assessed and, where possible, quantified. This research would need to provide a holistic appraisal of the benefits to improving freight transit across national borders.

Technological solutions at borders, such as developing an ICT infrastructure and smart border control measures, such as non-intrusive inspection and fully paperless systems (e-border) systems, will need to be considered. A further key research objective is to enhance freight movements between ports and hinterlands, particularly for Small and Medium Enterprises (SMEs) and others requiring effective freight consolidation services, by formulating a coherent and integrated intermodal strategy. This strategy will involve coordination between transport modes and strategies for the development of 'dry ports' (inland freight depots), free industrial zones, inland customs clearance arrangements, and bonded facilities. Consideration around how to harmonise regulations (such as restrictions on container transport by road) is needed, as are requirements for the employment of competent, trained personnel, which will reduce skill shortages and unofficial payments.

Thirdly, the failure of railway concessions will lead to escalating costs, low-quality service provisions, and, in extreme cases, the termination of services, with substantial societal and economic disruptions. An essential requirement for sustainable rail services in LICs is the design of concessions and establishing optimal regulatory and contracting models and solutions, as well as improving the mode share of passenger transport support and many UN Strategic Development Goals. Railways are inherently safer and, therefore, lead to fewer fatalities and accidents (SDG Target 3.6) and cause less environmental damage (SDG Target 3.9). Providing a viable passenger service as an alternative to road transport will increase the reliability and reduce the environmental impacts of the transport system (SDG Target 9.1).

Research should determine when it is appropriate to separate the management of railway services and infrastructure in LICs and to determine the most appropriate separation models. This would include considering the impact of different institutional and organisational structures for railways and, specifically, vertical separation versus vertical integration and intermediate structures. It should also consider which forms of competition (concessions, open access, or franchising) can best improve the efficiency and societal welfare of railways.

**Author Contributions:** Conceptualisation and methodology, P.W.; writing—original draft preparation, A.D.S. and Y.H.; writing—review and editing, A.S.

**Funding:** This research was funded by the UK Department for International Development, grant number HVT/005.

**Acknowledgments:** This research was funded via UK AID through the UK Department for International Development under the High Volume Transport (HVT) Applied Research Programme, managed by IMC Worldwide. The authors are particularly grateful for the advice of, Bruce Thompson the leader of Theme 1, Long Distance Strategic Road and Rail Transport, of the HVT Programme.

**Conflicts of Interest:** The authors declare no conflict of interest. The funders had no role in the design of the study; in the collection, analyses, or interpretation of data; in the writing of the manuscript, or in the decision to publish the results.

## References

1. Banister, D.; Berechman, Y. Transport investment and the promotion of economic growth. *J. Trans. Geogr.* **2001**, *9*, 209–218. [CrossRef]
2. Venables, A.; Laird, J.J.; Overman, H.G. *Transport Investment and Economic Performance: Implications for Project Appraisal*; Department for Transport: London, UK, 2014.
3. Farhadi, M. Transport infrastructure and long-run economic growth in OECD countries. *Trans. Res. Part A Policy Pract.* **2015**, *74*, 73–90. [CrossRef]
4. Arvis, J.-F.; Ojala, L.; Wiederer, C.; Shepherd, B.; Raj, A.; Dairabayeva, K.; Kiiski, T. *Connecting to Compete 2018*; World Bank: Washington, DC, USA, 2018.
5. McDonald, N. Safety and regulations restricting the hours of driving of goods vehicle drivers. *Ergonomics* **1981**, *24*, 475–485. [CrossRef]
6. Baas, P.H.; Charlton, S.G.; Bastin, G.T. Survey of New Zealand truck driver fatigue and fitness for duty. *Trans. Res. Part F Traffic Psychol. Behav.* **2000**, *3*, 185–193. [CrossRef]
7. Amundsen, A.H.; Sagberg, F. *Hours of Service Regulations and the Risk of Fatigue- and Sleep-Related Road Accidents*; Transportøkonomisk Institutt (Institute of Transport Economics) Report 659/2003; Institute of Transport Economics: Oslo, Norway, 2003.

8. Poulter, D.R.; Poulter, D.R.; Chapman, P.; Bibby, P.A.; Clarke, D.D.; Crundall, D. An application of the theory of planned behaviour to truck driving behaviour and compliance with regulations. *Accid. Anal. Prev.* **2008**, *40*, 2058–2064. [CrossRef]
9. Hall, R.W.; Mukherjee, A. Bounds on effectiveness of driver hours-of-service regulations for freight motor carriers. *Trans. Res. Part E Logist. Trans. Rev.* **2008**, *44*, 298–312. [CrossRef]
10. Jones, C.B.; Dorrian, J.; Rajaratnam, S.M.W.; Dawson, D. Working hours regulations and fatigue in transportation: A comparative analysis. *Saf. Sci.* **2005**, *43*, 225–252. [CrossRef]
11. Nantulya, V.M.; Reich, M.R. The neglected epidemic: Road traffic injuries in developing countries. *BMJ* **2002**, *324*, 1139. [CrossRef]
12. Kopits, E.; Cropper, M. Traffic fatalities and economic growth. *Accid. Anal. Prev.* **2005**, *37*, 169–178. [CrossRef]
13. Assum, T. *Road Safety in Africa: Appraisal of Road Safety Initiatives in Five African Countries*; Sub-Saharn Africa Transport Policy Program Working Paper No. 33; The World Bank: Washington, DC, USA, 1998.
14. Khayesi, M.; Peden, M. road safety in Africa. *BMJ* **2005**, *331*, 710. [CrossRef]
15. Chen, G. Road traffic safety in African countries–status, trend, contributing factors, countermeasures and challenges. *Int. J. Inj. Control Saf. Promot.* **2010**, *17*, 247–255. [CrossRef] [PubMed]
16. Sumaila, A.F. Road crashes trends and safety management in Nigeria. *J. Geogr. Reg. Plan.* **2013**, *6*, 43–62.
17. Abegaz, T.; Berhane, Y.; Worku, A.; Assrat, A. Effectiveness of an improved road safety policy in Ethiopia: An interrupted time series study. *BMC Public Health* **2014**, *14*, 539. [CrossRef]
18. Anbarci, N.; Escaleras, M.; Register, C. Traffic fatalities and public sector corruption. *Kyklos* **2006**, *59*, 327–344. [CrossRef]
19. Becker, G.S.; Stigler, G.J. Law enforcement, malfeasance, and compensation of enforcers. *J. Leg. Stud.* **1974**, *3*, 1–18. [CrossRef]
20. Van Rijckeghem, C.; Weder, B. Bureaucratic corruption and the rate of temptation: Do wages in the civil service affect corruption, and by how much? *J. Dev. Econ.* **2001**, *65*, 307–331. [CrossRef]
21. Rys, D.; Judycki, J.; Jaskula, P. Analysis of effect of overloaded vehicles on fatigue life of flexible pavements based on weigh in motion (WIM) data. *Int. J. Pavement Eng.* **2016**, *17*, 716–726. [CrossRef]
22. Kolo, S.S.; Jimoh, Y.A.; Ndoke, P.N.; James, O.; Amada, A.Y.; Bala, A. Analysis of Axle Loadings on a Rural Road in Nigeria. *Publ. Int. J. Eng. Sci. Invent.* **2014**, *3*, 8–14.
23. Bagui, S.; Das, A.; Bapanapalli, C. Controlling Vehicle Overloading in BOT Projects. *Procedia Soc. Behav. Sci.* **2013**, *104*, 962–971. [CrossRef]
24. Chan, Y.C.M. Truck Overloading Study in Developing Countries and Strategies to Minimize its Impact. Master's Thesis, Queensland University of Technology, Brisbane, Australia, 2008.
25. Toll Infrastructure Services. *Overloading Control Strategies in Africa*; Toll Infrastructure Services: Pretoria, South Africa, 2017.
26. Pinard, M.I. *Guidelines on Vehicle Overload Control in Eastern and Southern Africa*; Sub-Saharan Africa Transport Policy Programme; The World Bank: Washington, DC, USA, 2010.
27. Pinard, M.I. *Emerging Good Practice in Overload Control in Eastern and Southern Africa*; Discussion Paper 12, Sub-Saharan Africa Transport Policy Program; The World Bank: Washington, DC, USA, 2011.
28. Cottineau, L.M.; Hornych, P.; Jacob, B.; Schmidt, F.; Dronneau, R.; Klein, E. Automated overload enforcement by WIM. In Proceedings of the 25th World Road Congress, Seoul, Korea, 2–6 November 2015.
29. Jacob, B.; Cottineau, L.-M. Weigh-in-motion for Direct Enforcement of Overloaded Commercial Vehicles. *Trans. Res. Procedia* **2016**, *14*, 1413–1422. [CrossRef]
30. Hu, T. A framework of truck overload intelligent monitoring system. In Proceedings of the 2011 Fourth International Symposium on Computational Intelligence and Design, Hangzhou, China, 28–30 October 2011.
31. Nash, C.A.; Crozet, Y.; Link, H.; Nilsson, J.; Smith, A. *Liberalisation of Passenger Rail Services: Project Report*; Centre on Regulation in Europe: Brussels, Belgium, 2016.
32. Smith, A.S.J. *Liberalisation of Passenger Rail Services—Case Study—Britain*; Centre on Regulation in Europe: Brussels, Belgium, 2016.
33. Nilsson, J.-E. *Liberalisation of Passenger Rail Services: Case Study—Sweden*; Centre on Regulation in Europe: Brussels, Belgium, 2016.
34. Crozet, Y. *Liberalisation of Passenger Rail Services: Case Study—France*; Centre on Regulation in Europe: Brussels, Belgium, 2016.

35. Link, H. *Liberalisation of Passenger Rail Services: Case Study—Germany*; Centre on Regulation in Europe: Brussels, Belgium, 2016.
36. Wheat, P.; Smith, A.S.J. Do the usual results of railway returns to scale and density hold in the case of heterogeneity in outputs? a hedonic cost function approach. *J. Trans. Econ. Policy* **2015**, *49*, 35–57.
37. Nash, C.; Matthews, B.; Smith, A.S.J. The impact of rail industry restructuring on incentives to adopt innovation: A case study of Britain. *Proc. Inst. Mech. Eng. Part F J. Rail Rapid Transit* **2018**. [CrossRef]
38. Affuso, L.; Newbery, D.M.G. *Investment, Reprocurement and Franchise Contract Length in the British Railway Industry*; C.E.P.R. Discussion Papers; University of Cambridge: Cambridge, UK, 2000.
39. Smith, A.S.J.; Wheat, P. Evaluating alternative policy responses to franchise failure: Evidence from the passenger rail sector in Britain. *J. Trans. Econ. Policy (JTEP)* **2012**, *46*, 25–49.
40. Smith, A.S.J.; Wheat, P.E.; Nash, C.A. Exploring the effects of passenger rail franchising in Britain: Evidence from the first two rounds of franchising (1997–2008). *Res. Trans. Econ.* **2010**, *29*, 72–79. [CrossRef]
41. Cruz Carlos, O.; Marques Rui, C. Exogenous determinants for renegotiating public infrastructure concessions: Evidence from Portugal. *J. Constr. Eng. Manag.* **2013**, *139*, 1082–1090. [CrossRef]
42. Carbajo, J.; Estache, A. *Railway concessions: Heading down the right track in Argentina*; World Bank Other Operational Studies; The World Bank: Washington, DC, USA, 1996.
43. Crampes, C.; Estache, A. Regulatory trade-offs in the design of concession contracts. *Util. Policy* **1998**, *7*, 1–13. [CrossRef]
44. Campos, J. Lessons from railway reforms in Brazil and Mexico. *Trans. Policy* **2001**, *8*, 85–95. [CrossRef]
45. Guasch, J.L. *Granting and Renegotiating Infrastructure Concessions*; The World Bank: Washington, DC, USA, 2004.
46. Guasch, J.L.; Straub, S. Renegotiation of infrastructure concessions: An overview. *Ann. Public Coop. Econ.* **2006**, *77*, 479–493. [CrossRef]
47. Guasch, J.L.; Laffont, J.J.; Straub, S. Concessions of infrastructure in Latin America: Government-led renegotiation. *J. Appl. Econ.* **2007**, *22*, 1267–1294. [CrossRef]
48. Estache, A.; Guasch, J.L.; Iimi, A.; Trujillo, L. Multidimensionality and renegotiation: Evidence from transport-sector public-private-partnership transactions in Latin America. *Rev. Ind. Organ.* **2009**, *35*, 41. [CrossRef]
49. Rodríguez, D. Expanding the urban transportation infrastructure through concession agreements: Lessons from Latin America. *Trans Res Rec. J. Trans. Res. Board* **1999**, *1659*, 3–10. [CrossRef]
50. Stern, J. The relationship between regulation and contracts in infrastructure industries: Regulation as ordered renegotiation. *Regul. Gov.* **2012**, *6*, 474–498. [CrossRef]
51. Tam, C.M. Build-operate-transfer model for infrastructure developments in Asia: Reasons for successes and failures. *Int. J. Proj. Manag.* **1999**, *17*, 377–382. [CrossRef]
52. Sharipov, T. Results so far and prospects of Kazakhstan passenger rail franchising. *Res. Trans. Econ.* **2010**, *29*, 19–35. [CrossRef]
53. Gwilliam, K.M. *Africa's Transport Infrastructure*; The World Bank: Washington, DC, USA, 2011.
54. Roy, M.-A.; Kieran, P. Rail privatisation rolls across the continent. *Int. Railw. J.* **2006**, *46*, 38–41.
55. Bullock, R. *Results of Railway Privatization in Africa*; Transport Papers, No. TP-8; The World Bank: Washington, DC, USA, 2005.
56. Budin, K.-J.; Mitchell, B. *The Abidjan-Ouagadougou Railway Concession*; Infrastructure Notes, No. RW-14; The World Bank: Washington, DC, USA, 1997.
57. Pozzo di Borgo, P.; Labeau, A.; Eskinazi, R.; Dehornoy, J.; Parte, A.; Ameziane, M. *Sub-Saharan Africa: Review of Selected Railway Concessions*; Africa Transport Sector, Report No. 36491; The World Bank: Washington, DC, USA, 2006.
58. Ndonye, H.N.; Anyika, E.; Gongera, G. Evaluation of public private partnership strategies on concession performance: Case of rift valley railways concession, Kenya. *Eur. J. Bus Manag.* **2014**, *6*, 145–166.
59. Jones, J.; Murphy, J. The risks are rising for Africa's private operators. *Int. Railw. J.* **2008**, *48*, 104–108.
60. Mizutani, F.; Smith, A.; Nash, C.; Uranishi, S. Comparing the costs of vertical separation, integration, and intermediate organisational structures in European and East Asian railways. *J. Trans. Econ. Policy (JTEP)* **2015**, *49*, 496–515.
61. Van de Velde, D.; Nash, C.; Smith, A.; Mizutani, F.; Uranishi, S.; Lijesen, M.; Zschoche, F. *EVESRail—Economic Effects of Vertical Separation in the Railway Sector*; inno-V: Amsterdam, The Netherlands, 2012.

62. Benedetto, V.; Smith, A.S.J.; Nash, C.A. Evaluating the roles and powers of rail regulatory bodies in Europe: A survey-based approach. *Trans. Policy* **2017**, *59*, 116–123. [CrossRef]
63. Smith, A.S.J. The application of stochastic frontier panel models in economic regulation: Experience from the European rail sector. *Trans. Res. Part E Logist. Trans. Rev.* **2012**, *48*, 503–515. [CrossRef]
64. Office of Rail and Road. *PR18 Econometric Top-Down Benchmarking of Network Rail: A Report*; Office of Rail and Road: London, UK, 2018.
65. Smith, A.S.J. *Use of Benchmarking by British Regulators: Presentation to a Seminar on Yardstick Competition in Transport*; The Italian Regulatory Authority: Turin, Italy, 2017.
66. Smith, A.S.J. Benchmarking Methodologies: With Examples from Rail. In *Presentation to a Seminar: The art of ART: Measuring Efficiency for Growth, Development, and Better Quality Transport*; The Italian Regulatory Authority: Turin, Italy, 2018.
67. Smith, A.S.J.; Benedetto, V.; Nash, C. The impact of economic regulation on the efficiency of european railway systems. *J. Trans. Econ. Policy (JTEP)* **2018**, *52*, 113–136.
68. Button, K. Transport regulation and the environment in low income countries. *Util. Policy* **1992**, *2*, 248–257. [CrossRef]
69. Lee, J.; Hine, J.L. *Preparing a National Transport Strategy: Suggestions for Government Agencies in Developing Countries*; Transport Papers No. TP-19; The World Bank: Washington, DC, USA, 2008.
70. Sohail, M.; Maunder, D.A.C.; Cavill, S. Effective regulation for sustainable public transport in developing countries. *Trans. Policy* **2006**, *13*, 177–190. [CrossRef]
71. International Transport Forum. *Establishing Mexico's Regulatory Agency for Rail Transport: Peer Review of Regulatory Capacity. Case-Specific Policy Analysis 2016*; Organisation for Economic Co-Operation and Development: Paris, France, 2016.
72. Strathman, J.G. *Economics of Overloading and the Effect of Weight Enforcement*; Center for Urban Studies College of Urban and Public Affairs Portland State University: Portland, OR, USA, 2001.
73. Chen, Q.; Guo, L. Development of truck overloading automatic detection system based on TPMS. In Proceedings of the 2010 International Conference on Measuring Technology and Mechatronics Automation (ICMTMA 2010), Piscataway, NJ, USA, 13–14 March 2010; pp. 19–21.
74. Engineering News. Available online: http://www.engineeringnews.co.za/print-version/road-transport-2015-07-24 (accessed on 7 August 2015).
75. Christ, N.; Ferrantino, M.J. Land transport for export: The effects of cost, time, and uncertainty in sub-saharan Africa. *World Dev.* **2011**, *39*, 1749–1759. [CrossRef]
76. de Jong, G. *Value of Freight Travel-Time Savings. Handbook of Transport Modelling*; Hensher, D.A., Button, K.J., Eds.; Elsevier: Amsterdam, The Netherlands, 2007; pp. 649–663.
77. Feo-Valero, M.; Garcia-Menendez, L.; Garrido-Hidalgo, R. Valuing freight transport time using transport demand modelling: A bibliographical review. *Trans. Rev.* **2011**, *31*, 625–651. [CrossRef]
78. Zamparini, L.; Reggiani, A. Freight transport and the value of travel time savings: A meta-analysis of empirical studies. *Trans. Rev.* **2007**, *27*, 621–636. [CrossRef]
79. Zamparini, L.; Reggiani, A. The value of reliability and its relevance in transport networks: A systems perspective. In *Integrated Transport: From Policy to Practice*; Givoni, M., Banister, D., Eds.; Routledge: Abingdon, UK, 2010; pp. 97–116.
80. Zamparini, L.; Reggiani, A. The value of travel time savings in passenger a freight transport: An overview. In *Policy Analysis of Transport Networks*; Van Geenhuizen, M., Rietveld, P., Eds.; Routledge: London, UK, 2016; pp. 161–178.
81. Beuthe, M.; Bouffioux, C. Analysing qualitative attributes of freight transport from stated orders of preference experiment. *J. Trans. Econ. Policy (JTEP)* **2008**, *42*, 105–128.
82. Kurri, J.; Sirkiä, A.; Mikola, J. Value of time in freight transport in Finland. *Trans. Res. Rec J. Trans. Res. Board* **2000**, *1725*, 26–30. [CrossRef]
83. Bergantino, A.S.; Bolis, S. An analysis of maritime ro-ro freight transport service attributes through adaptive stated preference: An application to a sample of freight forwarders. *Eur. Trans.* **2004**, *25–26*, 33–51.
84. Fowkes, A.S.; Firmin, P.E.; Tweddle, G.; Whiteing, A.E. How highly does the freight transport industry value journey time reliability—And for what reasons? *Int. J. Logistics Res. Appl.* **2004**, *7*, 33–43. [CrossRef]
85. Hillestad, R.; Van Roo, B.D.; Yoho, K.D. *Fast-Forward: Key Issues in Modernizing the US Freight-Transportation System for Future Economic Growth*; RAND Corporation: Santa Monica, CA, USA, 2009.

86. Shinghal, N.; Fowkes, T. Freight mode choice and adaptive stated preferences. *Trans. Res. Part E Logistics Trans. Rev.* **2002**, *38*, 367–378. [CrossRef]
87. Larranaga, A.M.; Arellana, J.; Senna, L.A. Encouraging intermodality: A stated preference analysis of freight mode choice in Rio Grande do Sul. *Trans. Res. Part A Policy Prac.* **2017**, *102*, 202–211. [CrossRef]
88. Norojono, O.; Young, W. A stated preference freight mode choice model. *Trans. Plan. Technol.* **2003**, *26*, 1. [CrossRef]
89. Arunotayanun, K.; Polak, J.W. Taste heterogeneity and market segmentation in freight shippers' mode choice behaviour. *Trans. Res. Part E Logistics Trans. Rev.* **2011**, *47*, 138–148. [CrossRef]
90. Zamparini, L.; Layaa, J.; Dullaert, W. Monetary values of freight transport quality attributes: A sample of Tanzanian firms. *J. Trans. Geogr.* **2011**, *19*, 1222–1234. [CrossRef]
91. Ogwude, I.C. The value of transit time in industrial freight transportation in Nigeria. *Int. J. Trans. Econ./Rivista internazionale di economia dei trasporti* **1993**, *20*, 325–337.
92. Cundill, M.A. *Road-Rail Competition for Freight Traffic in Kenya*; Transport and Road Research Laboratory Research Report, No. RR 41; Transport Research Laboratory: Wokingham, UK, 1986.
93. Thomas, S. *The Value of Time Savings in West Malaysia: Commercial Vehicles*; Transport and Road Research Laboratory Research Report, No. HS-036 881; Transport Research Laboratory: Berkshire, UK, 1983.
94. Faye, M.L.; McArthur, J.W.; Sachs, J.D.; Snow, T. The challenges facing landlocked developing countries. *J. Hum. Dev.* **2004**, *5*, 31–68. [CrossRef]
95. Zhang, J.; Kawasaki, T.; Hanaoka, S. Freight transport value of time estimation for border crossing route: Case study of Tianjin Port-Ulan Bator route. In Proceedings of the 3rd International Conference on Transport and Logisitics, Malé, Republic of Maldives, 15–17 December 2011.
96. Banomyong, R.; Beresford, A.K.C. Multimodal transport: The case of Laotian garment exporters. *Int. J. Phys. Distrib. Logistics Manag.* **2001**, *31*, 663–685. [CrossRef]
97. De Souza, R.; Goh, M.; Gupta, S.; Lei, L. *An Investigation into the Measures Affecting the Integration of ASEAN's Priority Sectors (Phase 2): The Case of Logistics*; ASEAN: Jakarta, Indonesia, 2018.
98. Gupta, S.; Desouza, M.G.R.; Garg, M. Assessing trade friendliness of logistics services in ASEAN. *Asia Pac. J. Mark. Logistics* **2011**, *23*, 773–792. [CrossRef]
99. Tongzon, J.; Cheong, I. The challenges of developing a competitive logistics industry in ASEAN countries. *Int. J. Logistics Res. Appl.* **2014**, *17*, 323–338. [CrossRef]
100. Adaba, G.B.; Rusu, L. E-trade facilitation in Ghana: A capability approach perspective. *Electron. J. Inf. Syst. Dev. Ctries.* **2014**, *63*, 1–13. [CrossRef]
101. Hinson, R.E.; Adjasi, C.K.D. The Internet and export: Some cross-country evidence from selected African countries. *J. Int. Commer.* **2009**, *8*, 309–324. [CrossRef]
102. Hoffman, A.; Bhero, E. A simulation approach to the optimized design of cross-border operations. In Proceedings of the 2015 IEEE 18th International Conference on Intelligent Transportation Systems, Las Palmas, Spain, 15–18 September 2015.
103. Arvis, J.-F.; Carruthers, R.C.; Smith, G.; Willoughby, C. *Connecting Landlocked Developing Countries to Markets*; The World Bank: Washington, DC, USA, 2011.

*Article*

# Transport Corridors for Wider Socio–Economic Development

**A S M Abdul Quium**

Former member of staff, United Nations Economic and Social Commission for Asia and the Pacific Secretariat, Bangkok 10200, Thailand; abdul.quium@gmail.com

Received: 31 July 2019; Accepted: 19 September 2019; Published: 25 September 2019

**Abstract:** There can be two broad objectives of transport corridor development: to improve efficiency in the transport and logistics processes in the corridor, and to generate economic development in the corridor region, capitalizing on improved connectivity and transport networks. This paper focuses on the second objective of corridor development. A transport corridor can become a tool for spatially balanced and more sustainable economic development and human well-being in the corridor region. Considering the promise of this approach, this paper undertakes a critical review of transport infrastructure development studies undertaken in Sub-Saharan and South Asian countries to find evidence of infrastructure development impacts. Evidence gathered from the review suggests that transport infrastructure development can have significant positive impacts on economic growth, income, poverty, employment, equity, and inclusion. However, there can be important trade-offs between economy and welfare and environmental quality, and the distribution of impacts can be uneven. The paper also considers how some of the transport corridor development issues are addressed and complementary interventions that may be required, and, finally, discusses lessons learned from the review and their policy implications which can be useful for future corridor designs, and provides suggestions of research studies to fill the current knowledge gaps.

**Keywords:** transport corridor; transport infrastructure; transport development impacts; wider economic benefits; corridor management; South Asia; Sub-Saharan Africa

---

## 1. Introduction

High Volume Transport (HVT) corridors and networks comprise arterial and main roads and railways to form the national transport backbone, which connects the smaller feeder road and rail links. HVT corridors and networks carry the major share of passenger and freight traffic and play a key role in the economic and social development of a country. For example, one transport corridor region in Bangladesh (Dhaka–Chittagong corridor) generates almost 50% of Gross Domestic Product (GDP) and handles about 85% of international maritime trade [1] (pp. 345–357).

The impacts of investments in transport corridors and networks can be substantial. At the macro level, the transport network is linked to national output, employment, and income. At the micro level, well-connected transport networks link producers and consumers and affect people's wellbeing, including poverty alleviation through higher production and wages, new jobs, and lower input and higher output prices. Transport networks also facilitate access to education, employment, health, and other social and cultural facilities.

A well-managed HVT corridor can help to improve the quality of transport and logistics services in the corridor, reduce the cost of transport, increase efficiency in the overall supply and distribution chain, and reduce the carbon footprint of freight transport. In addition, an HVT corridor can bring together infrastructure facilities, policies, institutions, and investments to spur wider socio–economic development.

Transport development is linked to many Sustainable Development Goals (SDGs) and can be used as an intervention tool to achieve some of them (see United Nations (UN) [2] for SDGs). For example, HVT corridors and networks can be a tool to support achieving SDG 9 (sustainable infrastructure: Targets 9.1 and 9.a), and SDG 10 (reduced inequalities: Targets 10.2, 10.3, 10.7, and 10.8).

There are many theoretical works linking the contribution of transport infrastructure to growth and welfare. Aschauer [3] presented an econometric model of the relationship between production and public investments. Krugman [4] examined the forces that concentrate and disperse economic activity across economic space (geography). The new economic geography theories advanced by Krugman [4], Duranton and Puga [5], and other researchers have sought to explain the agglomeration of economic activity and its implications. Venables and Gasiorek [6], Department for Transport [7], Graham and Gibbons [8], Venables [9], and Vickerman [10] have considered the welfare implications of transport improvements in case of market failures and imperfect competition. They argue that conventional cost–benefit analysis (CBA) must be extended to include wider economic benefits (WEBs). WEBs are considered to be additional benefits, as they derive from sources of market failure and imperfect competition.

Discussions on some of the important theoretical works can be found in References [11,12] and elsewhere. It is generally agreed by academics, development practitioners, and policymakers that transport infrastructure is vital for economic development and human wellbeing. Transport development works through several mechanisms and various intermediate outcomes, such as decrease in trade costs, and increase in trade, investment, land value, and assets to induce WEBs in the long run.

An AECOM [13] study explains these effects of transport development in three stages. The first-order benefits (or direct benefits) are related to improvements in travel time, reductions in transport costs, increased reliability, and the introduction of new services, which result in cost reductions to transport users and transport service providers.

The second-order benefits arise as transport improvements enable access to larger markets and to wider facilities and services. The availability of better services and reduced cost of transport influence the location/relocation of firms, volume of trade, and higher agricultural production. An important characteristic of this stage is the visible signs of development: more traffic, shops, buildings, factories, etc.

The third-order effects arise in the long run from structural changes due to economy-wide dynamic processes activated by the second order effects. These changes ultimately lead to positive impacts on people's wellbeing, including poverty alleviation through an increase in employment, higher agricultural production and wages, and higher output prices. These changes, however, may also lead to externalities, such as negative impacts on the environment and some social aspects.

A recent study by Asian Development Bank (ADB) et al. [14] also describes a similar transformational process which leads to a set of final outcomes or wider economic benefits and costs. The chain of effects of transport development can be summarized in Figure 1.

It is important to mention here that the magnitude of such changes in different stages depends on responses of firms and households to effects in the first round, pre-existing conditions, institutional environment, and policy interventions of the government [15,16].

Considering the promise of transport corridors to be an important policy intervention tool, this paper aims to review the current knowledge on wider economic benefits and costs of transport corridor development. The specific objectives include:

- to review and analyze the literature on wider economic benefits and costs of major transport network/corridor development;
- to examine how some of the corridor development issues are addressed;
- to identify the current knowledge gaps related to transport corridor development and potential research studies that can be considered to fill these gaps.

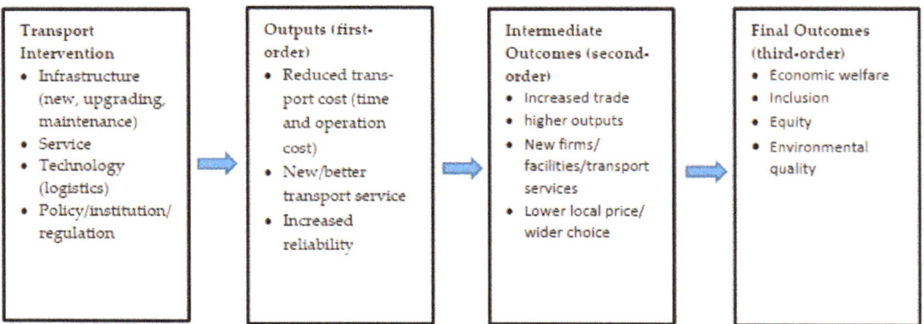


**Figure 1.** Effects of transport improvement/intervention. Source: Adapted from Asian Development Bank et al. [14] and based on preceding discussion.

To fulfil these objectives, the paper focusses on evidence from low- and middle-income countries from Sub-Saharan Africa (SSA) and South Asia, but relevant evidence from other developing countries are also considered.

Section 2 provides methodology and data sources. In Section 3, the paper provides a review of the development impacts of transport corridors and networks gathered from both peer-reviewed and grey literature.

The discussion in Section 4 is based on the main findings and lessons learned from the literature review. This section also considers issues related to corridor development and management and their policy implications, and future research to fill some current knowledge gaps.

## 2. Methodology

The findings and discussion in this paper are based on desktop literature review.

For this study, both peer-reviewed and grey literature have been used. Using search engines (Google Scholar, Google), an extensive search of online databases, namely, CrossRef, JStor, Research Gate, and Science Direct was undertaken, from which the cited papers were accessed.

The grey literature came from development banks and their institutes (World Bank, Asian Development Bank, African Development Bank, Asian Development Bank Institute, etc.), UN organisations, and research organisations, such as the Economic Research Institute for ASEAN and East Asia (ERIA), and Asian Institute of Transport Development (AITD). The grey literature was accessed mostly from the respective organization's official websites or e-Library.

## 3. Findings from Literature Review

### *3.1. Transport Infrastructure Investment and Economy*

This sub-section first reviews the literature for Sub-Saharan and South Asian countries, followed by a review of some important studies in developed countries. This has been done to find differences of the research in developed and developing countries.

Many studies have found positive effects of transport infrastructure investment on the economy. Although the measured growth effects vary among studies, the positive correlation between transport infrastructure investment and economic development is commonly accepted [17].

A number of studies have provided evidence of the substantial economic and social benefits of transport projects [14,18]. Estimates from multiple studies have suggested cumulative gains ranging from less than 1% to more than 10% of GDP [19–22]. The variation in estimates and other findings is thought to be due mainly to differences in characteristics of individual studies and the difference in context, and related to different phenomena (economic sectors, transport infrastructure, etc.) being measured [23].

A review of 78 studies, including 18 studies in Africa (from six countries, namely, Cameroon, Democratic Republic of Congo (DRC), Egypt, Ethiopia, Nigeria, Tanzania, and Uganda, and six regional studies) provides evidence of the substantial economic and social benefits of transport projects [16]. A meta-analysis of the results has revealed statistically significant benefits of transport networks for real and nominal income, consumption, gender, education, and job creation.

A major study, covering 16 countries in North Africa and 24 countries in SSA, found that infrastructure (transport and other infrastructure) accounts for more than half of Africa's recent economic growth and has the potential to contribute even more in the future [24].

Simulation-based estimates have demonstrated that transport infrastructure and services make substantial contributions to GDP in the long run [21,25–29]. Hahm and Raihan [25] used a Computable General Equilibrium (CGE) model to estimate the total economic gains from the six Belt and Road Initiative (BRI) economic corridors. The estimated gains in terms of percentage of GDP for three BRI countries in South Asia were 7% in Bangladesh, 4% in Pakistan, about 3% in India, and about 6% in Myanmar. For landlocked countries (Kazakhstan, Kyrgyzstan, Lao People's Democratic Republic, Mongolia, Tajikistan, Turkmenistan, and Uzbekistan), estimates vary between 5 and 10% of GDP. However, some other low-income countries, such as Cambodia, may also experience growth in GDP of more than 10%.

Zhai [26] investigated the real income gains from investment in expanded regional transport infrastructure in developing countries in Asia. The analysis has suggested that developing countries in Asia as a whole would gain about USD967.7 billion in 2020, which is equivalent to 6.0% of their baseline income in 2008.

Gilbert and Banik [21] estimated the impacts of South Asia Sub-regional Economic Cooperation (SASEC) transport infrastructure to connect Bangladesh, Bhutan, northeastern India, and Nepal. The simulation study indicated that the cumulative impacts as a percentage of GDP would vary between 0.7% and 14.8% (India 0.7%, Pakistan 2.7%, Bangladesh 4.1%, Sri Lanka 4.6%, and Nepal 14.8%). In absolute value, the gain was the highest for India (USD4330.3 million) followed by Pakistan (USD2600.8 million), Bangladesh (USD2295.1 million), Sri Lanka (USD933.8 million), and Nepal (USD2057.1 million).

Similar potential gains from increased sub-regional infrastructure investments were found in other studies. Stone and Strutt [27] valued social impacts at USD8.1 billion, resulting from moderate improvements to road infrastructure and trade facilitation in the Greater Mekong Sub-region (GMS). A major transport infrastructure project can have similar impacts on regional GDP. Anas et al. [20] used an Input–Output (IO) model to examine the impacts of a major toll way in West Java, Indonesia. Freight costs were found to decrease by 17%, and GDP in Bandung District to increase by 1.2%.

Hlotywa and Ndaguba [30] assessed the impact of road infrastructure investment on economic development in South Africa. The results of the study demonstrate that road transport investment variables account for approximately 86.7% of the variation in economic development in South Africa. Another study in South Africa examined the macroeconomic impact of labor-based road construction in undeveloped rural areas. Input–output analysis of 11 projects found that the GDP multiplier ranges between 1.34 and 1.53 (average 1.45) [31].

Using a cross-country analysis of 60 countries for the period 1980–2010, Ng et al. [32] compared improvement in mobility (by investment in access-controlled highways, such as motorways and freeways), accessibility (investment in local roads providing more direct links to destination), and economic growth. Their study found that improvement in mobility facilitated export-led growth in countries of medium and high-level development. On the other hand, it was more important to facilitate local business and trade activities investment in local roads to improve accessibility in countries of low-level development.

Pradhan and Bagchi [33] examined the effect of road and rail infrastructure on economic growth in India over the period 1970–2010. Using a Vector Error Correction Model (VECM), they found bidirectional causality between road transport investment and economic growth, and unidirectional causality between rail transport investment and economic growth. They also found bidirectional causality between road transport and gross capital formation and unidirectional causality from rail transport to gross capital formation. Considering the findings, they suggested that expansion of transport infrastructure, along with gross capital formation, can lead to substantial economic growth.

Hong et al. [34] examined the linkage between transport infrastructure and regional economic growth in China using a panel data model of 31 provinces from 1998 to 2007. Transport infrastructure was shown to play an important role in economic growth, and road infrastructure contributed more to economic growth in locations with poor roads. A retrospective analysis showed that the uneven distribution of transport infrastructure is an important reason for disparities between regions.

The major highway networks can significantly contribute to the manufacturing output growth of a country. Ghani et al. [18] found that the Golden Quadrilateral (GQ) highway network upgrades led to a substantial increase in Indian manufacturing activity. They found that the largest growth came from the non-nodal districts within 0–10 km of the network, which accounted for 34% of the initial levels. Their estimates credit about 43% of the observed increase to GQ upgrades and the rest to other factors. The N–5 highway in Vietnam had similar high output growth effects [14].

Evidence from the reviewed empirical studies suggests that transport infrastructure investment generally has strong positive growth effects. However, a common limitation of the reviewed studies is that they are restricted in scope, do not analyze how the growth effects of transport investment may have affected other aspects of economy, and, except for one study, do not show how growth effects may vary in different situations and for different groups. Some of the above studies indicate that other policies may have also contributed to the impacts of transport infrastructure. However, the review did not find any study that analyzed how the other policies may have influenced the impacts of transport infrastructure.

Numerous studies were undertaken in developed countries to find empirical evidence that transport infrastructure investment boosts overall economic growth. The Standing Advisory Committee on Trunk Road Appraisal (SACTRA) in the United Kingdom (U.K.) examined the linkages between transport and economy [35]. They observed that empirical evidence of the scale and significance of such linkages was weak and disputed, and concluded " ... that the theoretical effects can exist in reality, but that none of them is guaranteed" [35] (p. 8). A review of evidence on the transport sector's contribution to GDP in the developed countries can be found in Reference [36]. The earlier studies indicate that for a 10% increase in capital stock, GDP increases by about 2%. However, later studies based on more complex modelling indicate a much lower value.

More recently, government agencies and researchers in the developed countries have undertaken studies focusing on estimation of WEBs for inclusion in CBA of transport projects. The Department for Transport (DfT) in the U.K. has been a pioneer in including WEBs in the CBA of transport projects [37,38]. The Department recommends three categories of WEBs, namely, agglomeration, increased or decreased output in imperfectly competitive markets, and labor market impacts, and provides detailed guidelines to estimate them.

Following the DfT, other countries such as Australia and New Zealand have considered the inclusion of WEBs in CBA of transport projects. However, the number and category of WEBs and their treatment in evaluation vary. In New Zealand, five categories of WEBs (agglomeration, imperfect competition, increased competition, labor supply and job relocation) are considered in the evaluation framework, whereas, Australia's framework requires the exclusion of WEBs in CBA. WEBs are reported separately and considered for sensitivity tests, but not as an element of evaluation [39].

An inter-agency Workstream report [40] provides a comparison of WEB estimates for nine urban rail projects in Australia, New Zealand, and the U.K. The estimates for each category of WEB, in proportion to direct benefits in CBAs, varied considerably by project. The highest estimate of WEBs was for the Crossrail project in the U.K. (56%) and the lowest for the Brisbane cross-river project in Australia (19%). A similar comparison of seven rail and road projects can be found in Douglas and O'Keeffe [41]. These estimates show that for many projects, estimated WEBs can be substantial.

A study by Wangsness et. al. [42] investigated how 22 developed countries (Nordic countries, 15 countries in the EU, the USA, Canada, Switzerland, Australia, New Zealand, and Japan) treat WEBs in transport project appraisals. They found that 15 countries recognized at least one category of WEB, and only 10 of them recommended methods for their assessment. The recommended methods generally differed across the countries. Agglomeration benefits, which were most widely recognized, and by far the largest component of WEBs in many studies, were recognized by 14 out of 22 countries. However, only five of them recommended their monetization for CBAs, while the others were in favor of their inclusion in other types of analysis.

The criticisms of WEBs are largely due to questionable assumptions used in estimation methods and application in specific evaluations. Researchers [41,43,44] observed that there were several factors which might lead to biased estimates. These valid criticisms suggest that although the theoretical links are strong, empirical evidence is weak, and estimated values for WEBs are disputed.

The above discussion shows that WEBs are not universally accepted, and researchers and government agencies have divergent views on them. It also suggests that the quantification of WEBs is a new field of research in which more work is required.

Generally, the estimated contribution of transport infrastructure to GDP in developing countries is much higher than in developed countries. This is plausible, as transport networks in developing countries are not well-connected and generally not in good condition. Further infrastructure investment to improve network connectivity or infrastructure quality can produce substantial economic growth. Some of the reviewed studies identified different types of impacts, but did not explicitly consider their underlying causes or separately estimate different categories of WEBs. The review also did not find any study in developing countries that considered estimation methods for WEBs or their consideration in an evaluation framework.

### 3.2. Trade and Investments

An empirical study by Buys et al. [12] found that isolation from regional and international markets contributed significantly to poverty in Sub-Saharan Africa. Poor transport infrastructure and border restrictions were identified as significant deterrents to trade expansion. In another study, Hummels [45] found that the volume of trade between countries that share a land border varied widely by regions. It was only between 1 and 5% of trade for Africa, the Middle East, and Asia, compared to 10–20% for Latin America, and 25–35% for Europe and North America.

The land-locked developing countries (LLDCs) in Asia and SSA suffer more than coastal countries because of their higher transport costs. In a study of transport costs and trade, Limao and Venables [46] found that poor infrastructure accounts for 60% of transport costs for landlocked countries, compared to 40% for coastal countries. The estimated elasticity of trade flows with respect to trade cost was around −3. The analysis of trade flows showed that the relatively low level of trade in African countries was due mainly to poor infrastructure.

The decrease in transport cost has been the major driver of the increase in international trade [45]. Freund and Rocha [47] investigated the effects of different components of trade time on export in SSA. They found that a one-day increase in overland travel time implies a nearly 7% decline in exports. Buys et al. [12] investigated the economics of upgrading a primary road network that connects the major urban areas in Sub-Saharan Africa. Simulation results of the study indicated that upgrading has the potential to increase overland trade between countries in Sub-Saharan Africa by about USD250 billion over 15 years. The upgrading programme would require an estimated USD20 billion for initial upgrading and USD1 billion annually for maintenance.

Another empirical study by Bosker and Garretsen [48] also suggested that market access matters for economic development. The study showed that improvement of market access for Sub-Saharan Africa by investing in intra-continent infrastructure or through increased continental integration can have substantial positive effects on future economic development.

Simulation results of a study by Hahm and Raihan [25] showed that all countries under the BRI initiative would experience a rise in exports of goods and services. For example, the increase for Bangladesh, India, and Myanmar would be 3–7%, and 14% for Pakistan. Several other countries would also experience a high increase in exports. Exports of agricultural commodities would increase more than manufactured products. The export increase of agricultural commodities from Bangladesh, Cambodia, Lao People's Democratic Republic, and Myanmar could contribute to poverty reduction in those countries.

Hahm and Raihan [25] also suggested that the increase in imports would be higher than exports. For example, the import increase in Bangladesh, India, and Myanmar would be about 8–14% (compared to 3–7% export increase) and for Pakistan 14% (same as imports), and, for some countries, well above 15%. The researchers noted that this would lead to deterioration in the trade balance in most countries and would pose a risk for the overall balance of payments, which in turn could adversely affect economic growth.

Papriev and Sodikov [49] used a gravity model to evaluate overland trade expansion in 28 countries, resulting from improvements to the Asian highway network. The study has indicated that the highway network offers major potential for overland trade expansion through upgrading and improvement of the surface condition of selected roads, costing an estimated USD6.5 billion. One scenario of improving road quality indices up to 50 suggests that total intra–regional trade would increase by about 20%, or USD48.7 billion annually. In the second scenario of improving road quality indices up to 75, the predicted increase would be about 35%, or USD89.5 billion annually.

A gravity model-based study for the GMS by Fujimura and Edmonds [50] suggested that the development of cross-border road infrastructure has had a positive effect on intraregional trade in major commodities, with its elasticity in the range of 0.6–1.4. Similar effects on growth in trade were also observed in a study in Eastern Europe [51]. The gravity model simulations suggested that road upgrade could increase trade in the region by 50% above baseline, which exceeded the expected gains from tariff reductions or trade facilitation programmes of comparable scope. In nominal terms, this was about USD45 billion of trade benefits, and the estimated cost for road upgrade was USD8 billion.

The reviewed studies show that improvement of cross-border transport infrastructure can substantially increase transnational trade. However, they do not provide insight into how the growth of trade may affect other aspects of the economy. Only one study refers to the risk for the overall balance of payments due to the deterioration of trade balance. Neither do some studies provide any indication of the transport investment that would be needed. Further studies with a more comprehensive scope would be required to understand the trade-offs among increase in trade, fiscal risks, economic welfare, and other aspects of the economy, including negative effects.

### 3.3. Rural Economy, Poverty Reduction, and Social Impacts

People living in rural areas may also gain from major transport networks, especially roads and strategic transport infrastructure, such as a major bridge. Several studies provide evidence of transport networks having positive social impacts on rural people through poverty reduction and increased employment in non-farm activities [14,52–54]. Two studies [14,54] found a structural shift in the rural economy in terms of increase in non-farm activities and more employment.

The NH–5 Highway corridor in Vietnam made substantial positive impacts in the corridor region. The number of households living in poverty dropped by 35% between 1995 and 2000. Cities closer to and further away from NH–5 both experienced higher income growth per capita, as well as faster reduction in poverty than the rest of the country. The poverty rate in Vietnam as a whole reduced by 27% during this period as a result of broader spill overs from NH–5 to other regions [14].

An empirical study by Blankespoor et al. [52] investigated the impact of the Jamuna bridge on Bangladesh. They found that cropping intensity increased by 3% and the area using chemical fertilizer by 7%. The large reduction in transport costs (about 50%) due to the bridge led to agricultural development in the newly connected hinterland as a result of technology adoption and a better match of land to crops. The long-term estimate (2005–2013) was positive and statistically significant, with an increase in rice yield of 5.2%.

Neupane and Calkins [53] examined the status of poverty and income inequality in Southern Thailand along the Asian Highway network route (AH18) in Songkhla province. They used descriptive statistics and Analysis of Variance (ANOVA) method that included poverty and income inequality indices to analyze the household survey data. They found that the average household income varied with location, and poverty was lower along the Asian Highway route.

The impacts of a major National Highway (NH2) in three states in India were analyzed in a study by AITD [54]. Compared with a baseline survey, literacy increased by 6%, female literacy by 12%, school enrolment by 7%, female school children by 12%, and population gaining access to medical facilities by 7%. The study observed improvements in women's participation in the labor force (9% increase), employment in non-agricultural activities (7% increase), and an increase in annual (deflated) per capita income by Rs 243.

Another study found that the welfare impacts on rural people along the GQ highway in India was not similar in all areas [14] (p. 247). The highway reduced poverty significantly in districts with a large agro-processing base, but poverty did not drop in the average district near the highway.

In their study, Fan and Chan-Kang [55] found a trade-off between growth and poverty reduction from road investments in different parts of China. Road investments gave the highest economic returns in the eastern and central regions of China, while contributions to poverty reduction were greatest in western China.

An empirical study in Ethiopia by Minten et al. [56] found that an increase in transaction and transport costs over a 35 km distance led to a 50% increase of the prices of fertilizer and a 75% reduction in its use. The cost to bring fertilizer over a distance of 10 km from the distribution centre was as high as the costs to bring fertilizer to the distribution centre from the port about 1000 km away. The study concluded that tackling the "last mile(s)" costs should thus be a priority.

A study by Omamo [57] in Kenya showed that improved rural road networks that reduce transport costs could reduce the motivation of small farmers to meet food needs through domestic production and promote specialization that raises farm incomes.

Dorosh et al. [58] examined the effects of increases in road investments on travel times and agricultural production. They found that improvements in road infrastructure can facilitate a substantial increase in agricultural production in Sub-Saharan Africa.

The findings of some of the above studies show that transport development leading to improvements in access in rural areas can have direct welfare impacts for the rural people. However, the impacts may vary in different situations and for different groups. On the other hand, transport developments involving HVT corridors and networks are of strategic significance to a national economy, but their direct benefits to the rural people may remain limited unless they are linked with a system of feeder roads providing access to remote areas.

### 3.4. *Equity/Inclusive Development, Employment*

Roberts et al. [16] found that transport networks had a beneficial effect on social inclusion in terms of education and gender in most of the studies reviewed. About 75% of studies showed benefits of equality in terms of spatial distribution, though all studies showed substantial negative effects in terms of overall income distribution.

A major study by Donaldson [59] investigated the impact of India's vast colonial railway network using archival data. Donaldson [41] (p. 931) found that "Railroads reduced the cost of trading and interregional price gaps, and increased trade volumes". He (p. 931) also found that when the network was extended to a typical district, real agricultural income in that district rose by approximately 16%.

The findings of Zhenhua and Haynes [60] confirmed that the high-speed railway network in China contributed to decreasing regional economic disparity and promoted regional economic convergence.

The review by Roberts et al. [16] found that roads have a beneficial effect on social inclusion in terms of job creation. More jobs, especially in non-farm activities, and greater participation of women in the labor force were also observed in the AITD study [54].

The rehabilitation and improvement of the Maputo corridor successfully boosted transit trade flows and bilateral trade between South Africa and Mozambique. The Maputo corridor led to more than USD5 billion worth of investments, and 15,000 direct jobs in the construction and operation of transport, logistics, energy, and industrial ventures along the corridor [14].

The NH–5 highway corridor in Vietnam has attracted investment and created jobs. In 2006, 83,453 and 134,846 jobs were generated along the corridor in Hung Yen and Hai Duong provinces, respectively [14].

An International Labor Organization (ILO) study in two states of India (Gujarat and West Bengal) found that investment in infrastructure created a substantial number of jobs [61]. The study found that a 10% increase in the investment in highways and urban road construction sectors (INR2345 million in Gujarat and INR1831 million in West Bengal) led to 83,401 more workers being hired in Gujarat and 178,181 more workers hired in West Bengal. This also led to INR13.52 billion growth in Gujarat and INR14.05 billion growth in West Bengal.

Two studies on the Jamuna bridge in Bangladesh have provided evidence of its impacts on rural employment and job transition patterns, even though they followed different methodologies [44,45]. Blankespoor et al. [62] found that besides increasing employment, the bridge construction facilitated a farm to non-farm shift in employment. About 40% of the employment increase in services came from the reduction of employment in the manufacturing sector, and the rest from agriculture. The results suggest a long-term structural change in the employment pattern.

Mahmud and Sawada [63] also found that the bridge led to an increase in local employment and facilitated a shift from farm to non-farm in both districts. The share of non-farm employment increased from 6.7 to 14% in one district and from 8.6 to 16% in the other district.

A shift in production and labor from agriculture was also found in a study done in Cameroon [64]. The study found that better road access led to a diversification of the economic activities within those households that were most isolated.

The findings of the above studies suggest that transport investment can be an important policy instrument to create jobs and may contribute to decreasing regional economic disparity. However, substantial negative effects on overall income distribution can be expected. The studies do not provide any analysis of how the negative effects on overall income distribution may affect different groups, either in relative or absolute terms. The results of some studies suggest that transport development in conjunction with appropriate complementary interventions, such as in Vietnam, can make substantial positive changes in the economy.

### 3.5. Location and Spatial Effects

While the estimated overall impacts of transport networks are generally beneficial, there are often negative impacts in some country regions and for some groups in society. Dzumbira et al. [65] found differences in the development impacts of the Maputo Development Corridor. Economic impacts, access to services, employment opportunities, income levels, and access to formal housing in nodes along and near the N4 spine were better than those away from the corridor.

In an empirical study on the impacts of railways in colonial India, Donaldson [66] found that the network extended to a typical district increased its real agricultural income, but reduced the real income of its neighboring district without rail access.

A study by the United Nations Economic and Social Commission for Asia and the Pacific (UN ESCAP) [22] used a CGE model to assess the development impacts of three Asian Highway routes from Kunming, China through Southeast Asia to South Asia. The results showed that, although most country regions would remain unaffected, some regions would have substantial gains in GDP of about 2.2–2.8%. For other regions, the average losses would be small, about 0.3–0.4% in GDP.

Other simulation-based studies have also found an uneven distribution of economic benefits along transport corridors, such as the Dhaka–Kolkata corridor in Bangladesh [14], and the Delhi Mumbai Industrial Corridor (DMIC) in India [67].

In China, the National Express Network (NEN) has increased real incomes in nearby prefectures by nearly 4% on average, but decreased real wages in many prefectures in the urban and rural sector [68]. Other studies also provide evidence of uneven distribution of the impacts of transport networks [14,18,60,69]. This finding from many studies has implications for the planning and design of transport projects.

The major transport networks can motivate businesses to relocate in areas better served by the network. An empirical study in Uganda found that businesses gain more from being located in areas that offer agglomeration economies, availability of skilled workforce, and better infrastructure conditions [70]. Public infrastructure investments in other locations are likely to attract fewer private investors.

To ensure more inclusive development, it is important to understand the distribution of impacts across population groups as well as across geographical areas. The studies discussed above clearly show that the distribution of impacts can be uneven across geographical areas. However, this review did not find any major study that explored the distribution of impacts across different segments of the population. More research would be required to examine the distribution of impacts across different segments of the population for each development outcome and in different situations.

### *3.6. Cross-Border Facilitation*

De [71] analyzed the effects of inefficient facilitation of trade flow and concluded that transaction costs and delays at borders affect trade flows in the same way as tariffs do. The higher the transaction costs, the less is trade between partners in neighboring countries. A 10% drop in transaction costs at borders increases exports by about 2%.

Intercountry trade in goods and services can be greatly improved with efficient facilitation at border points and improved transit procedures [47,71]. These improvements would boost trade between landlocked countries [72]. However, the gains to countries may not be equal in either relative or absolute terms [21,22].

Stone and Strutt [27] have suggested that welfare gains of USD8.1 billion could be attained from moderate improvements in road infrastructure and trade facilitation in the Greater Mekong Subregion (GMS).

UN ESCAP has developed a corridor performance method that provides information on the relative importance and variability of time and cost at each interface point in a corridor [73]. This method was used to analyze the performance of trade corridors in East and Central Asia, and showed cost details for transport modes and transit time at each border post in the corridors.

Arvis et al. [72] did case studies on cost, time, and reliability of exports on some corridors in LLDCs, focusing on transit traffic for landlocked countries. The researchers concluded that the transit procedures regulating goods were poorly designed and implemented, which discouraged competition and high-quality logistics services.

In recent years, some countries have taken initiatives to streamline their border control and clearance procedures. These include Integrated Check Posts (ICP) by India and a single window system for southeast Asian countries which allows synchronized submission and processing of data, as well as faster clearance and release of shipments. The clearance process at border posts in the Maputo Corridor in southern Africa has also been streamlined. However, the literature does not provide any major studies on these initiatives or their impacts on trade flows and other aspects. Studies on these initiatives can provide important insights into how cross-border facilitation arrangements can be improved and adapted for other situations.

### 3.7. Corridor Governance

Transnational corridor development and operation is complex because of their wide reach and scope, and involvement of a large variety of stakeholders. Corridor management can be unique for various reasons, including the historical development of the corridor, initial conditions, and political objectives and institutions in countries along the corridor [72]. As a result, several management structures have emerged, see for example, References [72,74–76]. These structures include:

- Public–private partnership management structures, such as Maputo Corridor Logistics Initiative (MCLI) for Maputo Development Corridor (MDC) [75,77,78];
- Consensus-building structure, such as the Dar-es-Salaam Corridor Committee for Dar-es-salaam corridor [75,77];
- Project coordination structure, such as the Central Asia Regional Economic Corridor (CAREC) corridors in Central Asia [74];
- Legislative management structure based on treaties between countries, such as the Northern Corridor Transit and Transport Coordination Authority (NCTTCA) [75,77].

Many national corridors also have a formal management structure for coordination between government authorities, especially if they are large, multi-sectoral, or multi-modal. Several management structures for national corridors have emerged in India, Malaysia, and other countries [79,80]. These management structures were established under respective national laws; however, their legal status and governance structures are different.

This review finds that most transnational corridor managements have a multi-layer structure, including an apex/umbrella body, an executive/coordination committee, and a secretariat. However, the details of their structures and institutional arrangements vary. National corridors in India, and Malaysia also have multi-layer management structures [79,80].

Legal instruments such as treaties, conventions, agreements, protocols, covenants, compacts, exchange of notes, memoranda of understanding, etc., govern corridor management and operations [77]. Legal instruments can be bilateral, covering two countries, or multilateral, covering many countries along a particular corridor, a subregion, region, or global.

As mentioned above, corridor management ranges from a private sector-led entity operating as a lobby group (such as MCLI) to an intergovernmental body or state-run authority (such as NCTTCA). Each management structure has its strengths and weaknesses, but this review did not find any study that assessed the current management structures. Neither did the review find any study on designing of corridor governance structure, which is important for efficient corridor operation.

Inefficient facilitation arrangements at border crossings is a major deterrent to trade expansion. Corridors in Africa and Asia operate mainly under bi-lateral or sub-regional agreements [77,81]. Some 42 sub-reginal agreements have been signed in Asia alone. However, only some of these are in force. Non-uniformity of these agreements is a major challenge for region-wide trade and movement of traffic. In Africa, for example, multiple corridors cross Tanzania and DRC; these corridors operate under different management structures and their governance instruments are also different, which can lead to inconsistent corridor outcomes and administrative complications. However, studies on these issues are yet to appear in the academic literature.

### 3.8. Financing and the Private Sector's Involvement

The public sector has been the main source of financing new transport infrastructure, followed by official development assistance (ODA). For example, in Africa, more than half of transport infrastructure investments were from the public sector; the remaining were through ODA (from the Organization for Economic Co-operation and Development (OECD) and non-OECD countries) and the private sector [24]. Low-income countries (LICs) have not been very successful in attracting private investment in infrastructure development. Only 4% of private-financed projects took place in LICs, while the majority of projects have taken place in developing countries with relatively higher incomes countries [82].

Large gaps between required and available funds are a major challenge for developing transport infrastructure in LICs. Some recent studies have identified substantial investment gaps in infrastructure development. For example, the estimated investment need in South Asia is about 8.8% of GDP, but the current level of investment of about 3.2% of GDP [83,84]. The estimated financing gap for Africa's infrastructure is in the range of USD67.6–USD107.5 billion [85].

More recent data on private investment in infrastructure shows that globally private investment has declined [86]. The same trend has been observed in Asian countries [83]. It is necessary to understand the reasons for the decline in private investment and what can be done to reverse this trend. Given the limited success in implementing private-financed projects, LICs may explore alternative financing options, such as commercial borrowing by the government, issuing of bonds, and other financing modalities that may be available.

### 3.9. Environment

There are direct costs of transport development to the environment, such as deforestation, loss of biodiversity, and general degradation of ecosystems [87]. Roberts et al. [16] found that transport networks and corridors have a harmful impact on the environment in terms of deforestation and carbon dioxide ($CO_2$) emission. A study by Laurance et al. [88] provided evidence of road clearings on tropical forests in Central Africa and other regions. A highway across the northwestern Congo Basin has promoted massive logging, poaching, and forest loss.

The transport sector is a major consumer of energy resources and also one of the major emitters of carbon dioxide. Globally, the road sector accounts for most of the energy consumption in the transport sector. In 2010, the transport sector in the UN ESCAP region consumed 27.4% (648 million tons of oil equivalent, or Mtoe) of the sector's total global energy consumption (2362 Mtoe). In the UN ESCAP region, the road sector accounted for more than 80% of the total energy consumption in the transport sector [89]. Compared with the global increase, energy consumption by the road sector in the region is rising more steeply.

In a report by the United States (U.S.) Energy Information Administration [90], it is projected that during the period 2012–2040, the annual growth of the transport sector's energy consumption in Africa (3.1%) and Asia (2.9%) (excluding China and India) will be higher than in other regions.

Globally, transport accounted for one-quarter of total emissions in 2016 (about 8 $GtCO_2$) which was 71% higher than in 1990 [91]. Of this, the share of road transport emissions was 74%. In line with the global increase, $CO_2$ emissions from the transport sector in Asia and Africa also show an upward trend. $CO_2$ emissions in the transport sector in Asia increased from 0.78 $GtCO_2$ in 1990 to 2.44 $GtCO_2$ in 2016, and in the same period, Africa's emission increased from 0.11 $GtCO_2$ to 0.46 $GtCO_2$ [91]. Compared with the global share (24.21%), the share of emission from the transport sector in Asia was much lower (about 14%) but its share in African countries was much higher (39.65%).

The steep rise in $CO_2$ emissions from the sector is expected to continue with further economic development in African and Asian countries. As $CO_2$ emissions is a major source of negative impacts on welfare, greater efforts will be required to reduce the current trend of emission increase from the transport sector.

It may be generally assumed that transport projects will have negative effects on environmental quality outcomes. However, there is a knowledge gap in that the current studies do not provide insight into the trade-offs between economic and social impact outcomes and environmental quality. It may also be pragmatic to consider the environmental impacts of some transport projects from a different perspective, for example, the Delhi–Mumbai dedicated rail freight corridor (DFC). The alternative to DFC would have far more detrimental impacts in terms of $CO_2$ emissions [92]. The authors of Reference [58] estimate that annual $CO_2$ emissions under the low-carbon scenario with DFC (0.28 million tons) are less than one-fortieth under the business-as-usual scenario without DFC (12.32 million tons).

### 3.10. Congestion and Road Safety

Traffic congestion is an important source of welfare loss in almost all major developing countries. An Asian Development Bank report suggests that road congestion costs countries in the region about 2–5% of GDP every year, due to lost time and higher transport costs [93]. Congestion also has other negative impacts on the welfare of people. The major cities in Asia suffer from the highest air pollution levels in the world, about 80% of which is from transport. Kuala Lumpur, the capital city of Malaysia, has serious traffic congestion. According to a World Bank report [94], the city wastes 1.2 billion L of fuel on traffic congestion, which is about 2% of GDP. The results of a study show that the traffic congestion cost in Beijing was about RMB 58 billion (4.22% of GDP) in 2010 [95]. The estimated annual congestion cost in Dhaka in Bangladesh was USD3868 million, which included an environmental externality cost of USD375 million [96]. Similar congestion cost estimates for some African cities are also available.

The TomTom traffic index, a measure of traffic congestion, shows that in 2018, some of the most congested cities in the world were in Asia (Mumbai, Delhi, Jakarta, and Kuala Lumpur, for example). In Africa, except Cairo and Lagos, the other most congested cities were in South Africa [97]. Measures to reduce the congestion level can greatly help to reduce its adverse impacts on the environment and people's welfare.

The road traffic death rate is highest in Africa (26.6/million people) followed by south and southeast Asian countries (20.7/million people). In both regions, the death rate has increased compared with 2013. The burden of road traffic injuries and deaths is disproportionately borne by vulnerable road users and those living in low- and middle-income countries [98]. In addition to a public health problem, road traffic injuries are a development issue. Low- and middle-income countries lose approximately 3% of GDP as a result of road traffic crashes [99].

The LICs have about 9% of the world population and their vehicle population accounts for only 1% of the total vehicles but they share 13% of total road deaths in the world. The LICs in Africa has the highest rate of road traffic deaths in the world of 29.3 per 100,000 population [98]. The low standard and poor condition of most roads and inadequate/lack of road infrastructure facilities in most low- and middle-income countries in Africa and Asia are among the causes of high road traffic fatalities.

An analysis of the 2010 road safety data covering 34,370 km of highways in 23 countries, available in the Asian Highway database, clearly shows that the higher class of roads are generally much safer than the lower class of roads [89]. A number of safe road demonstration corridor projects are being implemented in many states of India and other countries. A preliminary analysis of some of these safety improvement projects finds similar results. For example, the Kadapa to Renigunta safety demonstration corridor project implemented under the Andhra Pradesh & Telangana Road Sector Project (APTRSP) in India shows impressive results. The locations where curves and junctions were improved saw a 53% reduction in road crashes and 42% reduction in fatalities [100]. These results show that significant improvement in road safety can be achieved through the upgrading of highways and safer road infrastructure design.

### 3.11. Trafficking, Spread of Disease, and Socio–Political Issues

Cross-border transport infrastructure is accompanied by a wide range of negative externalities, such as the spread of HIV/AIDS, trafficking of vulnerable groups, particularly women and girls, illegal trading of narcotics and other items, effects on local farmers and businesses, and erosion of social values and cultural identities. As a result, the perceptions of local people in border areas may not always be favorable to cross-border infrastructure [101–103].

Trafficking of women and girls is a serious problem in border areas of countries in South and South-East Asia. Deane [104] examined cross-border trafficking of women and girls from Nepal to India. This study cited different sources to estimate that 7000 to 10,000 girls between the ages of 9 to 16 years are trafficked each month from Nepal to India. The trafficking problem between other countries has been examined in other studies [105,106].

Cross-border transport infrastructure can also have other adverse social impacts on the local people. Günther Slesak et al. [107] reported alarming vulnerability rates in ethnic minorities to sexually transmitted diseases (STDs) and HIV/AIDS along a new major intercountry road in south-east Asia. A review by Regondi et al. [108] found evidence of the spread of HIV/AIDS along the road network in Southern Africa. The number of HIV-positive persons and AIDS patients increased sharply in Savannakhet in Lao People's Democratic Republic during the construction of the Second Mekong Bridge [103].

Some of the above externalities are deeply rooted in the problem of widespread poverty, especially in remote border areas. Along with direct intervention measures, other measures to reduce poverty in the border regions would also be necessary. However, apart from trafficking and the spread of diseases, the literature does not provide insights into other social issues. Displacement and marginalization of local communities, including indigenous people, are possible due to land appropriation and grabbing, and changing social structure along the corridors, especially in border areas. However, the review did not find any study on these social issues. The review also did not find any study on mitigation practices to counter the negative impacts of cross-border transport.

## 4. Discussion

For the convenience of discussion, a summary of the main impacts of transport infrastructure investment, discussed in Section 3, is presented in Table 1.

The outcomes or impacts of transport infrastructure development are generally positive on economy, income, poverty, employment, equity, and inclusion. However, there can be important trade-offs between economy and welfare and impacts on environmental quality. In some situations, there can even be trade-offs between economic growth and poverty [55]. In addition, the distribution of impacts can be uneven. To ensure more sustainable and inclusive development, the potential gains to economy and welfare must be balanced against the potential adverse impacts, and the gains should be more equitably distributed.

An important limitation of the current studies is that they are not comprehensive in scope, generally focused on some specific outcome, for example, economic growth, trade, etc. Comprehensive studies which consider the impacts of specific transport projects on all aspects of outcomes, including negative externalities, simultaneously would be useful to gain insight into the nature of trade-offs between different outcomes. Further research is also required to understand the uneven impacts across geographical locations and population groups for different types of infrastructure, and complementary policies, interventions, and institutions that can be considered to alleviate the adverse effects of uneven impacts. Findings of these studies can help when considering the choice of appropriate interventions and institutions which may then become part of the planning and design of HVT corridors and networks.

Transport development is necessary but may not be sufficient to generate wider economic benefits. Additional complementary interventions may also be required. For example, in the case of GQ Highways in India, non-availability of land for non-farm uses and low education and skills of the local labor force were found to be the main constraints for wider sharing of socio–economic benefits of the highways in some districts [16]. It also shows that transport investments and complementary policies should be based on a better understanding of the underlying mechanisms and the initial conditions that may affect the development outcomes. Otherwise, there is a risk that transport investments may not always produce the expected outcomes.

**Table 1.** Summary of main impacts from the reviewed literature.

| Area/Aspect of Impact (Section) | Summary of Main Impacts from the Reviewed Literature |
| --- | --- |
| Transport Infrastructure Investment and Economy (Sub-Saharan Africa and South Asia) (Section 3.1) | • substantial positive impacts on economy; gains from investment from less than 1% to more than 10% of Gross Domestic Product (GDP)<br>• Asia: estimated gains from regional transport infrastructure USD967.7 billion in 2020, equivalent to 6.0% of baseline income in 2008<br>• estimated gains for most Belt and Road Initiative (BRI) countries 3 to 10% of GDP<br>• infrastructure led to more than half of Africa's recent economic growth<br>• Africa: significant benefits to real and nominal income, consumption, gender, education, and job creation<br>• generation of wider economic benefits (WEBs) may need complementary investment, policy, and other measures; depends on initial condition and other factors |
| Transport Infrastructure Investment and Economy (developed countries) [1] (Section 3.1) | • economic impacts lower than developing countries (generally contribution to GDP 1% or less)<br>• WEBs generally recognized; induce higher productivity for firms and workers but applications in evaluation not universal<br>• WEBs can be substantial (20–50% of conventional benefits in cost–benefit analysis (CBA)); considered in evaluation framework of some countries but treatments vary; U.K. considers three, New Zealand five, Australia reports separately<br>• agglomeration benefits most widely recognized, and the largest component of WEBs in many studies |
| Trade and Investments (Section 3.2) | • improved cross-border transport infrastructure can substantially increase transnational trade<br>• poor infrastructure accounts for 60% of transport costs for landlocked countries compared to 40% for coastal countries<br>• estimated elasticity of trade flows with respect to trade cost about −3<br>• one day increase in transit time implies nearly 7% decline in exports in Sub-Saharan Africa (SSA)<br>• upgrading of roads can increase overland trade in SSA by about USD250 billion over 15 years<br>• BRI corridors in South Asia can increase imports between 8–14% compared to export by 3–4%<br>• increased trade can deteriorate trade balance in many countries; risk for the overall balance of payments |
| Rural Economy, Poverty Reduction and Social Impacts (Section 3.3) | • access improvements have direct welfare impacts for rural people<br>• positive impacts on rural people through poverty reduction and more jobs in non-farm sector<br>• NH–5 Highway corridor in Vietnam: poverty dropped by 35% between 1995 and 2000<br>• NH–2 corridor in India: increase of literacy by 6%, female literacy by 12%, school enrolment by 7%, female school children by 12%, access to medical facilities by 7%, women's participation in labor force 9%, employment in non-agricultural activities 7%<br>• increase in agricultural production and technology adoption |
| Equity/Inclusive Development, Employment (Section 3.4) | • beneficial effect in terms of substantial number of new jobs<br>• more jobs especially in non-farm activities<br>• greater participation of women in labor force<br>• benefits of more equality in terms of spatial distribution<br>• substantial negative effects in terms of overall income distribution |
| Location and Spatial Effects (Section 3.5) | • uneven impacts on population groups and geographical areas<br>• increase of real income in an area can decrease in other areas<br>• businesses can relocate to gain from agglomeration economies; some areas may lose businesses |
| Cross-Border Facilitation (Section 3.6) | • inefficient facilitation arrangements deter trade expansion<br>• cost reduction at borders increases exports and welfare; 10% drop in costs increases exports by about 2% in South Asia<br>• moderate improvements in infrastructure and trade facilitation can increase welfare gains of USD8.1 billion in Greater Mekong Sub-region (GMS)<br>• gains to countries are not equal in either relative or absolute terms |

**Table 1.** *Cont.*

| Area/Aspect of Impact (Section) | Summary of Main Impacts from the Reviewed Literature |
|---|---|
| Corridor Governance (Section 3.7) | • transnational corridor management can be unique<br>• several management structures have emerged<br>• most corridor managements have a multi-layer structure<br>• a wide array of legal instruments governs corridor management and operations; most agreements are bi-lateral and sub-regional<br>• non-uniformity of agreements is a challenge for region-wide trade expansion in Africa and Asia<br>• some domestic corridors have formal management structures |
| Financing and Private Sector's Involvement (Section 3.8) | • public sector is the main source of infrastructure financing<br>• only 4% private-financed projects took place in low-income countries (LICs)<br>• large funding gaps in LICs for infrastructure development<br>• globally private investment has declined in recent years |
| Environment (Section 3.9) | • harmful impact on the environment in terms of deforestation and loss of biodiversity<br>• transport sector is a major consumer of energy resources and emitter of $CO_2$<br>• globally transport sector emitted one-quarter of total emissions in 2016; share of road transport was 74%<br>• road sector accounts for 83.3% of $CO_2$ emissions in ESCAP region<br>• $CO_2$ emissions in Asia increased from 0.78 $GtCO_2$ in 1990 to 2.44 $GtCO_2$ in 2016 (312% increase) and in Africa from 0.11 $GtCO_2$ to 0.35 $GtCO_2$ (318% increase) |
| Road congestion and road safety (Section 3.10) | • congestion costs Asian countries 2 to 5% of GDP every year; Kuala Lumpur wastes 1.2 billion L of fuel, about 2% of GDP<br>• about 80% of air pollution in Asian major cities is from transport<br>• road congestion is an important source of welfare loss in African and Asian developing countries<br>• road traffic death rates are among the highest in Africa (26.6/million people) and south and southeast Asia (20.7/million people)<br>• low- and middle-income countries lose about 3% of GDP from road traffic crashes<br>• evidence shows that safety improvement measures and better infrastructure can reduce road traffic deaths and injuries |
| Trafficking, Spread of Disease and Socio–Political Issues (Section 3.11) | • trafficking of women is a serious problem in many border areas<br>• adverse impacts on local people, including displacement and marginalization, effects on local firms and businesses<br>• transport development can help spread of HIV/AIDS and other diseases<br>• illegal trading of narcotics and other items in border areas |

Source: Based on discussion presented in Section 3. [1] Comparison with developed countries has been made for this aspect only.

The NH–5 highway corridor in Vietnam is a good example of complementary interventions. The highway development was implemented in conjunction with complementary policies, including human resource development. The availability of an educated labor force allowed quick transfer of labor from agriculture to manufacturing and helped a major transformational shift in the economy [14]. The Maputo Corridor in Southern Africa is another good example where infrastructure rehabilitation and upgrading of transport infrastructure, along with improved facilitation measures at border posts, made the corridor successful in boosting transnational trade.

The review showed that HVT corridors and networks can create agglomeration effects in some locations. Businesses gain more from being in areas that offer agglomeration economies. Public infrastructure investments in other locations are likely to attract fewer private investors. Because of agglomeration benefits in established main centres, investment only in transport infrastructure has limitations to attract businesses to secondary centres outside the established main centres. Other intervention measures, such as public investment and policies, may be needed to induce growth in less attractive secondary centres or regions.

The burden of environmental and social costs of transport can be substantially reduced through various measures. For example, road safety is an important development issue in both SSA and South Asian countries, and should not be left as an "afterthought" in road infrastructure development projects. Although road safety is a cross-sectoral issue, evidence from the review suggests that the incidence of road crashes can be substantially reduced through road development with proper road safety audits at the road design stage.

Even though Environmental Impact Assessment (EIA) is now customary, further research may be necessary to guide the planning of transport infrastructure projects in environmentally sensitive areas. Other measures, such as the development of multi-modal transport systems, where possible, can substantially reduce the negative effects (see, for example, References [92,109]). Price instruments (congestion and pollution charges, for example) and regulations (such as emission and fuel standards) are useful tools to change the behavior of individuals and firms and address environmental externalities. However, the use of these tools is not common in LICs. They should be considered to reduce the burden of negative externalities, where feasible.

To improve transport project evaluation in developing countries, it would be worthwhile finding out if the same categories of WEBs, as in developed countries, are also important for developing countries. Further studies with such a focus, as well as developing suitable estimation methods for WEBs, would be needed.

Researchers have used a variety of models and methodologies to study the impacts of transport development, including CGE-based simulation studies, multi-regional input–output (IO) model, and difference-in-difference and other statistical and econometric models. Models based on CGE and IO methodologies are promising for estimating transport infrastructure investments and their distributional effects. Theoretically, the CGE and structural models are superior to other models. However, the structure and application of such models are still in the developmental stage. They are more complex, data-intensive, and require considerable expertise. Further efforts would be required to develop operational models for practical applications.

Research studies may also be considered to assess the effectiveness and suitability of the currently available analytical tools/models to understand the distributional impacts of regional and national transport projects/networks at the subnational level. Research is also needed to examine how such tools/models may be adapted for policy analysis and policy formulation, including designing of complementary intervention measures. A related issue in developing practical models for impact assessment is the availability of required data. The suggested research may also examine how this problem can be overcome.

Several corridor management structures have emerged. The review did not find any study on the assessment of the current management structures. An assessment study can provide valuable insights into the current structures and contribute to designing new and improving the current management structures. Therefore, a study on this matter is recommended.

There is another issue linked to management structures: the governance of transport corridors. Currently, there is no general framework for designing governance structures for transport corridors. Research may be considered on how a transport corridor governance structure can be organized, structurally and procedurally, so that multiple stakeholders in corridor development, management, and operation can play their roles and interact effectively. Allied to this, the research may also consider how governments can promote institutions to build partnerships, collaboration with such actors, and facilitate their action.

The establishment of functional linkages between local and rural communities and the urban/national economy by using major highways and railways is a major challenge. In addition to rural feeder road networks, some countries have considered rural logistics and market centres along the major transport networks and other intervention measures to improve efficiency in rural supply and distribution chains, serve as a direct market outlet for local produce, and generate non-farm local employment [110]. However, these initiatives are not widespread and do not follow a coordinated approach to establish effective rural–urban linkages. A study can be considered to develop case studies on such measures, and assessment for their adaptation in other countries.

The negative externalities of cross-border transport need closer attention. Human trafficking, illegal trade in narcotics and other items, and the spread of diseases are some of the major challenges that need to be tackled through appropriate interventions. Other social issues related to corridor development, such as displacement and marginalization of local communities, changing social structure, etc. need the attention of researchers. Research may be undertaken to study these problems in operational corridors. The study may also consider the effectiveness of the current mitigation measures, develop a general framework to formulate action plans for remedial measures, and examine how these measures may be incorporated in corridor project design, as well as in legal and regulatory instruments for border-crossing procedures.

Railways are expected to have a greater role in the future to meet the growing demand for transport infrastructure services. In recent years, many countries in South Asia and SSA have been building new railway lines. In addition, freight movement by railway is expected to be a key design feature of multimodal transport corridor projects in the future. However, evidence of the impacts of rail transport in the literature is not rich. A better understanding would be needed for designing economically and environmentally efficient and socially inclusive railway projects. Studies are suggested on impact evaluation of recent railway projects, including both passenger and railway freight corridors, which some countries are currently constructing.

**Funding:** This research was funded by UK AID through the UK Department for International Development under the High Volume Transport (HVT) Applied Research Programme, managed by IMC Worldwide.

**Acknowledgments:** This research was funded by UK AID through the UK Department for International Development under the High Volume Transport (HVT) Applied Research Programme, managed by IMC Worldwide. The author is particularly grateful for the helpful comments and advice of Bruce Thompson, the leader of Theme 1, Long Distance Strategic Road and Rail Transport, of the HVT Programme.

**Conflicts of Interest:** The author declares no conflict of interest.

## References

1. Planning Commission. *7th Five Year Plan-FY2016-FY2020*; Government of Bangladesh: Dhaka, Bangladesh, 2015; pp. 345–357.
2. United Nations. *Transforming Our World: The 2030 Agenda for Sustainable Development*; United Nations: New York, NY, USA, 2017; ISBN 978-0-8261-9011-6.
3. Aschauer, D. Is Public Expenditure Productive? *J. Monet. Econ.* **1989**, *23*, 177–200. [CrossRef]
4. Krugman, P. Increasing Returns and Economic Geography. *J. Polit. Econ.* **1991**, *99*, 483–499. [CrossRef]

5. Duranton, G.; Puga, D. Micro-foundations of Urban Agglomeration Economies. In *Handbook of Regional and Urban Economics*; Henderson, J.V., Thisse, J.-F., Eds.; Elsevier: Amsterdam, The Netherlands, 2004; Volume 4, pp. 2063–2117.
6. Venables, A.J.; Gasiorek, M. *The Welfare Implications of Transport Improvements in the Presence of Market Failure—Part 1 Report to Standing Advisory Committee on Trunk Road Assessment*; DETR: London, UK, 1999.
7. Department for Transport. *Transport, Wider Economic Benefits, and Impacts on GDP*; Department for Transport: London, UK, 2005.
8. Graham, D.J.; Gibbons, S. Quantifying Wider Economic Impacts of Agglomeration for Transport Appraisal: Existing evidence and future directions. *Econ. Transp.* **2019**, *19*, 100121. [CrossRef]
9. Venables, A.J. Evaluating Urban Transport Improvements: Cost-Benefit Analysis in the presence of agglomeration and income taxation. *J. Transp. Econ. Policy* **2007**, *41*, 173–188.
10. Vickerman, R. The Boundaries of Welfare Economics: Transport Appraisal in the UK. In *Transport Project Evaluation*; Haezendonck, E., Ed.; Edward Elgar Publishing: Cheltenham, UK, 2007; p. 12787. ISBN 978-1-84720-868-2.
11. Calderón, C.; Servén, L. Infrastructure and Economic Development in Sub-Saharan Africa. *J. Afr. Econ.* **2010**, *19*, i13–i87. [CrossRef]
12. Buys, P.; Deichmann, U.; Wheeler, D. Road Network Upgrading and Overland Trade Expansion in Sub-Saharan Africa. *J. Afr. Econ.* **2010**, *19*, 399–432. [CrossRef]
13. AECOM. *Freight Benefit/Cost Study: Capturing the Full Benefits of Freight Transportation Improvements: A Non-Technical Review of Linkages of the Benefit-Cost Analysis Framework*; Federal Highway Administration: Washington, DC, USA, 2001.
14. Asian Development Bank; Department for International Development; Japan International Cooperation Agency; World Bank. *The WEB of Transport Corridors in South Asia*; World Bank: Washington, DC, USA, 2018; ISBN 978-1-4648-1216-3.
15. Berg, C.N.; Deichmann, U.; Liu, Y.; Selod, H. Transport Policies and Development. *J. Dev. Stud.* **2017**, *53*, 465–480. [CrossRef]
16. Roberts, M.; Melecky, M.; Bougna, T.; Xu, Y. *Transport Corridors and Their Wider Economic Benefits: A Critical Review of the Literature*; Policy Research Working Paper; The World Bank: Washington, DC, USA, 2018.
17. Berechman, J.; Ozmen, D.; Ozbay, K. Empirical Analysis of Transportation Investment and Economic Development at State, County and Municipality Levels. *Transportation* **2006**, *33*, 537–551. [CrossRef]
18. Ghani, E.; Goswami, A.G.; Kerr, W.R. Highway to Success: The Impact of the Golden Quadrilateral Project for the Location and Performance of Indian Manufacturing. *Econ. J.* **2016**, *126*, 317–357. [CrossRef]
19. Asian Development Bank. *Bangladesh Quarterly Economic Update (December 2007) 2007*; Asian Development Bank Bangladesh Resident Mission: Dhaka, Bangladesh, 2007.
20. Anas, R.; Tamin, O.Z.; Wibowo, S.S. Applying Input-Output Model to Estimate the Broader Economic Benefits of Cipularang Tollroad Investment to Bandung District. *Procedia Eng.* **2015**, *125*, 489–497. [CrossRef]
21. Gilbert, J.; Banik, N. Socioeconomic Impacts of Cross-Border Transport Infrastructure Development in South Asia. *SSRN Electron. J.* **2010**. Available online: http://dx.doi.org/10.2139/ssrn.1586747 (accessed on 31 January 2019). [CrossRef]
22. UN ESCAP. *Growing Together: Economic Integration for an Inclusive and Sustainable Asia-Pacific Century*; United Nations Economic and Social Commission for Asia and the Pacific: Bangkok, Thailand, 2012; ISBN 978-974-680-332-8.
23. Deng, T. Impacts of Transport Infrastructure on Productivity and Economic Growth: Recent Advances and Research Challenges. *Transp. Rev.* **2013**, *33*, 686–699. [CrossRef]
24. Foster, V.; Briceño-Garmendia, C.M. *Africa's Infrastructure: A Time for Transformation*; The World Bank: Washington, DC, USA, 2009; ISBN 978-0-8213-8041-3.
25. Hahm, H.; Raihan, S. The Belt and Road Initiative: Maximizing Benefits, Managing Risks—A Computable General Equilibrium Approach. *J. Infrastruct. Policy Dev.* **2018**, *2*, 97–115. [CrossRef]
26. Zhai, F. Benefits of Infrastructure Investment: An Empirical Analysis. In *Infrastructure for Asian Connectivity*; Bhattacharya, B.N., Kawai, M., Nag, R., Eds.; ADBI, ADB with Edward Elgar Publishing: Cheltenham, UK, 2012; ISBN 978-1-78100-312-1.
27. Stone, S.; Strutt, A. *Transport Infrastructure and Trade Facilitation in the Greater Mekong Subregion*; ADBI Working Paper No. 130; Asian Development Bank Institute: Tokyo, Japan, 2009.

28. Francois, J.; Wignaraja, G. *Economic Implications of Deeper Asian Integration*; Johannes Kepler University: Linz, Austria, 2008.
29. Esfahani, H.S.; Ramírez, M.T. Institutions, Infrastructure, and Economic Growth. *J. Dev. Econ.* **2003**, *70*, 443–477. [CrossRef]
30. Hlotywa, A.; Ndaguba, E.A. Assessing the Impact of Road Transport Infrastructure Investment on Economic Development in South Africa. *J. Transp. Supply Chain Manag.* **2017**, *11*, 12. [CrossRef]
31. Standish, B.; Boting, A. The Macroeconomic Impact of Road Construction in Rural Areas of South Africa. *J. S. Afr. Inst. Civ. Eng. Johannesbg.* **2004**, *46*, 14–19.
32. Ng, C.P.; Law, T.H.; Wong, S.V.; Kulanthayan, S. Relative Improvements in Road Mobility as Compared to Improvements in Road Accessibility and Economic Growth: A Cross-Country Analysis. *Transp. Policy* **2017**, *60*, 24–33. [CrossRef]
33. Pradhan, R.P.; Bagchi, T. Effect of Transportation Infrastructure on Economic Growth in India: The VECM Approach. *Res. Transp. Econ.* **2013**, *38*, 139–148. [CrossRef]
34. Hong, J.; Chu, Z.; Wang, Q. Transport Infrastructure and Regional Economic Growth: Evidence from China. *Transportation* **2011**, *38*, 737–752. [CrossRef]
35. Standing Advisory Committee on Trunk Road Assessment (SACTRA). *Transport and the Economy*; DETR: London, UK, 1999.
36. Ministry of Transport. *Contribution of Transport to Economic Development*; Ministry of Transport: Wellington, New Zealand, 2014.
37. Department for Transport. *The Wider Impacts Sub-Objective, TAG Unit 3.5.14*; Department for Transport: London, UK, 2012.
38. Department for Transport. *Wider Economic Impacts Appraisal, TAG Unit A2.1*; Department for Transport: London, UK, 2018.
39. Commonwealth Department of Infrastructure. *Australian Transport Assessment and Planning Guidelines: T3 Wider economic benefits*; Transport and Infrastructure Council: Canberra, Australia, 2016.
40. Inter-agency Working Group. *Auckland CBD Rail Tunnel Assessment: Assessing Wider Economic Benefits*; Workstream Report Draft; 2011. Available online: https://www.transport.govt.nz/assets/Import/Documents/067cb8b7a6/Auckland-CBD-Rail-Workstream-report-Review-of-Construction-Costs.pdf (accessed on 23 August 2019).
41. Douglas, N.; O'Keeffe, B. Wider Economic Benefits—When and if they should be used in evaluation of transport projects. In Proceedings of the Australasian Transport Research Forum 2016 Proceedings, Melbourne, Australia, 16–18 November 2016.
42. Wangsness, P.B.; Rødseth, K.L.; Hansen, W. A review of guidelines for including wider economic impacts in transport appraisal. *Transp. Rev.* **2017**, *37*, 94–115. [CrossRef]
43. Kanemoto, Y. *Pitfalls in Estimating "Wider Economic Benefits" of Transportation Projects*; National Graduate Institute for Policy Studies: Tokyo, Japan, 2013.
44. Rye, T. Transport and economic growth. Presented at SCOTS Annual Conference, Pitlochry, Scotland, 19 May 2017.
45. Hummels, D. Transportation Costs and International Trade in the Second Era of Globalization. *J. Econ. Perspect.* **2007**, *21*, 131–154. [CrossRef]
46. Limão, N.; Venables, A.J. Infrastructure, Geographical Disadvantage, Transport Costs, and Trade. *World Bank Econ. Rev.* **2001**, *15*, 451–479. [CrossRef]
47. Freund, C.; Rocha, N. What Constrains Africa's Exports? *World Bank Econ. Rev.* **2011**, *25*, 361–386. [CrossRef]
48. Bosker, M.; Garretsen, H. Economic Geography and Economic Development in Sub-Saharan Africa. *World Bank Econ. Rev.* **2012**, *26*, 443–485. [CrossRef]
49. Papriev, Z.; Sodikov, J. The Effect of Road Upgrading to Overland Trade in Asian Highway Network. *Eurasian J. Bus. Econ.* **2008**, *12*, 85–101.
50. Fujimura, M.; Edmonds, C. *Impact of Cross-Border Road Infrastructure on Trade and Investment in the Greater Mekong Subregion*; ADBI Discussion Paper No.48; Asian Development Bank Institute: Tokyo, Japan, 2006.
51. Shepherd, B.; Wilson, J.S. Trade, Infrastructure, and Roadways in Europe and Central Asia: New Empirical Evidence. *J. Econ. Integr.* **2007**, *22*, 723–747. [CrossRef]

52. Blankespoor, B.; Emran, M.S.; Shilpi, F.; Xu, L. *Transport Costs, Comparative Advantage, and Agricultural Development: Evidence from Jamuna Bridge in Bangladesh*; Policy Research Working Paper 8509; Development Research Group, World Bank: Washington, DC, USA, 2018.
53. Neupane, H.S.; Calkins, P. Poverty, Income Inequality and Livelihood Diversification: A Case of Asia Highway in Songkhla Province of Thailand. In *Asian Economic Reconstruction and Development under New Challenges*; Huang, W., Leeahtam, P., Eds.; CMSE Press: Chiang Mai, Thailand, 2012.
54. Asian Institute of Transport Development. *Socio-Economic Impact of National Highway on Rural Population*; Asian Institute of Transport Development: New Delhi, India, 2011.
55. Fan, S.; Chan-Kang, C. Regional Road Development, Rural and Urban Poverty: Evidence from China. *Transp. Policy* **2008**, *15*, 305–314. [CrossRef]
56. Minten, B.; Koro, B.; Stifel, D. *The Last Mile(s) in Modern Input Distribution: Evidence from Northwestern Ethiopia*; ESSP Working Paper 51; International Food Policy Research Institute: Washington, DC, USA, 2013.
57. Omamo, S.W. Transport Costs and Smallholder Cropping Choices: An Application to Siaya District, Kenya. *Am. J. Agric. Econ.* **1998**, *80*, 116–123. [CrossRef]
58. Dorosh, P.; Wang, H.G.; You, L.; Schmidt, E. Road connectivity, population, and crop production in Sub-Saharan Africa. *Agric. Econ.* **2012**, *43*, 89–103. [CrossRef]
59. Donaldson, D. Railroads of the Raj: Estimating the Impact of Transportation Infrastructure. *Am. Econ. Rev.* **2018**, *108*, 899–934. [CrossRef]
60. Zhenhua, C.; Haynes, K.E. Impact of High-Speed Rail on Regional Economic Disparity in China. *J. Transp. Geogr.* **2017**, *65*, 80–91.
61. Sinha, A.; Prabhakar, A.; Jaiswal, R. *Employment Dimension of Infrastructure Investment: State Level Input-Output Analysis*; Employment Working Paper; ILO: Geneva, Switzerland, 2015; Volume 168.
62. Blankespoor, B.; Emran, M.S.; Shilpi, F.; Xu, L. Bridge to Bigpush or Backwash? Market Integration, Reallocation, and Productivity Effects of Jamuna Bridge in Bangladesh. *SSRN Electron. J.* **2018**. Available online: http://dx.doi.org/10.2139/ssrn.3162451 (accessed on 30 January 2019). [CrossRef]
63. Mahmud, M.; Sawada, Y. Infrastructure and Well-Being: Employment Effects of Jamuna Bridge in Bangladesh. *J. Dev. Eff.* **2018**, *10*, 327–340. [CrossRef]
64. Gachassin, M.C.; Najman, B.; Raballand, G. Roads and Diversification of Activities in Rural Areas: A Cameroon Case Study. *Dev. Policy Rev.* **2015**, *33*, 355–372. [CrossRef]
65. Dzumbira, W.; Geyer, H.S., Jr.; Geyer, H.S. Measuring the Spatial Economic Impact of the Maputo Development Corridor. *Dev. South. Afr.* **2017**, *34*, 635–651. [CrossRef]
66. Donaldson, D. *Railroads of the Raj: Estimating the Impact of Transportation Infrastructure*; Asia Research Centre Working Paper; London School of Economics: London, UK, 2010.
67. Kumagai, S.; Hayakawa, K.; Isono, I.; Keola, S.; Tsubota, K. Geographical Simulation Analysis for Logistics Enhancement in Asia. *Econ. Model.* **2013**, *34*, 145–153. [CrossRef]
68. Roberts, M.; Deichmann, U.; Fingleton, B.; Shi, T. Evaluating Road to Prosperity: A New Economic Geographic Approach. *Reg. Sci. Urban Econ.* **2012**, *42*, 580–594. [CrossRef]
69. Sun, F.; Mansury, Y.S. Economic Impact of High-Speed Rail on Household Income in China. *Transp. Res. Rec. J. Transp. Res. Board* **2016**, *2581*, 71–78. [CrossRef]
70. Lall, S.V.; Schroeder, E.; Schmidt, E. Identifying Spatial Efficiency–Equity Trade-offs in Territorial Development Policies: Evidence from Uganda. *J. Dev. Stud.* **2014**, *50*, 1717–1733. [CrossRef]
71. De, P. Why Is Trade at Borders a Costly Affair in South Asia? An Empirical Investigation. *Contemp. South Asia* **2011**, *19*, 441–464. [CrossRef]
72. Arvis, J.-F.; Carruthers, R.; Smith, G.; Willoughby, C. *Connecting Landlocked Developing Countries to Markets—Trade Corridors in the 21st Century*; The World Bank: Washington, DC, USA, 2011; ISBN 978-0-8213-8416-9.
73. UN ESCAP; KOTI. *Integrated International Transport and Logistics System for North-East Asia*; United Nations: New York, NY, USA, 2006.
74. Asian Development Bank. *Central Asia Regional Economic Cooperation Corridor Performance Measurement and Monitoring: A Forward-Looking Retrospective*; Asian Development Bank: Manila, Philippines, 2014; ISBN 978-92-9254-689-2.

75. Adzigbey, Y.; Kunaka, C.; Mitku, T.N. Institutional Arrangements for Transport Corridor Management in Sub-Saharan Africa. SSATP Working Paper No.86. *SSRN Electron. J.* **2007**. Available online: http://www.ssrn.com/abstract=1393864 (accessed on 1 January 2019).
76. Arnold, J. *Best Practices in Management of International Trade Corridors*; World Bank Transport Paper; World Bank: Washington, DC, USA, 2006.
77. Grosdidier de Matons, J. *A Review of International Legal Instruments*, 2nd ed.; SSATP, World Bank: Washington, DC, USA, 2014.
78. Sequeira, S.; Hartmann, O.; Kunaka, C. *Reviving Trade Routes Evidence from Maputo Corridor*; SSATP Discussion Paper No. 14; SSATP, World Bank: Washington, DC, USA, 2014.
79. Bhalaki, V. *The Delhi Mumbai Industrial Corridor: Salient Features and Recent Developments*; National University of Singapore: Singapore, 2013.
80. Athukorala, P.; Narayanan, S. Economic Corridors and Regional Development: The Malaysian Experience. *World Dev.* **2018**, *106*, 1–14. [CrossRef]
81. UN ESCAP. *Review of Developments in Transport in Asia and the Pacific 2013*; United Nations: Bangkok, Thailand, 2013; ISBN 978-92-1-054206-7.
82. Trebilcock, M.; Rosenstock, M. Infrastructure Public–Private Partnerships in the Developing World: Lessons from Recent Experience. *J. Dev. Stud.* **2015**, *51*, 335–354. [CrossRef]
83. UN ESCAP. *Financing for Development in Asia and the Pacific—Highlights in the Context of the Adis Ababa Action Agenda*; United Nations Economic and Social Commission for Asia and the Pacific: Bangkok, Thailand, 2018.
84. Asian Development Bank. *Meeting Asia's Infrastructure Needs*; Asian Development Bank: Manila, Philippines, 2017; ISBN 978-92-9257-753-7.
85. African Development Bank. *African Development Outlook 2018*; African Development Bank: Abidjan, Ivory Coast, 2018; ISBN 978-9938-882-43-8.
86. Harris, C.; Chao, J. Declining Private Investment in Infrastructure—A Trend or an Outlier? Available online: https://blogs.worldbank.org/ppps/declining-private-investment-infrastructure-trend-or-outlier (accessed on 7 April 2019).
87. Damania, R.; Russ, J.; Wheeler, D.; Barra, A.F. The Road to Growth: Measuring the Tradeoffs Between Economic Growth and Ecological Destruction. *World Dev.* **2018**, *101*, 351–376. [CrossRef]
88. Laurance, W.F.; Goosem, M.; Laurance, S.G.W. Impacts of Roads and Linear Clearings on Tropical Forests. *Trends Ecol. Evol.* **2009**, *24*, 659–669. [CrossRef] [PubMed]
89. UN ESCAP. *Statistical Yearbook for Asia and the Pacific 2013*; United Nations: Bangkok, Thailand, 2014; ISBN 978-92-1-120659-3.
90. US Energy Information Administration. *International Energy Outlook 2016*; US Energy Information Administration: Washington, DC, USA, 2016.
91. IEA. $CO_2$ Emissions. Available online: https://www.iea.org/statistics/co2emissions/ (accessed on 29 August 2019).
92. Pangotra, P.; Shukla, P.R. *Infrastructure for Low-Carbon Transport in India: A Case Study of the Delhi-Mumbai Dedicated Freight Corridor*; Indian Institute of Management: Ahmedabad, India, 2012.
93. Asian Development Bank. Urban Transport. Available online: https://www.adb.org/sectors/transport/key-priorities/urban-transport (accessed on 1 September 2019).
94. Sanders, G.; Mendivil, B.; Luis, C.; Reindert, W. *Malaysia Economic Monitor—Transforming Urban Transport*; World Bank: Washington, DC, USA, 2015.
95. Mao, L.-Z.; Zhu, H.-G.; Duan, L.-R. The Social Cost of Traffic Congestion and Countermeasures in Beijing. In Proceedings of the Ninth Asia Pacific Transportation Development Conference, Chongqing, China, 29 June–1 July 2012; pp. 68–76.
96. Khan, T.; Islam, R.I. Estimating Costs of Traffic Congestion in Dhaka City. *Int. J. Eng. Sci. Innov. Technol.* **2013**, *2*, 9.
97. TomTom—International BV. Traffic Index 2018. Available online: https://tomtom.com/en_gb/traffic-index/ranking/ (accessed on 5 September 2019).
98. World Health Organization. *Global Status Report on Road Safety 2018*; World Health Organization: Geneva, Switzerland, 2019; ISBN 978-92-4-156568-4.
99. World Health Organization. *Global Status Report on Road Safety 2015*; World Health Organisation: Geneva, Switzerland, 2015; ISBN 978-92-4-156506-6.

100. Gupta, N. Road Safety Action Pays off, and "Demonstration Corridors" Are here to Prove it. Available online: https://blogs.worldbank.org/transport/road-safety-action-pays-and-demonstration-corridors-are-here-prove-it (accessed on 1 September 2019).
101. Lin, S.; Grundy-Warr, C. One Bridge, Two Towns and Three Countries: Anticipatory Geopolitics in the Greater Mekong Subregion. *Geopolitics* **2012**, *17*, 952–979. [CrossRef]
102. Warr, P.; Menon, J.; Yusuf, A.A. Regional Economic Impacts of Large Projects: A General Equilibrium Application to Cross-Border Infrastructure. *Asian Dev. Rev.* **2010**, *271*, 104–134.
103. Japan International Cooperation Agency; ALMEC Corporation. *Cross-Border Transportation Infrastructure*; Japan International Cooperation Agency: Tokyo, Japan, 2007.
104. Deane, T. Cross-Border Trafficking in Nepal and India—Violating Women's Rights. *Hum. Rights Rev.* **2010**, *11*, 491–513. [CrossRef]
105. Molland, S. "The Perfect Business": Human Trafficking and Lao-Tai Cross-Border Migration. *Dev. Change* **2010**, *41*, 831–855. [CrossRef]
106. Data and Research on Human Trafficking: A Global Survey. *Int. Migr.* **2005**, *43*, 99–128.
107. Slesak, G.; Inthalad, S.; Kim, J.; Manhapadit, S.; Somsavad, S.; Sisouphanh, B.; Bouttavong, S.; Phengsavanh, A.; Barennes, H. High HIV Vulnerability of Ethnic Minorities After a Trans-Asian Highway Construction in Remote Northern Laos. *Int. J. STD AIDS* **2012**, *238*, 570–575. [CrossRef] [PubMed]
108. Regondi, I.; George, G.; Pillay, N. Hiv/Aids in the Transport Sector of Southern Africa: Operational Challenges, Research Gaps and Policy Recommendations. *Dev. South. Afr.* **2013**, *30*, 616–628. [CrossRef]
109. Hanaoka, S.; Regmi, M.B. Promoting Intermodal Freight Transport Through the Development of Dry Ports in Asia: An Environmental Perspective. *IATSS Res.* **2011**, *35*, 16–23. [CrossRef]
110. Yokota, T. *Guideliness for Road Side Stations "Michinoeki"*; World Bank: Washington, DC, USA, Undated.

*Article*

# Low-Carbon Quick Wins: Integrating Short-Term Sustainable Transport Options in Climate Policy in Low-Income Countries

**Stefan Bakker [1,*], Gary Haq [2], Karl Peet [1], Sudhir Gota [1], Nikola Medimorec [1], Alice Yiu [1], Gail Jennings [1] and John Rogers [1]**

1 Partnership on Sustainable, Low-Carbon Transport, Rijnlaan 66, 3522 BO Utrecht, The Netherlands
2 Stockholm Environment Institute, Department of Environment and Geography, Environment Building, Wentworth Way, University of York, York YO10 5NG, UK
* Correspondence: sjabakker@gmail.com

Received: 5 June 2019; Accepted: 8 August 2019; Published: 12 August 2019

**Abstract:** In low income countries (LICs) in Africa and Asia per capita transport greenhouse gas emissions are relatively low but are expected to grow. Therefore, a substantial reduction in projected increases is required to bring emissions in line with long-term global climate objectives. Literature on how LICs are integrating climate change mitigation and sustainable transport strategies is limited. Key drivers of transport policy include improving accessibility, congestion, air quality, energy security, with reducing greenhouse gas emissions being of lower priority. This paper assesses the current status, feasibility and potential of selected low-carbon transport measures with high sustainable development benefits that can be implemented in the short to medium term, so- called 'quick wins'. It examines to what extent ten such quick wins are integrated in climate change strategies in nine low- and middle-income countries in Africa and South Asia. The research method comprises expert interviews, an online questionnaire survey of experts and policymakers in the focus countries, and a review of literature and government plans. Results indicate that sustainable urban transport policies and measures are considered high priority, with vehicle-related measures such as fuel quality and fuel economy standards and electric two- and three-wheelers being of key relevance. In existing national climate change strategies, these quick wins are integrated to a certain extent; however, with better coordination between transport and energy and environment agencies such strategies can be improved. A general conclusion of this paper is that for LICs, quick wins can connect a 'top-down' climate perspective with a 'bottom-up' transport sector perspective. A knowledge gap exists as to the mitigation potential and sustainable development benefits of these quick wins in the local context of LICs.

**Keywords:** low-income countries; low-carbon transport; sustainable mobility; climate change strategies; transport policy; Paris Agreement

---

## 1. Introduction

Greenhouse gas (GHG) emissions from transport are rising faster than any other economic sector [1]. Under a business-as-usual (BAU) scenario, global transport GHG emissions are expected to increase from 8 to 16 billion tonnes (t) by 2050. This poses a significant challenge to achieving long-term objectives outlined in the United Nations Paris Agreement, which set out a global action plan to avoid dangerous climate change [2]. Limiting global warming to below 1.5 °C will require GHG emissions being reduced to 2 to 3 gigatonnes (Gt) by 2050 and to up to 6 Gt for a 2 °C scenario [3].

Per capita transport GHG emissions in low-income countries (LICs) are approximately 0.1 to 0.5 t compared to high-income countries (HICs) at 2 to 5 t. The contribution of LICs to total global transport

GHG emissions is 0.5% [4,5]. Rapid motorisation in LICs is expected to increase future transport GHG emissions which threaten the long-term global climate objective of 0.2 to 0.6 t per capita for transport (assuming 10 billion in 2050).

LICs recognise the need to reduce transport emissions as part of international commitments, but this is not the primary driver behind many transport policy interventions. Other development concerns are higher on the political agenda such as improving urban mobility and accessibility, rural connectivity, efficient logistics, and sometimes energy security (i.e., reducing of oil imports) and health (e.g., improving in air quality and physical activity levels). For example, in Accra (Ghana) climate change was not prioritised in sustainable urban transport project assessment although air and noise pollution were included [6]. In Dhaka (Bangladesh) environmental criteria for sustainable urban transport included noise and air pollution but GHG emissions were not explicitly stated [7]. Within this context, low-carbon transport (LCT) can be seen as a co-benefit of sustainable transport policies (Figure 1).

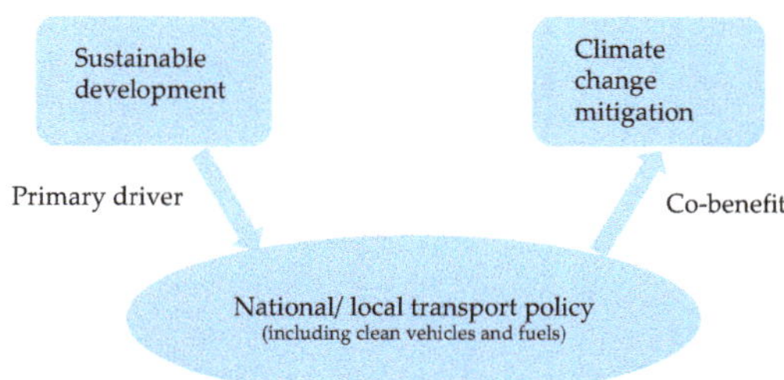


**Figure 1.** Sustainable development and low-carbon transport (source: authors, based on [8]).

LCT can be defined as transport sector developments that emit less GHG emissions than in a business-as-usual (BAU) scenario. LCT interventions are those that contribute to climate change mitigation no matter what the primary policy goals are. This framing is consistent with 'low-carbon development' in which national and local development objectives are the starting point, and opportunities are identified to achieve these in a manner which emits the least GHG emissions [9].

Under the United Nations Framework Convention on Climate Change (UNFCCC), countries are required to report national climate change strategies in the form of nationally determined contributions (climate action plans) (NDCs), national communications (NCs) and biennial update reports (BURs) [2]. These strategies could be seen as reflecting a 'top-down' view of climate change mitigation at the national level driven by international agendas. However, climate policy objectives need to be implemented in each sector at the national and local level in the form of policies and measures related to transport development, vehicles and energy systems.

Reaching long-term climate targets requires short-, medium- and long-term actions which are described in recent literature, a synthesis of which can be found in [1], which also shows the benefits for sustainable development objectives. More detailed global climate change mitigation scenarios for the transport sector are described in [3,10]. Key measures include fuel efficiency improvements for passenger and freight vehicles; renewable energy in transport; electrification of passenger cars, motorcycles, three-wheelers, buses and trucks; as well as transport demand management, logistics efficiency and shifting from private vehicles to more efficient modes.

Global progress on these measures is mixed and falls short of what is required for a well-below-2-degrees scenario, as shown in the Transport and Climate Change Global Status Report (TCC-GSR) [4]. For example, while fuel economy of new cars is improving globally, city-level sustainable mobility planning is becoming more common, more cities are implementing low-emission zones, and the stock of electric two-wheelers is increasing rapidly (mainly in China and Europe), there remains limited progress on electrifying cars and trucks [11], implementing more efficient road freight and logistics measures [12], realising envisioned shifts to more efficient transport modes, and increasing the share of renewables in the transport sector [4].

While there is a substantial body of literature on LCT at the global level and for developed countries, existing research on developing country mitigation pathways for transport is more limited and general. Figueroa et al. [13], for example, highlight the key role of informal transport and land-use planning, and Dhar et al. [14] link transport policies in India to the NDC. However, a research gap exists as to which solutions exist for LICs that can be implemented quickly, have GHG mitigation potential, and can contribute to climate strategies. Quick wins (QWs) are low-carbon transport policies and measures that can be implemented in the short to medium term, are relatively low-cost, and have high sustainable development benefits. This paper therefore aims to address the following research questions:

- What are the most suitable low-carbon transport QWs for LICs?
- Which QWs are priorities for stakeholders in LICs?
- What is the current status of implementation of QWs?
- To what extent are QWs integrated in national climate change strategies?

The research method and data sources used in the study are outlined in Section 2. This is followed by a discussion of the climate mitigation potential of QWs and how they are integrated in climate strategies (Section 3), i.e., the top-down perspective. A bottom-up analysis is undertaken of QWs in nine focus countries: six in Africa and three in Asia (Section 4), followed by a discussion of the results, main conclusions and recommendations.

## 2. Materials and Methods

The research method and data sources are summarised in the flow diagram in Figure 2 and described below. The corresponding sections are indicated in brackets.

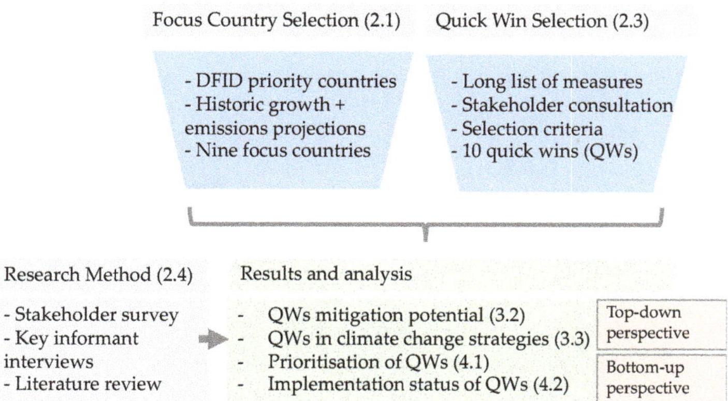


**Figure 2.** Study method (with section numbers).

### 2.1. Country Selection Criteria

The study focused on the UK Department of International Development's (DFID) priority countries in Africa and Asia: Bangladesh, Ghana, India, Indonesia, Kenya, Nigeria, Rwanda, South Africa, and Uganda. These nine countries were selected through a robust, quantitative process based on their need for low-carbon transport interventions. This was assessed using the following criteria: motorisation rates, transport emissions currently, and projected BAU transport emissions growth in the future.

### 2.2. Low-Carbon Transport Quick Wins Definition

Transport QWs are defined here as actions which can be taken in the short to medium term and contribute to the low carbon transformation of transport (i.e., a well below 2 degrees or even 1.5-degree scenario). They provide different benefits (e.g., access, efficiency, safety and environment benefits) to local and national stakeholders and are main policy drivers with climate change mitigation as a co-benefit. QWs have been tested at scale and are replicable with the possibility for large-scale impact. They are technically and economically feasible in both developed and developing countries using available technologies. Finally, QWs address both passenger and freight transport and should be relevant to the Avoid-Shift-Improve (ASI) concept [15]. This is a sustainable transport policy framework that emphasises avoiding and reducing trips (A), shifting to environmentally-friendly modes of transport (S), and improving (I) vehicle energy and carbon efficiency of each mode [16].

### 2.3. Selection of Transport Quick Wins for LICs

From March to July 2016, a six-step process (see [15]) was undertaken to develop a list of QWs. An email invitation was sent to more than 100 international organisations working on sustainable transport and resulted in the compilation of a long list of over 100 measures. Through grouping and balancing measures across themes, modes and world regions, the list was reduced to approximately 40 options. These measures were further evaluated against sustainable development benefits, implementation barriers and coverage of 'Avoid-Shift-Improve' measures, and both passenger and freight transport. Expert feedback was received on the remaining 23 measures while an on-line survey of more than 100 stakeholders provided feedback on the definition and feasibility of the QWs. Finally, a more detailed assessment against the following criteria provided a final list of 10 QWs specifically for the purpose and context of the analysis:

- Sustainable development benefits, such as air quality and improved accessibility;
- Applicability in DFID priority countries that reflect policy needs of the local transport system context;
- Climate change mitigation potential; and
- Passenger and freight including avoid-shift-improve options.

The final 10 QWs for this analysis are:

1. Accelerate phase-out of fossil fuel subsidies;
2. Formulate Sustainable Urban Mobility Plans (SUMPs) in primary and secondary cities, supported by a National Urban Mobility Policy or programme;
3. Promote electric two- and three wheelers, including e-vehicle sharing systems in primary and secondary cities as well as rural areas, for both passenger mobility as well as last-mile urban freight;
4. Limit import of inefficient and polluting secondhand trucks;
5. Implement (ultra-) low emission zones, including car-free zones in city centres;
6. Introduce and scale up pricing for car-related travel options (e.g., congestion/road charging, parking pricing, workplace parking levy) in primary and secondary cities;
7. Tighten fuel economy standards for passenger cars, coupled with labelling schemes and fiscal incentives such as carbon dioxide ($CO_2$) based vehicle taxation;
8. Provide and improve walking and cycling infrastructure (e.g., connected walking paths, protected cycle lanes, safe intersections), reallocating road space where necessary;
9. Improve freight efficiency (e.g., reduce empty load running by freight trucks) through route optimisation, asset sharing between companies, and increased use of information and communication technology (ICT) solutions;
10. Accelerate deployment of tighter diesel fuel quality standards to reduce emissions of black carbon (BC) and other short-lived climate pollutants.

### 2.4. Method for Quick Wins Assessment

Primary data for the QW assessment were collected from interviews and a questionnaire survey. From October to December 2018, semi-structured key informant interviews were undertaken with 23 experts from the nine countries (2–3 per country). Experts were from national and local transport government agencies, research institutions and implementing agencies. Among other questions, they were asked to rate each QW option on a scale from 1 to 5 on their appropriateness in the short to medium term for their country. Respondents were also able to suggest additional QWs.

From September to October 2018, a survey of low carbon, high volume transport knowledge and capacity in Africa and South Asia was undertaken of the stakeholder network of the Partnership for Sustainable, Low-carbon Transport. The online tool SurveyMonkey was used. The survey included qualitative and quantitative questions exploring how LCT knowledge is gained, how capacity is achieved, and how transport users and other affected parties are engaged. As part of one question in the survey, respondents were requested to rank the 10 QWs from 1 to 10 according to their appropriateness for the respondent's country. The survey was sent to 816 experts and received 67 respondents with 49 completing the question on QWs. The 67 respondents were equally represented across government agencies, private sector, research institutions and implementing agencies. However, most of the respondents were from African countries.

Secondary data on QWs implementation, integration in climate plans and mitigation potential was collected from existing peer-reviewed and grey literature as well as technical reports, government plans and reports, and additional sources such as online media. See below for further details on this data collection process. The findings from the literature review were validated and elaborated by stakeholder interviews.

## 3. Results and Analysis Part 1: A Top-Down Climate Change Perspective on Transport Quick Wins

This section briefly discusses general literature on low-carbon transport in the nine focus countries and then looks at the mitigation potential of quick wins and to what extent these are present in climate strategies in the focus countries.

### 3.1. Mitigation Potential of Quick Wins

Literature on transport and climate change mitigation exists for some of the focus countries, including economy-wide and transport sector-specific low-carbon scenarios. Gota et al. [5] compiled a comprehensive global database of mitigation studies and reports for 81 countries including six out of the nine countries. Transport was one of the sectors included in those economy-wide mitigation potential modelling studies; as these are based on energy-economy models, the level of detail is limited. These studies show that the reduction potential (averaged across multiple studies for one country) below BAU for the transport sector in 2050 is 62% for Bangladesh, 65% for India, 61% for Indonesia, 82% for Nigeria and 61% for South Africa, while for Kenya the potential is 37% for 2030.

Transport sector-specific literature shows more detail on the mitigation potential of quick wins. Among other low-carbon options, these show the importance of some of the quick wins as well. In India, Dhar et al. [17] modelled low-carbon scenarios for transport for a 1.5 degree scenario, with vehicle fuel efficiency, transport demand management (in passenger and freight), biofuels, modal shift (in passenger and freight), and electric vehicles (including two-wheelers) playing a key role. For Bangladesh, Gota & Anthapur [18] developed low-carbon freight scenarios, also estimating black carbon emissions, that consider broad Avoid, Shift and Improve strategies. Siagian et al. [19] modelled an economy-wide scenario for Indonesia, in which energy efficiency and biofuel use in transport can help in achieving NDC targets. In Kenya, e-scooters and vehicle efficiency for cars and trucks are included in the transport sector low-carbon scenarios [20]. Finally, Stone et al. [21] developed a detailed model to estimate historical fuel demand for road freight and passenger vehicles in South Africa. In short, there is sector-wide low-carbon transport literature for a few of the nine countries; however, the scenarios include few short-term policy options and limited detail on these options, and the freight sector in particular is covered only to a limited extent.

### 3.2. Mitigation Potential of Quick Wins

A literature review was undertaken on the climate mitigation potential of the 10 QWs including global studies as well as research on the nine focus countries. The main methodology was a keyword search in Scopus and Google Scholar. Keywords included the 10 QWs, as well as "low-carbon transport", in combination with the names of the nine countries. In addition, relevant references from these articles were used. This resulted in 24 peer-reviewed articles, of which 14 cover climate change potential and 10 are indirectly relevant. In addition, 17 grey literature sources such as technical reports from international organisations were found by searching in Google using similar keywords. Literature was from the past five years (2013–2018) with a few sources from before this period if more recent sources were unavailable.

Table 1 summarises literature on transport sector GHG mitigation and shows the importance and potential of all 10 QWs in the nine selected countries.

For India, most mitigation options are included in low-carbon scenarios or studies for specific measures. For Indonesia, several studies are available as well. For Bangladesh and the six African countries, literature is limited, except for options related to fuel efficiency and fuel standards, where international organisations have done extensive analysis (even if not specifically on GHG mitigation potential). For some specific options (e.g., limiting imports of used trucks and introducing low emission zones) more analysis would be beneficial.

**Table 1.** Mitigation potential literature for 10 LCT (low-carbon transport) quick wins.

| Quick Win | Key Points from Mitigation Potential Literature (Global) | Mitigation Literature Country-Specific |
|---|---|---|
| 1. Fossil fuel subsidy phase out | Removal of fossil fuel consumption subsidies could lead to global GHG emission reductions of 2–4% by 2020, rising to 8–12% by 2050 [22]. Removal of fossil fuel subsidies is a prerequisite to carbon taxation, which is required to achieve a beyond 2-degree scenario [3]. | For India and Indonesia, fuel subsidy reform could lead to between 1 and 9% GHG savings in 2030 [23]. For Ghana, removal of subsidies could result in negative impact on household welfare [24]. Phasing out energy subsidies could reduce Indonesian $CO_2$ emissions from fuel combustion by 11–13% in 2020 [25] |
| 2. Sustainable Urban Mobility Plans, National Urban Mobility Programme | Urban passenger transport emits about 25% of total transport sector emissions. SUMPs mainly focus on non-technology options, i.e., 'avoid' and 'shift', which contribute 2–40% of emission reductions in the 2050 low-carbon scenario [10]. Implementation of a SUMP in Burgos (Spain) resulted in 17% lower $CO_2$ emissions [26]. Pisoni et al. model impact of SUMPs on air quality [27]. | For the 1.5-degree scenario in India, demand-side urban transport measures are essential [17], and $CO_2$ emissions is one of the key indicators in comprehensive mobility plans in India [28]. Urban transport measures in 7 Indonesian cities, supported by a national urban transport framework, can save 0.1–0.2 $tCO_2$ per capita in 2030 [29] |
| 3. Electric two- and three-wheelers | Over 80% of the 29 Mt $CO_2$ savings in 2017 by all types of EVs globally are due to e-bikes in China [11]. Full decarbonisation of two- and three-wheelers is necessary for the beyond 2 degrees scenario [3]. For Vietnam, e-bikes are the mitigation option with the second-largest potential in the transport sector [30]. In Thailand, deploying electric motorcycles could reduce two-wheeler life cycle $CO_2$-eq emissions by 42–46% [31]. | For a 2-degree scenario in India, over 90% of two-wheelers should be electric [14]; however, energy-use of two- and three-wheelers varies depending on driving conditions [32]. There are cost and $CO_2$ emission savings for electric tricycles in Nigeria [33]. A study shows benefits of electric two-wheelers for Africa (no $CO_2$ estimates) [34] Benefits are shown for ojek (motorcycle taxi) drivers by switching to electric vehicles in Indonesia [35]. There is a rapidly expanding market in Africa for motorcycles and boda boda (motorcycle taxis) in particular [36]. |
| 4. Limit import of inefficient and polluting secondhand trucks | Import restrictions for secondhand vehicles as a key part of the policy package in a global low-sulphur scenario [37]. Fuel efficiency of vehicles declines rapidly after 15 years of use, up to 50% by 25 years [38]. | Import policies are considered in the fuel efficiency scenario in a green freight study for Bangladesh [18]. A low-carbon scenario for Kigali (Rwanda) considers vehicle age restrictions [39]. |

**Table 1.** *Cont.*

| | Quick Win | Key Points from Mitigation Potential Literature (Global) | Mitigation Literature Country-Specific |
|---|---|---|---|
| 5. | Low-emission zones | Impact assessments show some impact of environmental zones in EU cities on the impact on PM/soot emissions [40]. Further climate benefits would accrue from more EV deployment; however, no study has been found estimating the GHG impact. | No sources found. |
| 6. | Pricing of car use | There is a relatively strong knowledge base of ex-post and ex-ante studies on road pricing and parking management [41,42]. In Singapore, a package of measures including congestion charging and $CO_2$-based vehicle taxation results in low transport emissions per capita [1]. | Study on congestion pricing in Delhi shows significant shift from private vehicles to public transport [43]. |
| 7. | Fuel economy standards and incentives | Regulation on the energy-use and lifecycle GHG emissions of vehicles is necessary for a beyond 2 degrees scenario [3] and progress is seen in many countries [44]. The IPCC presents emission intensity reduction potentials for different types of diesel and petrol vehicles [45]. | Progress in fuel economy policies in major markets, including India, Indonesia and South Africa is reviewed [44]. Fuel efficiency strategies are included in low-carbon transport scenario for India [14]. |
| 8. | Non-motorised transport (NMT) infrastructure | Globally, it is estimated that in 2050, 22% of urban passenger travel can be by (e)bike, compared to 6% in the base case. This results in 300 $MtCO_2$ reductions in 2050 and USD 1 trillion in savings from vehicle purchase and operation and construction and maintenance of infrastructure [46]. For walking, no specific mitigation potential estimates have been found; however, it is acknowledged for its key role in mitigation [45] and reaching public transport modal shift targets. | Sustainable urban transport scenario for Bangalore includes NMT and transit-oriented development (TOD) and $CO_2$ estimates [47]. Low-carbon scenario for Kigali (Rwanda) quantifies $CO_2$ savings from bike lane investments [39]. In Ghana, three out of four would not cycle to social events for public image; health is driver [48]. Research on walking and cycling in African cities, including in Kenya and South-Africa [49]. |
| 9. | Logistics optimization/ freight efficiency | The IEA [12] analyses 15 measures, e.g., urban consolidation centres, platooning, co-modality, backhauling, retiming of deliveries, etc. Most of these could have a best-case impact of up to about 5% emission reductions, while some measures may have a reduction potential over 10%. Implementation of these measures, including in developed countries, is still in an early stage. | A study $CO_2$ scenario with freight efficiency for Bangladesh [18]. A report highlights many measures (no mitigation potential calculation) and note that logistics is 7% of India's total $CO_2$ emissions and 67% of transport PM [50]. Green freight programme for Northern Corridor (including Rwanda, Kenya) with measures, in context of mitigation and air pollution (objective of 10% reduction in $CO_2$ per ton-km) [51]. |
| 10. | Diesel quality standards | A mitigation scenario for black carbon (BC) reduces such emissions by about half, corresponding to about 4 $GtCO_2$-eq (GWP100) in 2050 [52]. Diesel road vehicles and ships are one of the main sources of BC emissions with 19% of global BC emissions [53]. Research shows that a global sulphur scenario reduces BC emission from diesel road transport by about 90% from the baseline in 2040 [37]. Such strategies will result substantial health benefits from reduced exposure to air pollution. | Report includes country level market analysis for low-sulphur diesel for Bangladesh, Ghana, India, Indonesia, Kenya, Nigeria, Rwanda, South Africa, and Uganda [54]. Diesel vehicles contribute 20–55% of total BC in South Asian cities [55]. |

### 3.3. *Integration of Quick Wins in Climate Change Strategies*

To assess how QWs are integrated in climate change strategies of the nine countries, an analysis of the Paris Agreement reporting mechanisms (NDCs, BURs and NCs) (a methodology used by for example Stead [56]) and other strategies, such as climate change strategies (CCSs), climate change action plan (CCAP), low-carbon development strategy (LCDS) and National Climate Change Policy (NCCP) was undertaken. In Table 2, 'Y' means the QW is in at least one of the plans/reports. In most cases a QW is not included in all of these. One reason for this is that NDCs are often short whereas BURs and NCs present measures in more detail. In addition, it should be noted that all countries have submitted an NDC and at least one NC, whereas five have submitted one or multiple BURs, and five have additional national strategies.

**Table 2.** Quick wins in climate change strategies and reports.

| Country | Bangladesh | Ghana | India | Indonesia | Kenya | Nigeria | Rwanda | South Africa | Uganda |
|---|---|---|---|---|---|---|---|---|---|
| Reports Analysed | NDC, NC2 (2012) | NDC, BUR2, NC3, CC policy | NDC, NC2, BUR1, LCS | NDC, NC3, BUR | NDC, NC, CCAP 2013 | NDC, NC2, BUR1 | NDC, NC2, LDCS 2011 | NDC, NC3, BUR2 | NDC, NC2, NCCP 2015 |
| 1. Phase-out fossil fuel subsidies | | Y | Y | | | Y | | | |
| 2. SUMPs and NUMP | | | | | | | | | |
| 3. Promote electric two- and three-wheelers | | | Y | | | | | | |
| 4. Limit import of 2nd hand trucks | | Y | | | | | Y | | Y |
| 5. Low emission zone (LEZ) | | | | | | | | | |
| 6. Pricing for car-related travel | | | Y | Y | | Y | | | |
| 7. Fuel economy policies | Y | | Y | Y | | Y | Y | Y | Y |
| 8. NMT infrastructure | | Y | Y | Y | | | Y | Y | Y |
| 9. Freight efficiency/logistics | | | | | | | | | |
| 10. Diesel quality standards | | Y | Y | | | | Y | | Y |

NDC: nationally determined contribution; NC: national communication; BUR: biennial update report; LC(D)S: low-carbon (development) strategy; CC: climate change; NCCP: national climate change policy.

Fuel economy policies and non-motorised transport (NMT) infrastructure are the QWs that are included most often: six or more countries do so. Fossil fuel subsidy reduction, import restrictions, pricing measures, and diesel quality standards are included by 3–5 countries. SUMP, electric two- and three-wheelers, low emission zone (LEZ) and freight efficiency by zero or one country.

## 4. Results and Analysis Part 2: A Bottom-Up Transport Perspective on Quick Wins for Focus Countries

### *4.1. Perceived Feasibility of Quick Wins for Focus Countries*

Figure 3 shows perceived feasibility of transport QWs according to the survey responses. Respondents regarded the following QWs as most feasible in their countries (noting that data are not sufficient to suggest a proposed ranking of these QWs on a country or regional basis):

- QW 2 Formulate Sustainable Urban Mobility Plans (SUMPs) in primary and secondary cities, supported by a National Urban Mobility Policy or programme.
- QW 8 Provide and improve walking and cycling infrastructure (e.g., connected walking paths, protected bicycle lanes), reallocating road space where necessary.
- QW 3 Limit imports of inefficient and polluting secondhand trucks, complemented by age limitations for the existing fleet.
- QW 4 Promote electric two- and three-wheelers (including shared e-vehicles) in primary and secondary cities.
- QW 1 Accelerate phase-out of fossil fuel subsidies.

Other key options from respondents include diesel quality standards, increasing freight efficiency, and paratransit reform and regulation. The paratransit reform and regulation option refers to changing from unregulated mini-buses and three-wheelers to a system where routes are organised by a regulator, potentially in support of high-quality public transport such as metro or bus rapid transport (BRT). Several respondents suggested this as an option to be added to the 10 QWs.

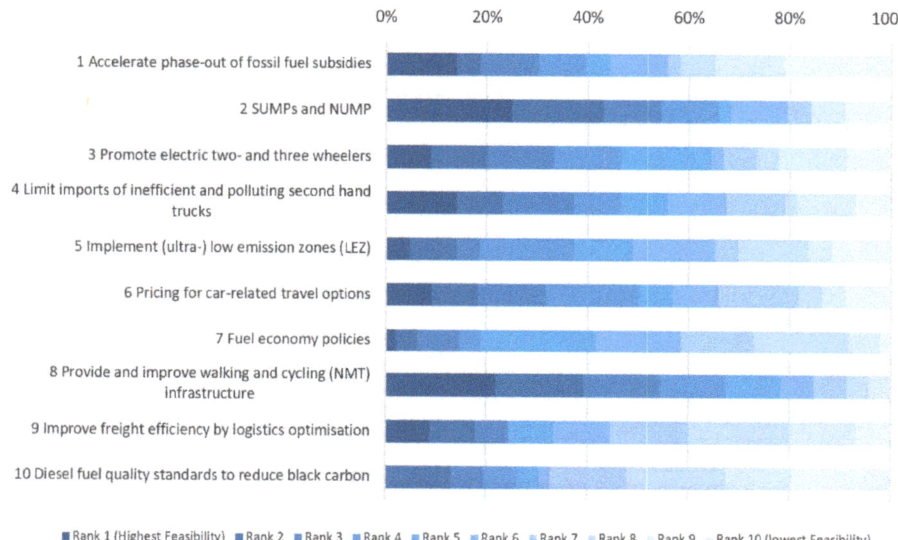


**Figure 3.** Perceived feasibility of the low-carbon transport quick wins.

### 4.2. Implementation Progress of QWs

A review of planning and implementation of the QWs in the nine selected countries was also undertaken. The analysis was based mainly on peer-reviewed and grey literature from 2017–2018 which included reports and policy briefs, government official documents, and if no other sources available, news articles from online media. All sources are included in a matrix (see Supplementary Materials S2) with Table 3 presenting the summary assessment.

Methodology for the rating of the implementation status:

0 No implementation or discussion of the options in the policy domain
* Measure in discussion or pilot implementation
** Policy partial in place or planned, or partial implementation
*** Full-scale implementation

Three researchers independently rated implementation based on the same literature.

The detailed application of this methodology varies depending on the nature of each QW. For example, evaluation of regulatory measures such as diesel quality standards or limiting imports of used trucks import are more straightforward than more diverse measures such as improving freight efficiency or NMT infrastructure, which require a multitude of smaller interventions (see S2 Matrix on quick win implementation in Supplementary Materials).

**Table 3.** Assessment of quick win implementation status in focus countries.

| | Country | Bangladesh | Ghana | India | Indonesia | Kenya | Nigeria | Rwanda | South Africa | Uganda |
|---|---|---|---|---|---|---|---|---|---|---|
| 1. | Phase-out fossil fuel subsidies | ** | ** | *** | ** | *** | * | ** | * | * |
| 2. | SUMPs and NUMP | ** | * | ** | ** | ** | 0 | * | ** | * |
| 3. | Promote electric two- and three-wheelers | ** | * | *** | * | ** | * | ** | ** | ** |
| 4. | Limit import of 2nd hand trucks | * | * | *** | *** | ** | ** | ** | *** | * |
| 5. | Low-emission zones | 0 | 0 | 0 | 0 | 0 | 0 | 0 | 0 | 0 |
| 6. | Pricing for car-related travel | * | 0 | ** | ** | * | ** | * | ** | * |
| 7. | Fuel economy policies | * | * | *** | ** | * | * | * | ** | * |
| 8. | NMT infrastructure | * | ** | ** | ** | ** | ** | ** | ** | ** |
| 9. | Freight efficiency/logistics | * | * | ** | ** | * | * | * | ** | * |
| 10. | Diesel quality standards | ** | ** | *** | ** | *** | ** | *** | ** | *** |

There is progress across most or all nine countries for QW 4 (Limiting imports of used trucks) and QW 10 (Diesel quality standards). This could indicate an increased awareness of air pollution issues with both these options playing a key role in addressing diesel emissions. In addition, a number of international organisations are also promoting these actions (e.g., the International Council on Clean Transportation and UN Environment).

Options where progress varies among the nine countries are the following:

- QW 1 (Fossil fuel subsidy reduction): this is a politically challenging option, where public opposition to policy changes can be expected [22];
- QW 2 (SUMPs and NUMP): India and South Africa have a nationally-guided programme for cities, while in other countries only a few cities develop SUMP-like mobility plans. Awareness of SUMPs and NUMPs as a key policy tool is picking up in recent years. NUMPs are challenging especially because the national government is often reluctant to allocate financial resources to cities, with the local governments in turn not seeing the benefit of planning guidelines from a higher-level authority;
- QW 3 (Promoting electric two and three wheelers): two-wheelers are not popular with policymakers [57], who see motorcycle drivers as reckless and often involved in traffic crashes, rather than a flexible, fast and space and energy-efficient mode of transport. However, electrification is increasingly acknowledged as part of air quality and energy security strategies;
- QW7 (Fuel economy policies) with for example India adopting relatively ambitious standards. Although fuel economy policies are beneficial to the national economy, implementing these could be politically challenging due to potential impacts on the car market and manufacturers;

- QW 8 (NMT infrastructure): many cities are examining this option and consider it important, yet implementation is limited, with unsafe and inconvenient conditions for walking and cycling. Allocation of government budget to the various transport modes is a key issue, with for example the political economy in Ghana favouring road investments over NMT and rail [58].

There is little progress on the following options: QW 5 (Low-emission zones), which is not discussed in any project country yet with no clear examples in other LICs and MICs; QW 6 (Pricing for car-related travel) is being considered and discussed but with little implementation; and QW 9 (Freight efficiency improvements) with progress limited to isolated projects in some countries. This may be due to governments considering freight as the domain of the private sector and the complexity of such projects and plans.

Tables 2 and 3 show that more QW policy activity is taking place than reported in climate plans. Table 2 shows that diesel quality standards and import restrictions are included in only a few climate strategies. In addition, electric two- and three-wheelers are missing in climate plans, even though there has been policy progress in recent years, for example in India's electric vehicle support scheme FAME [59]. To some extent the same is true for SUMPs and freight efficiency measures. On the other hand, car-related pricing measures are mentioned in climate plans but with limited progress in policy development.

Although the reasons for these reported differences are beyond the scope of this paper, possible explanations could be that for some measures (e.g., vehicle import restrictions, diesel quality standards and SUMPs) the climate benefits, which are partially based on the warming potential of black carbon, are not explicitly recognised. It may also be that in climate plans, urban passenger transport and alternative fuels are better represented than freight and two- and three-wheelers [60].

It may also be noted that climate change reports are often from 2016 or earlier, whereas recent literature from 2017 and 2018 was used in Table 1. Electric two- and three-wheelers and SUMPs may have only recently received more attention and are therefore better reflected after 2017.

In summary, from an examination of the QW mitigation potential, QW in climate plans (top-down perspective, Section 3), QW implementation status, and QW perceived feasibility (bottom-up perspective, Section 4), the following observations can be drawn:

- Freight efficiency measures appear to be less prioritised in literature, climate strategies, policy implementation and by stakeholders.
- Fuel efficiency policies are acknowledged as key in literature, but lack in implementation and priority by stakeholders.
- SUMPs and NUMPs, as well as NMT infrastructure, are seen as a key option but implementation is lagging behind.
- Attention by stakeholders and literature coverage for electric two- and three-wheelers is increasing, especially in recent years in Asian and African countries.
- Improving diesel quality standards is considered important in literature and by stakeholders from a local air pollution and health perspective, with substantial co-benefits due to the climate warming potential of black carbon.
- Little attention is given to LEZs in the climate change context, even though these may play a key role in promoting electric vehicles [61].

It should also be noted that some of the highly rated QWs are measures that are typical for LICs and MICs and are often less relevant for HICs, for example diesel quality standards, truck import limitations, electric two- and three-wheelers, and paratransit reform.

## 5. Discussion

### 5.1. Interpretation of Results and Limitations of Data and Method

The results presented here can be considered fairly robust. Nevertheless, the following caveats are noted. In the primary data collection (online questionnaire survey and interviews), nearly all of the respondents are from the transport sector. They may have knowledge about energy and environmental issues, yet their perspective and priorities may differ from experts that are primarily from the energy, industry and environmental sectors. This may impact the QW prioritisation outcome to some extent. In addition, most of the survey and interview respondents are from Africa.

The literature review covered a range of sources from peer-reviewed literature, grey literature, government reports and media articles, mostly from the last two years and nearly all more recent than 2012. It covered all focus countries and QWs. Although non-exhaustive, it provides an up-to-date picture of the literature on QW implementation and mitigation potential in the nine countries. The scoring of the implementation status involved a certain degree of subjectivity, both from a methodological perspective in assessing level of policy development and scope. For example, a large country such as India is not comparable to a small country like Ghana when it comes to implementation of NMT at the local level. Finally, the mentioning of a policy measure in climate plans does not necessarily mean it will be implemented and in what way (i.e., scope and level of ambition).

### 5.2. Climate Change Strategy Development Process and Quick Wins

Climate policy is still a relatively new phenomenon. Since the adoption of the Paris Agreement in 2015 it is higher on the political agenda, and developing countries realise that the climate change policy framework is here to stay. The Paris Agreement climate reporting mechanisms (NDC/BUR/NC) have resulted in developing countries thinking about aspects of low-carbon transport, as shown in [8] for Southeast-Asian countries.

LCT strategies require actions from a variety of government agencies. Transport-related agencies such as national or local transport or planning bodies develop and implement Avoid and Shift interventions while Improve measures related to vehicles are often in the mandate of other agencies such as national ministries of environment, energy, finance and industry [62]. Some options may involve all these stakeholders. Therefore, to develop a comprehensive climate strategy for the transport sector, collaboration and coordination between these various agencies is required. The NDC process has been shown to help inter-ministerial coordination for LCT in some Southeast-Asian countries, including in Indonesia [8]. However, various agencies may have different and sometimes conflicting objectives, for example, producing cars and reducing congestion, as was raised by an Indonesian respondent.

QWs may be an opportunity to improve integration of transport in climate change plans. However, more actions are being developed and taken than currently included in climate change plans. These can be discussed in future inter-ministerial processes for LCT strategies. In this way, QWs can act as an important bridge between the top-down and bottom-up perspectives on LCT, as described in Sections 3 and 4 above (Figure 4).

To further promote development of QWs as well as their integration with climate strategies, more information and analysis about how to implement these is required. This also requires more information about the mitigation potential of QWs for each local context, as well as distributional effects on various social groups and sustainable development benefits. International organisations such as the International Energy Agency, UN Environment and multilateral development banks also have a role to play as various global or regional initiatives are supporting QWs, in particular diesel quality standards, fuel economy policy, SUMPs/NUMP, NMT, electric vehicles (though not focusing specifically on two- and three-wheelers) and green freight [15]. Examples of such initiatives include the Global Fuel Economy Initiative, Mobilise Your City, Global Green Freight Action Plan and the Electric Vehicles Initiative.

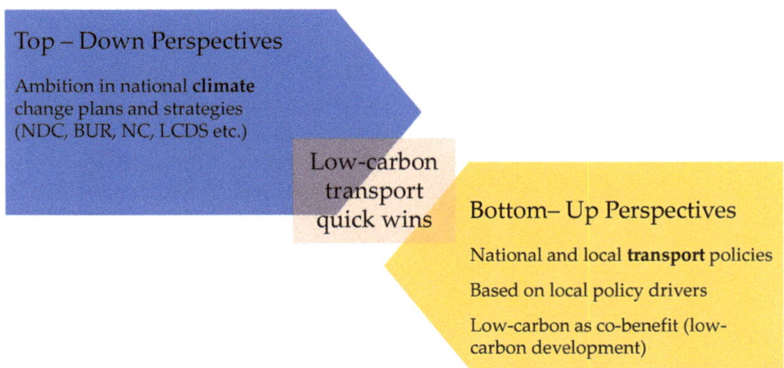


**Figure 4.** Quick wins can connect top-down climate and bottom-up transport perspectives on low-carbon transport. (Source: authors).

### 5.3. Broader Context of Long-Term Decarbonisation

QWs have a key role to play in climate change mitigation; however, they are only part of the package of measures that is required to bring the transport sector onto a 1.5-degree pathway. Moreover, QWs do not substitute large-scale investments in low-carbon infrastructure (e.g., rail, waterways, and electric vehicle charging) infrastructure and vehicles that have to be taken in the short to medium term in order to avoid a long-term lock-in into high-carbon transport systems. Given the long lead time of planning required for such infrastructure to be operational, decision-making processes require a focus on low-carbon infrastructure investment.

## 6. Conclusions

For transport policy makers in LICs, climate change mitigation is generally not a high priority. Therefore, GHG reduction can be seen as a co-benefit of sustainable mobility. LCT interventions are those that are implemented to improve access, reduce congestion, increase equity, improve air quality and energy security, yet that also contribute to climate change mitigation. In this way, the transport sector can contribute to achieving climate objectives stipulated in the NDCs.

LCT QWs are interventions with high sustainable development benefits and can be implemented in the short-term. They can be seen as a bottom-up approach to integrate sustainable mobility options in climate change strategies, or in other words, can operationalise low-carbon development in the transport sector. This paper presents 10 such QWs. As the climate change mitigation potential of these is significant, they can play an important role in short-term decarbonisation of the transport sector and help to achieve long-term targets required under the Paris Agreement.

Analysis of nine LICs in Africa and Asia shows that fuel economy policies and NMT infrastructure are the QWs that are included most often in climate change strategies: six or more countries do so. Fossil fuel subsidy reduction, secondhand vehicle import restrictions, pricing measures, and diesel quality standards are included by 3–5 countries. SUMPs, electric two and three-wheelers, LEZ and freight efficiency by zero or one country. However, when investigating policy development in the nine countries, it appears more is happening than the climate change reports are showing. For example, most countries are pursuing diesel quality standards, import restriction, and, to a lesser extent, promotion of electric two- and three-wheelers, even though the climate reports do not show this. It should also be noted that these three options together with, for example, paratransit reform, are measures that are typical for LICs and MICs and are less relevant for high-income countries. In addition to these options, stakeholders in the nine countries prioritised SUMPs, NMT infrastructure, and fossil fuel subsidy reduction. However, low-carbon freight options are emphasised less by stakeholders, literature, and current policy development, yet freight is responsible for 36% of carbon dioxide emissions from the land transport [12].

Based on the analysis of LCT QWs for LICs, the following recommendations can be made. Firstly, more country-level analysis on the various costs and benefits, including distributional impacts, mitigation potential and sustainable development impacts of QWs, would be beneficial in enhancing the knowledge base and awareness. Secondly, each QW requires a detailed analysis of design options to implement the measure, specific for each local or national context. Finally, improved coordination and collaboration between transport authorities (mainly Avoid and Shift measures) and agencies focusing on energy, industrial and environmental aspects of vehicles (mainly Improve measures) is key in bridging the top-down and bottom-up gap and achieving low carbon transport in LICs in Asia and Africa.

This paper contributes to a growing body of literature that connects transport and climate policy. It particularly increases the understanding of how transport-related policies and measures can support climate strategies in LICs while also contributing to sustainable development.

**Supplementary Materials:** The following materials are available online at http://www.mdpi.com/2071-1050/11/16/4369/s1, Matrix on low-carbon transport measures in climate plans, and http://www.mdpi.com/2071-1050/11/16/4369/s2, Quick wins implementation status matrix.

**Author Contributions:** Conceptualization, S.B., K.P.; methodology, S.B., K.P., S.G., N.M., G.J., G.H.; investigation, S.B., J.R., G.J., N.M., S.G.; formal analysis, S.B., N.M.; writing—original draft preparation, S.B.; visualization, S.B., N.M.; supervision, K.P., G.H.; writing—review and editing, G.H., K.P., G.J.; project administration, A.Y.; funding acquisition, A.Y., K.P. All authors have reviewed the entire manuscript.

**Funding:** This research was funded by the UK Department for International Development's (DFID) Applied Research Programme in High Volume Transport, Theme 3—Low Carbon Transport by DFID.

**Acknowledgments:** Feedback on a draft version by DFID and the comments by the anonymous reviewers are greatly appreciated.

**Conflicts of Interest:** The authors declare no conflicts of interest.

## References

1. IPCC. *Global Warming of 1.5 °C. An IPCC Special Report on the Impacts of Global Warming of 1.5 °C Above Pre-Industrial Levels and Related Global Greenhouse Gas Emission Pathways, in the Context of Strengthening the Global Response to the Threat of Climate Change, Sustainable Development and Efforts to Eradicate Poverty*; Masson-Delmotte, P., Zhai, H.O., Pörtner, D., Roberts, J., Skea, P.R., Shukla, A., Pirani, W., Moufouma-Okia, C., Péan, R., Pidcock, S., et al., Eds.; World Meteorological Organization: Geneva, Switzerland, 2018.
2. United Nations 2015 Paris Agreement. Available online: http://unfccc.int/files/essential_background/convention/application/pdf/english_paris_agreement.pdf (accessed on 20 May 2019).
3. International Energy Agency. *Energy Technology Perspectives 2017, Catalysing Energy Technology Transformations*; OECD/IEA: Paris, France, 2017.
4. Peet, K.; Gota, S.; Huizenga, C.; Medimorec, N.; Enriquez, A.; Yiu, A. *Transport and Climate Change 2018 Global Status Repor*; Partnership on Asutainable Low Carbon Transport, 2018; 184p. Available online: http://slocat.net/tcc-gsr (accessed on 20 May 2019).
5. Gota, S.; Huizenga, C.; Peet, K.; Medimorec, N.; Bakker, S. Decarbonising transport to achieve Paris Agreement targets. *Energy Effic.* **2019**, *12*, 363–386. [CrossRef]
6. Jones, S.; Tefe, M.; Appiah-Opuku, S. Proposed framework for sustainability screening of urban transport projects in developing countries: A case study of Accra, Ghana. *Trans. Res. Part A* **2013**, *49*, 21–34. [CrossRef]
7. Munira, S.; San Santoso, D. Examining public perception over outcome indicators of sustainable urban transport in Dhaka city. *Case Stud. Trans. Policy* **2017**, *5*, 169–178. [CrossRef]
8. Bakker, S. Shifting to Low-Carbon Transport in ASEAN: Policy Development in a Rapidly Motorising Region. Ph.D. Thesis, University of Twente, Enschede, The Netherlands, 29 March 2018. Available online: https://ris.utwente.nl/ws/portalfiles/portal/104186485/bakker_2.pdf (accessed on 20 May 2019).

9. Van Tilburg, X.; Würtenberger, L.; de Coninck, H.; Bakker, S. Paving the Way for Low-Carbon Development Strategies. Report ECN-E-11059. Petten: Energy Research Centre of The Netherlands, 2011. Available online: https://www.researchgate.net/publication/236594280_Paving_the_way_for_Lowcarbon_development_strategies (accessed on 20 May 2019).
10. International Transport Forum. *ITF Transport Outlook 2017*; OECD Publishing: Paris, France, 2017. Available online: https://doi.org/10.1787/9789282108000-en (accessed on 20 May 2019).
11. International Energy Agency. *Global EV Outlook 2018. Towards Cross-Modal Electrification*; OECD Publishing: Paris, France, 2018.
12. International Energy Agency. *The Future of Trucks. Implications for Energy and the Environment*; OECD Publishing: Paris, France, 2017; 164p. Available online: https://webstore.iea.org/the-future-of-trucks (accessed on 20 May 2019).
13. Figueroa, M.; Fulton, L.; Tiwari, G. Avoiding, transforming, transitioning: pathways to sustainable low carbon passenger transport in developing countries. *Curr. Opin. Environ. Sustain.* **2014**, *5*, 184–190. [CrossRef]
14. Dhar, S.; Shukla, P.; Pathak, M. India's INDC for Transport and 2 °C Stabilization Target. *Chem. Eng. Trans.* **2017**, *56*, 31–36. [CrossRef]
15. Quick Wins on Transport, Sustainable Development and Climate Change. Kick-Starting the Transformation of the Transport Sector. Available online: http://www.ppmc-transport.org/wp-content/uploads/2016/11/SLoCaT-Quick-Wins-Report-1.pdf (accessed on 20 May 2019).
16. Bakker, S.; Zuidgeest, M.; de Coninck, H.; Huizenga, C. Transport, development and climate change mitigation: Towards and integrated approach. *Trans. Rev. A Trans. Transdiscipl. J.* **2014**, *34*, 335–355. [CrossRef]
17. Dhar, S.; Pathak, M.; Shukla, P. Transformation of India's transport sector under global warming of 2 °C and 1.5 °C scenario. *J. Clean. Prod.* **2018**, *172*, 417–427. [CrossRef]
18. Gota, S.; Anthapur, S. *Advancing Green Freight in Bangladesh: A Background Paper*; Clean Air Asia: Manila, Philippines, 2016. Available online: http://www.ccacoalition.org/es/node/1315 (accessed on 20 May 2019).
19. Siagian, U.; Yuwono, B.; Fujimori, S.; Masui, T. Low-Carbon Energy Development in Indonesia in Alignment with Intended Nationally Determined Contribution (INDC) by 2030. *Energies* **2017**, *10*, 52. [CrossRef]
20. Notter, B.; Weber, F.; Füssler, J. *Greenhouse Gas Emissions from the Transport Sector: Mitigation Options for Kenya, Methodology and Results*; Federal for The Environment, Nature Conversation and Nuclear Safety: Zurich, Swizterland, 2018; 25p. Available online: https://www.changing-transport.org/wp-content/uploads/2018_GIZ_INFRAS_Transport_Mitigation_Options_Kenya.pdf (accessed on 20 May 2019).
21. Stone, A.; Merven, B.; Maseela, T.; Moonsamy, R. Providing a foundation for road transport energy demand analysis: A vehicle parc model for South Africa. *J. Energy South. Afr.* **2018**, *29*, 29–42. [CrossRef]
22. Burniaux, J.M.; Chateau, J. Greenhouse gases mitigation potential and economic efficiency of phasing-out fossil fuel subsidies. *Int. Econ.* **2014**, *140*, 71–88. [CrossRef]
23. Asian Development Bank. *Fossil Fuel Subsidies in Asia: Trends, Impacts and Reforms—Integrative Report*; Asian Development Bank: Mandaluyong City, Philippines, 2016.
24. Cooke, E.; Hague, S.; Tiberti, L.; Cockburn, J.; El Lahga, A.R. Estimating the impact on poverty of Ghana's fuel subsidy reform and a mitigating response. *J. Dev. Eff.* **2016**, *8*, 105–128. [CrossRef]
25. Durand-Lasserve, O.; Campagnolo, L.; Chateau, J.; Dellink, R. *Modelling of Distributional Impacts of Energy Subsidy Reforms: An Illustration with Indonesia*; OECD Publishing: Paris, France, 2015. Available online: https://www.oecd-ilibrary.org/docserver/5js4k0scrqq5-en.pdf (accessed on 20 May 2019).
26. Diez, J.; Lopez-Lambas, M.; Gonzalo, H.; Rojo, M.; Garcia-Martinez, A. Methodology for assessing the cost effectiveness of Sustainable Urban Mobility Plans (SUMPs). The case of the city of Burgos. *J. Trans. Geo.* **2018**, *68*, 22–30. [CrossRef]
27. Pisoni, E.; Christidis, P.; Thunis, P.; Trombetti, M. Evaluating the impact of "Sustainable Urban Mobility Plans" on urban background air quality. *J. Environ. Man.* **2019**, *231*, 249–255. [CrossRef]
28. Preparing a Comprehensive Mobility Plan (CMP)—A Toolkit. Institute of Urban Transport. Available online: http://www.iutindia.org/downloads/Documents.aspx (accessed on 20 May 2019).
29. Sustainable Urban Transport Programme Indonesia (NAMA SUTRI) Pilot Phase. Available online: http://transferproject.org/wp-content/uploads/2015/02/Indonesia_NAMA-SUTRI_Full-NAMA-Concept-Document.pdf (accessed on 20 May 2019).

30. Asian Development Bank. *Pathways to Low-Carbon Development for Viet Nam. Asian Development Bank*; Asian Development Bank: Mandaluong City, Philippines, 2017. Available online: https://www.adb.org/sites/default/files/publication/389826/pathways-low-carbon-devt-viet-nam.pdf (accessed on 20 May 2019).
31. Kerdlap, P.; Gheewala, S.H. Electric Motorcycles in Thailand: A Life Cycle Perspective. *J. Ind. Ecol.* **2016**, *20*, 1399–1411. [CrossRef]
32. Saxena, S.; Gopal, A.; Phadke, A. Electrical consumption of two-, three- and four-wheel light-duty electric vehicles in India. *Appl. Energy* **2014**, *115*, 582–590. [CrossRef]
33. David, A.; Adelakun, A.; Etukudor, C.; Femi, A. Electric tricycle for commercial transportation. In Proceedings of the 3rd International Conference on Africa Development Issues, Ota, Nigeria, 9–11 May 2016; Covenant University Press: Ota, Nigeria, 2016. Available online: https://www.researchgate.net/publication/316470835_Electric_Tricycle_for_Commercial_Transportation (accessed on 20 May 2019).
34. Black, A.; Barnes, J.; Makundi, B.; Ritter, T. Electric Two-Wheelers in Africa? Markets, Production and Policy. 2018. Available online: https://www.die-gdi.de/fileadmin/user_upload/pdfs/veranstaltungen/2018/20180618_green_transformation/Electric_two-wheelers_in_Africa.pdf (accessed on 20 May 2019).
35. Nugroho, S.; Zusman, E. Low carbon paratransit in Jakarta, Indonesia: Using econometric models to T improve the enabling environment. *Case Stud. Trans. Policy* **2018**, *6*, 342–347. [CrossRef]
36. Sietchiping, R.; Permezel, M.; Ngomsi, C. Transport and mobility in sub-Saharan African cities: An overview of practices, lessons and options for improvements. *Cities* **2012**, *29*, 183–189. [CrossRef]
37. Miller, J.; Jin, L. *Global Progress Toward Soot-Free Diesel Vehicles in 2018*; ICCT: Washington, DC, USA, 2018; 52p. Available online: https://www.theicct.org/publications/global-progress-toward-soot-free-diesel-vehicles-2018 (accessed on 20 May 2019).
38. Macias, J.; Aguilar, A.; Schmid, G.; Francke, E. *Policy Handbook for the Regulation of Imported Second-Hand Vehicles*; Report No.: 7; Global Fuel Economy Initiative: Mexico City, Mexico; 86p. Available online: https://www.globalfueleconomy.org/media/45362/wp7_regulation_for_2nd-hand_vehicles-lr.pdf (accessed on 20 May 2019).
39. Sudmant, A.; Colenbrander, S.; Gouldson, A.; Chilundika, N. Private opportunities, public benefits? The scope for private finance to deliver low-carbon transport systems in Kigali, Rwanda. *Urban Clim.* **2017**, *20*, 59–74. [CrossRef]
40. Calvert, T. The EVIDENCE project: Measure no.6-Environmental zones. *World Trans. Policy Pract.* **2016**, *22*, 56–65.
41. Nash, C.; Whitelegg, J. Key research themes on regulation, pricing and sustainable urban mobility. *Int. J. Sustain. Trans.* **2016**, *10*, 33–39. [CrossRef]
42. Cavallaro, F.; Giaretta, F.; Nocera, S. The potential of road pricing schemes to reduce carbon emissions. *Trans. Policy* **2018**, *67*, 85–92. [CrossRef]
43. Swamy, S. Congestion pricing: A case of Delhi. In Proceedings of the Urban Mobility India 2016 conference, Gandhinagar, India, 8–11 November 2016. Available online: http://www.urbanmobilityindia.in/Upload/Conference/4ef8e29d-4c90-448a-a5d5-37b2c1cac9e3.pdf (accessed on 20 May 2019).
44. Global Fuel Economy Initiative. *International Comparison of Light-Duty Vehicle Fuel Economy 2005–2015: Ten Years of Fuel Economy Benchmarking, Report No.: 15*; GFEI: Mexico City, Mexico, 2017. Available online: https://www.globalfueleconomy.org/media/418761/wp15-ldv-comparison.pdf (accessed on 20 May 2019).
45. Sims, R.; Schaeffer, F.; Creutzig, X.; Cruz-Núñez, M.; D'Agosto, D.; Dimitriu, M.J.; Figueroa Meza, L.; Fulton, S.; Kobayashi, O.; Lah, A.; et al. 2014: Transport. In *Proceedings of the Climate Change 2014: Mitigation of Climate Change. Contribution of Working Group III to the Fifth Assessment Report of the Intergovern-Mental Panel on Climate Change*; Edenhofer, O., Pichs-Madruga, Y., Sokona, E., Farahani, S., Kadner, K., Seyboth, A., Adler, I., Baum, S., Brunner, P., Eickemeier, B., et al., Eds.; Cambridge University Press: Cambridge, UK; New York, NY, USA, 2014.
46. Mason, J.; Fulton, L.; McDonald, Z. *A Global High Shift Cycling Scenario: The Potential for Dramatically Increasing Bicycle and E-bike Use in Cities Around the World with Estimated Energy, $CO_2$ and Cost Impacts*; Institute for Transport and Development Policy, 2015; 42p. Available online: https://www.itdp.org/wp-content/uploads/2015/11/A-Global-High-Shift-Cycling-Scenario-_-Nov-12-2015.pdf (accessed on 20 May 2019).

47. Shastry, S.; Madhav, P. The Role of Transportation in the Future of Urban Developing Asia: A Case Study of India. Pacific Energy Summit Working Paper. 2016. Available online: https://www.nbr.org/publication/the-role-of-transportation-in-the-future-of-urban-developing-asia-a-case-study-of-india/ (accessed on 20 May 2019).
48. Acheampong, R. Cycling for Sustainable Transportation in Urban Ghana: Exploring Attitudes and Perceptions among Adults with Different Cycling Experience. *J. Sustain. Dev.* **2016**, *9*, 110–124. [CrossRef]
49. Mitullah, W.; Verschuren, M.; Khayesi, M. *Non-Motorized Transport Integration into Urban Transport Planning in Africa*, 1st ed.; Routledge: New York, NY, USA, 2017; 240p.
50. NITI Aayog and Rocky Mountain Institute. *Goods on the Move: Efficiency and Sustainability in Indian Logistic*; NITI Aayog and Rocky Mountain Institute, 2018; 44p. Available online: https://niti.gov.in/writereaddata/files/document_publication/Freight_report.pdf (accessed on 20 May 2019).
51. Northern Corridor Transit and Transport Co-ordination Authority. *Northern Corridor Green Freight Programme. For a Competitive and Sustainable Economic Corrido*; Northern Corridor Transit and Transport Co-ordination Authority: Mombasa, Kenya; 44p. Available online: http://www.ccacoalition.org/sites/default/files/resources/2017_northern-corridor-green-freight_NCTTCA.pdf (accessed on 20 May 2019).
52. Klimont, Z.; Shindell, D. Bridging the gap – The role of short-lived climate pollutants. In *The Emissions Gap Report 2017—Bridging the Gap—Phasing out Coa*; United Nations Environment Programme: Nairobi, Kenya, 2017. Available online: https://www.unenvironment.org/resources/emissions-gap-report-2017 (accessed on 20 May 2019).
53. Sims, R.; Gorsevski, V.; Anenberg, S. *Black Carbon Mitigation and the Role of the Global Environment Facility: A STAP Advisory Document*; Global Environment Facility: Washington, DC, USA, 2015; 113p. Available online: https://www.thegef.org/sites/default/files/publications/Black-Carbon-Web-Single_1.pdf (accessed on 20 May 2019).
54. Malins, C.; Kodjak, D.; Galarza, S.; Chambliss, S.; Minjares, R. *Clean up the on-Road Diesel Fleet. A Global Strategy to Introduce Low-Sulfur Fuels and Cleaner Diesel Vehicles*; Climate and Clean Air Coalition: Paris, France, 2016; 77p. Available online: https://wedocs.unep.org/bitstream/handle/20.500.11822/21552/Cleaning_up_Global_diesel_fleet.pdf (accessed on 20 May 2019).
55. U.S. Environmental Protection Agency. *Reducing Black Carbon Emissions in South Asia: Low Cost Opportunities*; U.S. Environmental Protection Agency, 2012; 83p. Available online: http://ccacoalition.org/en/resources/reducing-black-carbon-emissions-south-asia-low-cost-opportunities (accessed on 20 May 2019).
56. Stead, D. Policy preferences and the diversity of instrument choice for mitigating climate change impacts in the transport sector. *J. Environ. Plan. Manag.* **2017**, *61*, 2445–2467. [CrossRef]
57. Bakker, S. Electric two-wheelers, sustainable mobility and the city. In *Sustainable Cities—Authenticity, Ambition and Dream*; Almusaed, A., Almssad, A., Eds.; IntechOpen: London, UK. Available online: https://www.intechopen.com/books/sustainable-cities-authenticity-ambition-and-dream/electric-two-wheelers-sustainable-mobility-and-the-city (accessed on 20 May 2019). [CrossRef]
58. Obeng-Odoom, F. Drive left, look right: The political economy of urban transport in Ghana. *Int. J. Urban Sustain. Dev.* **2010**, *1*, 33–48. [CrossRef]
59. Dhar, S.; Pathak, M.; Shukla, P.R. Electric vehicles and India's low carbon passenger transport: A long-term co-benefits assessment. *J. Clean Prod.* **2017**, *146*, 139–148. [CrossRef]
60. Gota, S.; Huizenga, C.; Peet, K.; Kaar, G. *Nationally-Determined Contributions (NDCs) Offer Opportunities for Ambitious Action on Transport and Climate Change*; PPMC/SLoCaT, 2016; 58p. Available online: http://www.ppmc-transport.org/wp-content/uploads/2015/06/NDCs-Offer-Opportunities-for-Ambitious-Action-Updated-October-2016.pdf (accessed on 20 May 2019).
61. Hall, D.; Cui, H.; Lutsey, N. *Electric Vehicle Capitals: Accelerating the Global Transition to Electric Drive*; ICCT Briefing: Washington, DC, USA, 2018; 15p. Available online: https://www.theicct.org/publications/ev-capitals-of-the-world-2018 (accessed on 20 May 2019).
62. Purwanto, J.; Karmini, K.; Kappiantari, M.; Sehlleier, F. *Indonesia Stocktaking Report on Sustainable Transport and Climate Change. Data, Policy, and Monitoring*; GIZ Indonesia: Jakarta, Indonesia, 2018; 93p.

*Article*

# Addressing the Linkages between Gender and Transport in Low- and Middle-Income Countries

**Tanu Priya Uteng [1,*] and Jeff Turner [2]**

1 Institute of Transport Economics (TOI), Gaustadalléen 21, NO 0349 Oslo, Norway
2 Gender, Inclusion & Vulnerable Groups, High Volume Transport Research Programme, IMC Worldwide Consultants, London RH1 1LG, UK
* Correspondence: tpu@toi.no

Received: 1 July 2019; Accepted: 8 August 2019; Published: 22 August 2019

**Abstract:** The Millennium Development Goals (MDGs) specifies gender equality and sustainable development as their two central priorities. An area of critical importance for sustainable and gender-fair development is mobility and transport, which has so far been neglected and downplayed in research and policy making both at the national and global levels. Rooted in the history of the topic and the emerging ideas on smart, green and integrated transport, this paper presents a literature review of on gender and transport in the low- and middle-income countries. The paper presents a host of cross-cutting topics with a concentrated focus on spatial and transport planning. The paper further identifies existing research gaps and comments on the new conceptualizations on smart cities and smart mobilities in the Global South. Due attention is paid to intersections and synergies that can be created between different development sectors, emerging transport modes, data and modeling exercises, gender equality and sustainability.

**Keywords:** gender; transport; accessibility; smart city; smart mobility; low- and middle-income countries

---

## 1. Women, Development and Transport

There is a growing recognition, both among the research and the practice communities, that societies across the world are undergoing rapid mutation processes due to convergence of various forms of mobilities. The physical and virtual mobilities are intersecting at an ever-increasing pace in sync with what was originally discussed under the 'sociology beyond societies' [1] and the 'new mobilities paradigm' [2]. Yet, access to mobilities remains fractured and is unevenly distributed. Here, we would like to clarify the basic position of this paper—transport is distinct from mobilities. Mobilities encompass dimensions like access to opportunities, quality of life and wellbeing of people. The fact that the transport sector is among the top three contributors to GHG emissions, the biggest consumer of non-renewable energy and has most negatively contributed towards climate change, makes it a suitable candidate for further analysis in light of the mobilities agenda, particularly for the low- and middle-income countries (LMICs; the paper uses the terms developing economies, Global South and low- and middle-income countries (LMICs) in an interchangeable manner).

Unabated urbanization and a voluminous growth in personal motorization (driven solely on non-renewable energies) remains one of the key challenges in developing economies. Motorization is expected to continue to increase at an unprecedented rate in the developing parts of the world [3]. Urban areas in these parts are experiencing a dramatic increase in air pollution, roadway congestion, noise, health issues and traffic accidents as a result of increased car ownership. This is occurring in conjunction with substantial shifts from active modes to motorized modes, and a categorical lack of focus on public transport. Thus, issues concerning the environment, climate and renewable energy gains paramount importance here. Increasing public transport usage and a behavioral shift towards reduced consumption seems to be the only realistic way to curb the negative effects on environment, climate and energy usage. For example, the urban population in India shot from 17% in 1951 to 32% in 2011 and is expected to rise to 35% in 2021 [4]. In absolute numbers, it is estimated that 91 million joined the ranks of urban dwellers in the 2000s. Regarding vehicular growth, 35% of the total vehicles in the country are concentrated in the metropolitan cities alone, which constitute just 11% of the population. Though public transport usage is high, the share of buses is negligible—two-wheelers and cars constitute 90% of the total vehicles on the road in contrast to buses, which constitute less than 1% of the motorized vehicles [4]. In the transport arena, the associated risk, uncertainty and irreversibility (RUI) issues gets exacerbated by the fact that decisions like constructing a highway or major road projects are both resource consuming, practically irreversible and generally operate on longer time horizons [5,6]. The dominance of road building exercises in the Global South, despite the majority of people walking or cycling, highlights that firstly, policies and investment decisions are based on imperfect and incomplete knowledge of the relationship between increased motorization and energy issues, climate change, etc. and secondly, there is a strong lack of a context-based planning methodologies/approach.

Simultaneously, for the first time in the history of policy-making and implementation, both Global North and South are immersed in resolving ways to restructure urban governance in light of Agenda 2030 as many of the areas targeted by the Agenda 2030's 17 objectives are linked to social sustainability and yet ways and means to achieve these objectives remain diffuse. This is particularly the case for gender-based analyses and policy-making. As Razavi [7] notes "In the end, while six of the 17 goals include gender-specific indicators, the indicator framework under five of the goals can be described as 'gender-sparse' (Goals 2, 10, 11, 13 and 17) and for the remaining six critical areas it is depressingly 'gender-blind' (Goals 6, 7, 9, 12, 14 and 15) [8]." The reasons for this apparent gender neglect are many and diverse but one of the simplest explanations is that gender as a category of analysis is difficult to constrain in simple indicator-based systems, more so for systemic issues like climate change urban planning and transport. In the field of urban development and transport, the issue gets further complicated in the Global South, which does not have a strong quantitative data collection and analysis tradition, and thus social sustainable perspectives on thematic issues like accessibility, universal design, gender and diversity mainstreaming, equity, power and influence on planning and decision making, seldom finds its way in discussions and analyses influencing policy making. The primacy of forecasts (based on existing travel patterns) and technical models rule the roost and even when certain trends like 'women exhibiting more sustainable travel behavior than men' [9,10] are established, the prevailing norms of the sector simply do not allow for alternate ways of planning to emerge [11]. Even in societies with a firm agenda on gender equality like Sweden, research suggests that decisions on infrastructure investments and processes followed to reach these decisions are rarely in sync with the broad goals specified in the official documents [12].

This paper builds on the preceding arguments and is positioned in the domain of development planning with particular reference to spatial and transport planning. The paper explores the topic of gender, transport and mobilities in low- and middle-income countries and reflects on critical research gaps emerging from this review. A further layering to the discussion on mobilities/accessibility is provided by the 'smart' city agendas and smart mobilities, which currently pervades discussions undertaken in urban and transport planning domains across the world. The paper structures its arguments with due regards to the digitalization and smart agendas being currently discussed.

Our aim has not been to discuss one particular agenda in detail and given the multiplicity of topics covered in this literature review, we have borrowed from a number of theories. Section 2 highlights some relevant theories while Section 3 presents an overview of the current research findings and gaps identified under a host of topics including access to various opportunities and the issue of safety. Section 3 also highlights the spatial issues of the urban, peri-urban and rural areas of the Global South. The elements of cross-border trading and its importance for women are also discussed in this section. Section 4 initiates a discussion on the methodologies and data needs for addressing the topic of gender, transport and mobilities, while Section 5 presents the identified research gaps to be addressed in future studies. Section 6 concludes the study.

## 2. Theoretical and Methodological Underpinnings

In this paper, we reviewed the current status of research knowledge available in the Global South and the research gaps to be addressed in future studies. Our aim was not to discuss one particular agenda in detail and given the multiplicity of topics covered in this literature review, we borrowed from a number of theories.

The complexity of the transport domain calls for a multi-theoretical approach, which considers macro level perspectives as well as the micro determinants for individual attitudes and behavior. Such theoretical foundations are not readily available and need to be constructed through bringing a host of theoretical standpoints to reflect on the theme of gender and transport in the Global South.

We present some relevant perspectives that hold the potential for taking the inclusivity agenda forward. Though the theories, we briefly touch upon here, have been discussed in the domain of travel behavior, they have not been processed to reflect on gendering of travel behavior. This lack gets further pronounced when the current mores of planning smart cities and smart mobilities are put under scrutiny. We propose that these theories merit further deliberation and have the potential to insert a thematic focus on gendered mobilities into discussions related to governance and innovations.

### 2.1. *Social Psychological and Feminist Theories*

Taking the element of perceptions forward, social psychological theories related to norms, emotions, attitudes and behavior could be further employed to illuminate peoples' opinions and sensitivity toward socio-technical conditions in the context of everyday travel and likely social and cultural trends. A key issue is also to map various 'cultures' and the gendered variations within different cultures for coping with changing mobility conditions. Relevant theories in this context are the theory of planned behavior [13], norm activation model [14] and theories including habits (e.g., [15]). Dijst et al. [16] provide a good discussion of different attitudinal models for understanding both the gendered nature of travel behavior and the kind of smart interventions, which promises most sustainable solutions.

Taking the agenda of prevalent norms, social and cultural practices, relations and organization of roles of institutions and the subsequent phenomenon of gendering [17,18], the feminist theory insists on studying gender both at the individual level but also at the organizational or institutional levels [11]. Feminist theory can be specifically employed in two particular ways to examine the field of transport [19,20]:

i. The separation of men and women in transport decision making—Hirdman [20] points out to the mechanisms of separation, which keep men and women restricted to separate domains even when women enter the transport sector. Men are mostly involved in transport both as an economic and policy sector, while women get relegated to the service functions.
ii. The dominant masculine norms of the transport sector—transport institutions, across the world, have historically been dominated by male bodies and thus the prevalent masculine norms exert normative powers over its agenda [19]. Economic and policy sectors have evolved to operate in the purview of masculine norms and thus its activities coded as masculine, while the service functions tend to be feminine coded.

### 2.2. *Socio-Technical Transition Theory*

Socio-technical transition theory [21–25] builds on the premise that environmental problems represent major societal challenges, whose solution requires structural changes in key areas of society. A socio-technical transition is defined as " ... a major shift or step change, in which an existing socio-technical system—a cluster of aligned elements including technology, regulations, consumer practices, cultural meanings, markets, infrastructure, scientific knowledge, supply and maintenance networks—is durably reconfigured" [26] (p. 1003). A key idea is that large-scale transformations are initiated by ongoing activities in small-scale networks (niches), which over time change prevailing practices, competences and knowledge that constitute the existing (transport) system.

The basic ethos of the socio-technical perspective is that transitions are non-linear processes that result from the interplay of multiple developments at three analytical levels: "Niches" (the locus for radical innovations), "socio-technical regimes" (the locus of established practices and associated rules) and an exogenous socio-technical "landscape" [24]. Within the transport system, various "green niches" have emerged like new mobility systems based on mobile ICT technology and electrical vehicles.

In discussing the gender component of mobilities, socio-technical transition theory is relevant since (i) it draws attention to a broad range of elements and actors and their interactions, (ii) analyses of past transitions tells about factors conductive to change and (iii) it shows the social interpretation of technology [26]. This conceptual framework can be applied in understanding both the social trends and driving forces, along with perceptions and access to solutions.

### 2.3. Mobility Biographies Theory

The biological markers of a woman's life have traditionally underpinned her freedom to negotiate the inside-outside boundaries. This is particularly true for adolescent girls and young mothers in the GS (Global South) affecting decisions to continue education, employment and seeking health interventions as accessing all the three domains remains interlinked with accessing spaces beyond the confines of home. In light of the importance of life-events, a useful theoretical departure is offered by the life course approaches or 'mobility biographies'. This strand of research in travel behavior studies underpins that both travel demand and needs change over the life course of individuals. Mobility biographies emerged as a reaction to the aggregate results emerging from analyses of travel surveys where travel demand on an individual level appeared to be relatively stable in the medium term. Once disaggregated by key turning points in life like changes in the places of residence, childbirth, education and employment [27], daily travel behavior underwent marked changes as the spatial distribution of activities and associated activity spaces altered.

The concept is further unpacked by Scheiner [28] as being embedded in other 'partial biographies', namely residential biography, employment biography and household biography. Studying 'mobility biographies' includes studying the tools, practices and context affecting daily travel behavior, and topics like ownership of mobility tools (such as cars, and access to public transport), factors influencing people to start, stop or significantly change their mobility behavior [29] become important sub-headings to be studied. Most of the studies employing this theory are from the Global North—for example, Chatterjee [30,31] confirm the effect of life course and turning events on the uptake of cycling, Clark [32] expands on the case of car-ownership while Priya Uteng et. al. [33] broach on the topic of car sharing. Thus topics like contextual change, intrinsic motivations, facilitating conditions and the interactions between structural factors and human agency take center stage in discussing travel behavior. The policy implications of studying when and how mobility behavior changes can open up a 'window of opportunity' to plan for and maintain (desired) behavioral changes [34].

From a methodological perspective, this approach urges us to go beyond the purview of quantitative studies and engage with the narrative-interpretative inquiries into the meaning and complexity of mobility biographies. This typology of engagement in the transport sector, even in the developed economies with a relatively matured transport research arena, remains scarce [35].

### 2.4. Social Practice Theory (SPT)

Social practice theory (SPT) attempts to bridge the gap between two primary approaches of treating human behavior—the *homo economicus approach* emphasizing that social order is a combination of individual purposes, intentions and interests and the *homo sociologicus approach*, which relies on the collective norms and values [36] (pp. 245–246).

Rather than treating these two entities of individual vs. collective as mutually exclusive domains, SPT urges us to conceptualize the body, mind, things, knowledge, discourse, structure/process and agent to localize the social within the practice as the main unit of analysis [36]. This essentially means that we focus on both the local or micro phenomena and large social phenomena [37].

SPT defines a practice as a routinized behavior consisting of a set of interconnected elements of *materials, skills and meanings* associated with a practice [36,38]. SPT provides a framework for analyzing the recruitment and retention mechanisms through seeking answers—how practices emerge, how they persist and how they are abandoned—and through analyzing both the product/service providers as well as their users and adopters.

Merging these different theories allowed us to combine results on in-depth analyses of day-to-day behavior with the broader perspectives of policy-making to comment on the fundamental question of "How do practices envisioned in policy-making and those who carry them actually intersect?".

## 3. Current Research Findings and Gaps

In the following points, we present some consistent findings emerging from studies focusing on gendered mobilities in the developing countries [9,10,39]:

1. Women's travels are multi-purpose, complex and resource-constrained (vs. the male norm). Accessibility to health, education and employment opportunities remains constrained for women due to a number of social, physical and economical reasons;
2. Walking, public transport (both formal and informal) and intermediate means of transport (IMTs) are the most used transport modes. Yet, public transport and IMTs inevitably get a lower focus than road/highways/bridge building projects, thus putting women's needs to a further disadvantage;
3. Safety issues are critical and fear of sexual harassment on public transport and public spaces remain widespread. Fear of sexual harassment and personal security remain great concerns in negotiating daily mobilities;
4. Women on low-income suffer a disproportionately high loss of employment opportunities in the face of slum eviction and relocation. The same holds true for rural women when avenues of employment like cross-border trading does not get enough focus in trade agreements;
5. Cultural restrictions placed on the mobility of girls and women in accessing public spaces influences the time, space and duration of women's movements. Extreme cultural restrictions necessitate a context-specific and inclusive approach to women's mobility and transport;
6. The issue of affordability restricts women to a great extent. Constraints on women's mobility keep them away from income sources and from services, with negative implications for both economic and social objectives of development;
7. Smart mobilities and smart cities do not necessarily help the agenda of creating inclusive cities (findings from the Global North so far indicate smart solutions have exclusively facilitated a particular group—young, educated, high-income, white male).

We looked up the following terms in literature search covering both published works and grey literature (from UN and other development organizations and consultancy reports) to consolidate the findings emerging from across the Global South: Gender and transport ; Women and transport; Gendered moblities; Women and safety; Urban Women and safety, Transport; Women and cross-border trade; Women, development schemes in low-income countries/LMICs; Access to education, health and employment in LMICs; Urban and rural accessibility, women; Transport and post-disaster rehabilitation; Gender and space; Transport, capacity building; Women, informal transport; Women, informal employment; Transport in developing countries—methodologies; Transport in developing countries—data needs.

In the following sections, we briefly present both the established and emerging issues to frame the problematic of gendered mobilities.

### 3.1. Access to Education

Linkages between transport, (im)mobility, spatial/social stagnation and resultant poverty in Asia and sub-Saharan Africa have been well established. Instead of being an isolated issue of cultural restraints imposed on girls or a lack of physical access to schools, these two often intersect towards girls' low educational achievement. A study from Morocco highlights that in girls attending primary school tripled to 54% in the area of influence of major paved rural roads, while similar jumps were evidently absent in areas where physical accessibility was still an impediment [40]. Porter [41] presents research findings from a three-country study (Ghana, Malawi and South Africa) where it was consistently found that girls living in remote rural areas with poor roads and poor or expensive transport services were unable to access schools due to an interplay of a variety of cultural, economic and social factors. One of the cross-cutting factors, especially for young girls, is their contribution in the household chores. In the highland village in Malawi, it was found that a significant majority of the students were absent from school on the market days, held twice a week in the nearby towns. Girls were expected to headload local products for sale to the town, which becomes a time-consuming weekly routine since the villages were approximately 8 km from the paved road, with an irregular and erratic supply of public transport. There are myriad examples available from across the developing world, which will confirm the same (refer Box 1). In such circumstances, the inevitable outcome is that girls in remote rural areas often do not obtain even a basic education and further opportunities to develop a livelihood are severely curtailed.

**Box 1.** Cycling to school: Increasing secondary school enrollment for rural girls.

An innovative program was launched in one of the poorest states of India, Bihar, with an aim to reduce the gender gap in secondary school enrollment by providing girls who continued to secondary school with a bicycle that would improve their physical access to school. Using data from a large representative household survey, Muralidharan and Prakash [42] find that being in a cohort that was exposed to the Cycle program increased girls' age-appropriate enrollment in secondary school by 30% and simultaneously reduced the gender gap in age-appropriate secondary school enrollment by 40%. Distance to the nearest secondary school was a crucial element, and increases in enrollment mostly took place in villages where the nearest secondary school was further away. This suggests that the program was most effective in reducing the time and safety costs of school attendance by providing a bicycle. The Cycle program was deemed as being more cost effective at increasing girls' enrolment than comparable conditional cash transfer programs in South Asia.

Coordinated provision of bicycles or other accessible modes (school bus for a cluster of villages etc.) to girls has the potential to generate externalities beyond the cash value of the program. The cycle program went beyond the mere provisioning of a transport mode, and included positive externalities like improved safety from girls cycling to school in groups, and changes in patriarchal social norms that typically discourages and condemns female mobility outside the village, inhibiting female education and employment at large.

An extensive review of the data across 24 rural, peri-urban and urban sites is now available [43] pointing to major issues around mobility for education, even in urban areas, associated with cost and availability of transport as well as the constraints imposed by demands for children's work within the household.

### 3.2. Access to Employment Opportunities

Employment in the informal sector dominates the livelihood landscape in the developing countries. Even though the official numbers are typically conservative estimates, they remain staggeringly high—48 percent in northern Africa; 51 percent in Latin America; 65 percent in Asia and 72 percent in sub-Saharan Africa [44]. In the transport domain as well, informal sector dominates both in terms of absolute numbers of vehicles on roads and in the number of people, mostly men, employed in this sector. The informal modes—para-transit and non-motorized transport modes (NMTs)—are primarily the main public transport modes available in a number of developing countries. They form the main carriers of both the vast majority of population and of informal economies. Despite their pivotal role in connecting people to different opportunities, para-transits and NMTs are either unrecognized in the transport plans or in some instances, rules and regulations insist on either removing them from circulation or barring their access on the main arterial roads.

Non-motorized modes like bicycles and rickshaws and para-transit play a significant role in the lives of women who are dependent on these modes to access employment and other opportunities [39,45]. In urban areas, where often zoning legislation separates commercial from residential areas, women remain the hardest hit if transport accessibility is affected in negative ways. The same is true for women in rural areas as well since they remain dependent on others to sell their products in the regional markets thereby minimizing their control over the profits. Further, availability, affordability and acceptability of transport remains contested. Most formal public transport supply caters to peak hours facilitating the formal sector workers. Additionally, the issue of pricing, and the physical safety of women on public transport also impact their freedom of movement.

The case of Bangladeshi garment workers invokes how a lack of focus on physical accessibility, particularly in the form of safe and reliable public transport, is counterproductive to the issue of making women financially independent and active participants of the society. The ready-made garments (RMG) industry of Bangladesh is a booming industry, which exports to over 30 countries in the world, employing about 1.8 million workers of which 1.5 million were women [46]. On one hand, this case can be celebrated as a major breakthrough of female employment in established sectors and yet a mere scratching of the surface of this success story reveals how both living conditions and capacity to save is severely affected by a categorical lack of affordable transportation facilities for workers on limited wages [47]. For example, planning decisions in Dhaka prohibits cycle rickshaws to drive on certain major roads where several garment factories are located thus putting a ban on the most viable transport mode for the female garment workers. Sharp differences in living conditions and saving potentials were further found among female workers who worked in factories proving bus transport as compared to workers who did not have access to such provisions (for detail discussion, refer to [10]).

Female employment in the Global South is also often within the premises of the household in a format popularly known as home-based manufacturing, which typically ranges from garments, consumable products to providing ancillary product creation for various industries. The savings potential of women employed in the home-based sectors vary greatly depending on physical accessibility to markets as illustrated through the case of home-based garment producers in Ahmedabad, India [48]. She concludes that development decisions need to include a focus on spatial mobility to improve livelihood outcomes of female producers and depending on the sector's market characteristics, an important intervention facilitating women producers could be improving access to the range and quality of markets available to them [48].

Microenterprise credit programs have received innumerable support in the past decades and a Nobel prize simply bolsters the effectiveness of this solution in addressing female empowerment. However, these mechanisms to support women's income generating opportunities and economic empowerment have been contested [49–56] and Omorodion [53] points to the inconclusive nature of the micro-credit programs in improving the economic situation of women. A reading on the topic of daily mobilities highlights that economic empowerment and mobilities remain interlocked. The first issue has already been raised in the previous point and concerns direct access to market. Women's inability to access markets and directly sell their products and make networks to access information greatly inhibits their saving potentials. The ability of this section of women to virtually access information and networks through mobile phones remains contested. Further, lending institutions need to take the accessibility criterion in consideration and locate their outlets for repayments close to markets, training centers and in communities involved in the program, enabling the women to make repayments without social and physical obstructions. The case of Esan women in Nigeria [53] highlight "The lack of financial institutions in rural areas meant traveling long distances to make loan repayments also contributed to the failure of the micro-credit schemes in Esan communities". Nigam [57] reinstates this point through assessing microcredit schemes in five countries—Nepal, Viet Nam, Egypt, India and Kenya. He states that the credit schemes can be truly effective in reducing the worst manifestations of poverty only when credit dole outs operate in combination with basic social services. Access to market, repayment nodes, basic education and training are among the most vital elements of such services (refer Box 2).

**Box 2.** Spatial mismatch.

The comparative analysis of working and non-working women's mobility in Navi Mumbai, India [58] reveals that economic empowerment coupled with improvement in literacy levels could result in three to four-fold increase in an average women's mobility. The time and activity pattern study of the working women reflects greater obligatory time requirements, which results in lesser time for travel in comparison to those observed in more developed societies. It was also observed that there are spatial mismatches between the distribution of low paid female jobs and locations where low-income women live resulting in longer commuting by low income women compared to high income women. The working women also are greatly dependent on the safe, reliable and affordable mass transport systems for their long work trips journeys.

*Resettlement/subsidized/affordable housing schemes.*

One of the most prominent responses of city governments, across the developing world, to the issue of slums and squatter settlements has been resettlement schemes, variants of which can be found under the name of subsidized and affordable housing schemes. These resettlement colonies are typically located in the peripheral edges of the city with poor or no public transport connection. The most immediate response to these resettlement programs has been the loss of livelihood for women, which were originally anchored in walking distance of the slums. The mesh created by distant relocation, inadequate transport services with respect to frequency, connectivity and affordability, unsafe public space designs (primarily bus tops, access pathways) invariably and continuously hits low-income women the hardest. Acknowledgement of these issues, and planning of residential housing areas, which are either mixed land uses or zoned with adequate provision for accessible services and employment opportunities continues to elude development authorities, donor organizations and development sector at large [10,59]).

### 3.3. Access to Health Services and Well-Being

Figures state that between 50%–60% of people in poor countries live more than 8 km from a healthcare facility [60]. Mortality rates for women in time-critical medical emergencies related to childbirth and infant illnesses continue to be high in a large part of the developing world simply due to a lack of availability of access to these health centers [61]. This is typically manifested in form of either lack of transport options, unaffordability or a combination of both (Mlay et. al. [62] illustrates this point through the case of Tanzania and similar case findings from Ethiopia are brought forth by Hamlin [63]). Similarly, to illustrate the prevalence of this incidence in all parts of the developing world, a study from Cebu in the Philippines quantified the (strong) association between infant, child and maternal mortality rates and distance to healthcare services [64] by calculating that a 10% increase in distance from a hospital was associated with a 2% increase in all three mortality rates.

The issue of head-loading continues to plague health deficits of women and is directly linked to their mobility burdens. For example, figures derived from 276 women fuel carriers sampled in Addis Ababa highlighted an average load of 36.2 kg (i.e., 75% of body weight) being carried for an average trip length of 11.7 km, and close to 17% of the women were carrying loads heavier than their body-weight [65]. With reference to the maximum carrying weight of 20 kg recommended by the ILO (International Labor Organization), it is not surprising that these women suffer from eye, chest and back pains coupled with high rates of miscarriage. Porter et al. [66] presents an analysis of load carrying impacts on children from 24 urban and rural research sites in Ghana, Malawi and South Africa, emphasizing substantial detriments to both children's health and their education. A full review of likely health impacts of head loading in sub-Saharan Africa highlights the need to build scientifically validated evidence base with health professionals.

The incidence of HIV has also been linked to transport availability and spatial concentration of medical services. Several cases from Africa highlight that comprehensive HIV services, primarily provided through hospitals, remain inaccessible to rural population. The cost of transport to these facilities is often high, and Amnesty International calls for a meaningful consideration to the transport needs of economically and socially marginalized people, especially rural women at risk of or living with HIV.

One of the research gaps identified so far is that there is little recorded evidence in the form of research studies to highlight the constrained access of urban low-income women to health services. Most of the studies in urban areas have focused on the quality, patient–provider relationships, accountability and affordability of the health services [67]. However, referral services like assisted or non-assisted transport services have been rarely studied in the urban areas of the developing economies, except for studies such as that by Murray et al. [68] that focused on Lusaka, Zambia and there is need for much more rigorous research work.

Essentially, even in the urban areas, women and girls experience maternal death and a myriad of health-related problems for reasons similar to what has been consistently found in the rural areas—lack of affordable and reliable transportation to clinics and hospitals. A combination of lack of ambulance services to the urban slums and unaffordable public transport often leave women in extremely dangerous conditions [69,70]. During a focus group conducted with female residents of Bwaise slum in Kampala, Uganda, it was revealed that women were restricted by financial resources and distance. "It took at least one hour to walk from Bwaise to the nearest health centre, and there are reports of women giving birth en route. The only vehicular access to health centres is by boda-boda (motorcycle taxis) because of the poor conditions on the surrounding roads" [71]. One of the concrete suggestions put forth by the female slum residents was provision of an ambulance station (or a designated pickup point) in close vicinity of the slum.

### 3.4. *Humanitarian Efforts, Post-Disaster and Post-Conflict Rehabilitation*

Massive humanitarian efforts and resources are put in the post-disaster/post-conflict rehabilitation processes. These processes of transitioning from relief to development offer unique opportunities to correct spatial development-related imbalances and reduce vulnerabilities to hazards [72]. However, decisions under these stressful circumstances are taken more as a response to establish immediate control rather than pre-mediated rehabilitation decisions with long-term consequences [73]. For example, the primary focus of post disaster investments in Honduras after Hurricane Mitch was on major arterial or secondary roads and not rural roads. Often, implementation of the transport sector's modern rhetoric of sustainability, gender and the environment are strangely absent. Post disaster reconstruction stages need to be streamlined to reconstruct with change thereby avoiding the creation of new vulnerabilities or exacerbating the existing ones. The International Forum for Rural Transport and Development [74] notes that 'prioritizing the rehabilitation of rural road networks to enable small farmers to access markets could potentially discourage post-disaster migration to vulnerable rural areas and urban slums. By continuing to listen to the needs of the poor in the post-disaster context, the transport sector has the potential to avoid creating new societies with even greater vulnerability.'. In light of the unprecedented urbanization taking place in the developing economies and issues like, which we expand on in the later sections of the paper, feminization of slums in the urban areas, cross-border trading in which women are heavily involved, inserting evaluation and implementation of accessibility modules in the rehabilitation processes becomes a necessity.

### 3.5. *Traffic Injuries and Women*

The World Health Organization [75] reports that 90% of the world's road fatalities occur in low- and middle-income countries, even though these countries have approximately only half of the world's vehicles. Majority share of fatal injuries consist of 'vulnerable road users' comprising pedestrians, cyclists and motorcyclists. It is also reported that globally, three out of four road deaths are among men [76,77]. However, when we start analyzing disaggregated data in the low- and middle-income countries, a different picture emerges but is often under discussed as both access to and analyses of disaggregated data remains problematic in the low- and middle-income countries.

For example, Ghanaian accident data highlights that over 40% of fatal road traffic accidents (RTA) involved pedestrians, and the majority of these were women [78].

As previous sections have highlighted, women constitute a major share of pedestrians in the LMICs owning to their livelihood options like street vending, hawking and unaffordability to use other modes of option than walking. Street vending across the LMICs operates through the modality of sharing road space with the motorized and non-motorized vehicles thus putting women in precarious situations. Further, Williams et al. [79] highlight the age dimension specific to the case of LMICs where the incidence of falls among elderly women remains much higher than men.

The intersectionality of traffic injuries remains an understudied topic, particularly in the LMICs. Reasons and the particular ways in which women are or become victims of road accidents needs further probing [80].

### 3.6. Safety, Smart Cities and Smart Mobilities

It is not new knowledge that women across the world are more fearful of crime than men. This fear gets accentuated by a combination of limitations to defend themselves in face of physical attack and a greater propensity to transfer past experiences and memories of victimization to present [81]. For example, in Chennai, India, 66 percent of female respondents stated that they have been harassed while commuting [82]. A series of studies document the prevalence of fears and concerns about safety in the developing world. They plot how such fears influence travel decisions taken by women and can have a substantial impact on both the volume and timing of ridership. In tandem with the main findings from a U.S. based study [83], studies from the developing economies come to similar conclusions—there is a significant mismatch between the safety and security needs and desires of female passengers and the types and locations of strategies public transport agencies use [59,84–89]. Studies focusing exclusively on safety concerns of women on their daily travels are scarce in the developing world. However, the ones that have studied this topic unanimously assert that safety concerns have strong contextual determinants—public lighting, characteristics of sidewalks, isolation and neighborhood characteristics [90]. Perceptions of insecurity in a variety of urban environments—at the bus and rail stations, on their way to and from the bus and rail stations, etc.—are a primal force restricting women from achieving their maximum potential not only in education, employment and health, but in their general well-being.

A major criticism of the smart cities approach is the gap between the technical/digital approach and quality-of-life approaches. Lauwers and Papa [91] claim the shift from conventional mobility planning towards smart mobility is primarily applying new technology to existing infrastructures instead of creating better solutions. For example, buses are being retrofitted with tracking devices rather than increasing public transport supply and checking outcome measures such as access to work, education, etc. In this sense, smart mobility concerns itself primarily with innovative technological or consumer-centric solutions rather than adopting a social sustainability lens to the entire mobility agenda.

Studies exploring the under-delivery of smart solutions for women are restricted to the Global North, but they have conclusively established the fact that smart solutions can be highly exclusive. Shaheen et al. [92] studied 23 bike-sharing programs in North America and found that the main obstacles identified for low-income groups were the need for smart devices, debit/credit cards, minimum bank balance or deposit to cover for vandalism or theft. Although there has been a considerable increase in smartphone users in most developing countries, and this continues to grow, the actual penetration level by population remains less than 50%. Further, there are examples of significant disparities in access to smart phones between men and women in some Asian and African countries. For example, women in India between the ages of 15–65 are 46% less likely than men to own a mobile phone, while in Bangladesh and Rwanda women of the same age group are 62% less likely than men to use the Internet (for details, refer to Box 3).

**Box 3.** AfterAccess: Uncovering the gender gap.

AfterAccess is a large-scale data collection initiative aimed at compiling comparable information and communication technology indicators for countries in the Global South. The survey has focused on creating comprehensive mobile and Internet use database for the Global South. Funded by IDRC and conducted jointly by DIRSI, LIRNEasia, and Research ICT Africa, AfterAccess surveys collect data on ICT access and use through household and individual surveys across 22 countries, covering a sample of 38,000 and counting. The sampling method allows for representation with a plus or minus 3% margin of error, and the sample sizes are large enough to allow for gender-disaggregation of the indicators collected. Among other factors, the surveys measure the gender gaps in mobile and Internet usage. The disparities reveal a sobering picture in some Asian and African countries. For example, women in India between the ages of 15–65 are 46% less likely than men to own a mobile phone, while in Bangladesh and Rwanda women of the same age group are 62% less likely than men to use the Internet.

Such analyses assist us in reflecting on the relevance of Smart Cities approach in current times. Before implementing smart solutions, which are primarily based on digital interface, we need to address the gender gap of women's (versus men's) access to and use of ICTs and gendered barriers of use to enable evidence-based policymaking.

Source: AfterAccess: Uncovring the gender gap (https://www.idrc.ca/en/research-in-action/afteraccess-uncovering-gender-gap)

Another issue concerns the lack of digital literacy, knowledge, comfort and confidence to use smartphones. In many emerging economies, disparities in digital literacy compound disparities in basic literacy and reduce people's access to smart solutions and services.

Ride-sourcing (also known as 'on-demand-rides' or 'ride-hailing') is one of the most popular forms of smart mobility. However, there have been some considerable safety setbacks because cases of drivers sexually assaulting female passengers have emerged from across the globe. Given the emerging demand for safe transport services for women, women-only ride-hailing services (exclusively women drivers and passengers) have been launched in many countries, such as Riding Pink in Malaysia, LadyDriver and FemiTaxi in Brazil, See-Jane-Go in the USA and almost a dozen similar services in India. In Indonesia, where people often hitch rides, two women-only, motorcycle ride-hailing services, LadyJek and Sister Jek, were launched. Ways to factor in safety in the smart solutions need immediate and urgent attention [89,93].

On the positive side, apps like *Safetipin* facilitate mapping exercises that can greatly enhance weeding out unsafe spaces, corridors and routes (refer Figure 1). In 2014, Safetipin was launched in India to help women safely navigate the city by identifying its safe and unsafe zones [94]. It is a location-based mobile app that collects safety-related information and conduct safety audits of different places by calculating a safety score. Users of the app can identify how safe certain areas are and can plan their travel routes and timings accordingly. The safety audit is based on nine parameters—*lighting, openness, visibility, security, crowd, public transport (connectivity), presence of women and children, feelings of safety and presence of footpaths or walkways*. Safety scores for Nairobi are illustrated in Figure 1 (for further discussion, check [93]).

Each of the parameters, except for feelings (of safety), is measured objectively using a rubric and scored on a scale of 1 to 4 (from 'poor' to 'good' conditions). The safety data is available in multiple forms, including maps, reports and csv files, which can support urban stakeholders in making judicious urban planning and monitoring decisions, such as identifying areas that need more lighting, security, CCTVs and/or public transport at night. Safetipin has now extended to more than 20 cities, including some outside India, such as Bogota, Manila-Quezon, Jakarta and Nairobi.

**Figure 1.** Safety score for Nairobi (source: Safetipin [95]).

Such exercises have become relatively easy, cheap and accessible to planning authorities across the world. We urge a replication and continuous mapping of safety as part of the smart cities/mobilities agenda (for further discussion, check [93,96]).

The safety scores that Safetipin has generated in the case cities have driven city leaders to take action to improve women's safety. In Bogotá, Safetipin has assisted policymakers in helping to create a data-driven approach to women's safety in public space. City officials have been able to combine data based on the safety audits with other data sources in order to better understand urban problems. For instance, the city was able to overlay maps indicating Safetipin's security parameter rating with maps of police station locations and incidences of crime. Based on Safetipin data, the city identified five priority locations for interventions to generate a broader dialogue about women's safety at night that included local operative councils for women and gender, local women's organizations and citizens. These public engagement activities are key to educating people and changing people's perceptions and attitudes about women, gender and gender-based violence.

Data-driven approaches to women's safety are increasingly being adopted in other cities as well (for example, web-based interactive map campaign 'Free to Be' [97]).

### 3.7. Feminization of Slums

This section builds on the increasing feminization of slums and spatial development rationales. It establishes links between transport and housing policies and the need for these two sectors to work together to facilitate women's mobility.

Across a sample of 51 developing countries from Southern Asia, Latin America and sub-Saharan Africa, women are more likely to live in slums than men [98]. For a majority of the countries, there are 108 women living in slums for every 100 men in the same category. Living in slum conditions means that residents lack access to access to safe water, adequate sanitation, durable housing, sufficient living space and/or security of tenure.

For Kenya, housing the world's fifth largest slum by population size, there are 13 more women for every 100 men living in slum conditions. The figure is 15 or more women aged 15–49 for every 100 men of the same age group in other countries, including Swaziland, Gabon, Ghana, Cameroon and

Senegal. In Colombia about 11 and in Dominican Republic, Bolivia and India, about five more women aged 15–49 respectively live in slum conditions than their male counterparts.

There is a further spatial layering of center vs. periphery. Lima et al. [99] underline that land use in Recife's Metropolitan Area does not obey the center-periphery occupation pattern (European) or the periphery-center occupation pattern (US). This observation, however, is applicable to a host of developing countries. A common denominator to urban development in the developing countries is the fact that slums and poor communities are scattered throughout the municipalities, in conurbation with the richest neighborhoods. There are trends in some developing countries towards in-situ establishment and legalization of the slums. For example, in the case of Recife [100], through establishing special zones of social interests (known as ZEIS) in 1980s, legal right to the urban land of previously informal low-income settlements was recognized. This paved way for a spate of formal services being provided to the area to improve urbanization standards, provide basic infrastructure (e.g., sewage, drainage, pavement and water supply) and legal tenure of these settlements [100]. Contrary to this case, there are myriad cases of slum relocation and allotment of low-income housing in the peripheries of urban areas with little or no public transport available to these areas. Women remain the hardest hit group in these reallocation schemes with loss of employment and further isolation from income-generating avenues. Majority of slum dwellers in Kolkata predominantly walk (56%) for their travel needs followed by bus travel (26%) [101]. In terms of travel distance, nearly 50% of slum dwellers commute within 0.8 km, while 75% travel within 1 km. The percentage share of income on transport expenditure tends to increase for slum dwellers residing away from city centre, which exhibits that slum dwellers tend to limit their travel distance in order to optimize their travel expenditure. It was observed that with improved accessibility, per capita income of slum dwellers tends to increase due to better access to work opportunities. The per capita average monthly savings of slum dwellers is more sensitive with transport accessibility than accessibility to employment opportunity highlighting the importance of transport in improving standards of living of urban poor. In light of 'feminization of slums' (refer Figure 2), which is a rapidly increasing phenomenon, there is a need to understand the gender differences in this experience and report gender disaggregated travel patterns [102] (refer Box 4).

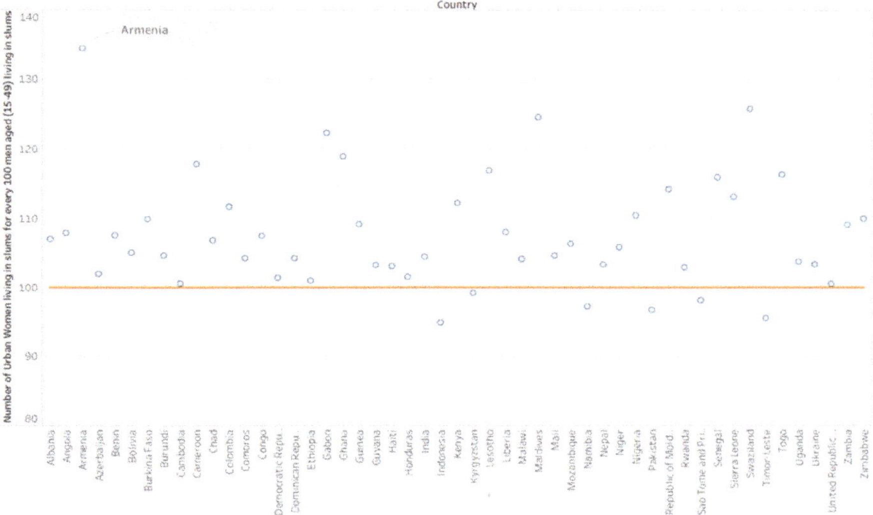


**Figure 2.** Urban females aged 15–49 living in slums for every 100 males aged 15–49 living in slums, 2007 or later (Source: UN Women calculations based on USAID 2018 [98]).

**Box 4.** Mobility, poverty and gender: Travel 'choices' of slum residents in Nairobi, Kenya.

Salon and Gulyani [103] present the gendered variations and the empowering potential of daily travels of female slum residents. Their findings are based on one of the most comprehensive travel data collected for slum population in the developing countries—survey of 4375 slum residents in Nairobi, Kenya. Analyses revealed that though majority of the slum population could not afford any of the motorized transport options in the city and coped by limiting their travel outside their settlement by often 'choosing' to walk, the burden of reduced mobility is borne disproportionately by women. Women faced distinct barriers to access, and though policies aiming to improve mobility and transport access for the slum population first and foremost need to address the issue affordability, but specific constraints faced by women needs additional support.

The story that emerges from our analysis is that both poverty and gender matter in explaining differences in the travel choices of working adults in Nairobi's slums. It is expected that poverty level would affect travel choices, but the gender effect that we find is surprisingly strong among adults. Characteristics of women's travel are systematically distinct from those of men in this population, even when controlling for poverty level.

Policy implications—four divergent policy implications emerge from their study, and other studies confirm that these are widely applicable in the developing contexts across the world:

i. Affordable and reliable public transport—if affordability increases—either through reduction in poverty or lower fares—women's use of public transport will rise sharply.
ii. Child care facility—for women, children in the household greatly reduces their chances of being employed, while men become more mobile. Gainful employment takes place either by working closer to home or through running a household microenterprise (HME). Provision of (affordable) childcare facility can enhance employment opportunities for women.
iii. Education and employment—women reap larger mobility gains from education than do men. The increase in the likelihood of using matatu (local informal public transport) is accompanied by a large decrease in the likelihood of being unemployed.

For rural areas, the location of market and health centers, and access to vocational training and higher education gains paramount importance in ascertaining how much of the benefits finally percolate down to women. Specific and targeted interventions like providing cycles to girls in high school (Cycle program in the state of Bihar, India) greatly reduce the 'distance cost' of attending schools and related opportunities. "Comparisons with conditional cash transfer programs in other South Asian contexts suggest that the Cycle program was much more cost effective at increasing girls' secondary school enrollment than an equivalent-valued cash transfer. Given the importance of increasing women's education attainment in developing countries like India (especially in its most under-developed regions) and the fiscally-constrained policy environment, these results are important and suggest that the Cycle program was not just politically popular but also much more cost-effective than the most frequently considered and implemented policy alternative to increase girls' secondary school enrollment in developing countries in the past couple of decades (CCT's)." [42] p. 26).

### *3.8. Space, Access and the Informal Economy*

"Engagement with informality is in many ways quite difficult for planners. Informal spaces seem to be the exception to planning, lying outside its realm of control" [104] (p. 155). What Roy further argues and builds in the same paper is that though informality is the exception, it is an outcome and product of the state. What is important to note that in the developing countries, informality encompasses all modes of production and access to these modes of production.

At the same time, informality continues to facilitate women. Women are either equal participants or, in some cases, dominate the informal markets, trades and usage of the informal transport modes. In urban areas, these three facets get typically expressed by:

i. Creation and adherence to land use plans without giving recognition or space to the informal use of land, thus treating land use, social development, women empowerment and distributive justice as separate blocks;
ii. Prioritizing car-based mobility or an imposed public transport system, which fails to both recognize the already existing informal modes and to cater to the majority of the population;

iii. Borrowing and implementing best practice models, which might not be conversant and fit for the local context.

Another area of neglect is cross-border trading, both formal and informal, which severely affects the lives of rural women. Informal cross-border trade (ICBT), a trade arrangement, which is informal, and thus precarious, remains dominated by women traders in the rural areas. Though this fact is more pronounced in Africa, increasing evidences can be traced in Asia as well. Kusababe [105] (p. 582) puts it "Very little literature exists on the gendered effect of international borders. Of the few extant studies, scholars have elaborated how geographical borders define and label women's work. Cheater [106] shows how cross-border 'shoppers' in Zimbabwe were labelled as a security threat to the country. Vila [107], in her work on the US-Mexico border, demonstrates how borderlands, as margins of society, are seen as 'dangerous' and full of 'vice', and that certain gendered behaviours and attitudes are seen as characteristic of Fronterisas/os. Similar findings are reported by Biemann [108], who concludes that, 'The border thus becomes a metaphor as well as an actual material institution that capitalizes on the differences between the economic and the sexual' (p. 108)".

ICBT or in the absence of formal regulations around small-scale trade what is often deemed to be 'smuggling' has had positive effects for women, which is highlighted in the scattered research studies and reports found on this topic. However, to date, this form of trade is not typically recorded in government statistics leading to an obfuscation of data and consequently the roles and needs of women traders. The gendered effects are significant given that 70%–80% of African traders involved in ICBT are women [109–112].

Kusakabe [105] (p. 581) highlights how "the formalization of the border trade has changed women's 'sense of space' and their relations with men and other women". Her study highlights how international borders create different scales of places [113], and women's ability to access these scales depends on both material (transport connectivity) and social (societal and household's definition of gendered work) definitions of access. This is significant in the context of increasing efforts to improve the 'efficiency' of regional land borders through the introduction of technologies and harmonization of bureaucrat processes.

UN Women [114] notes that poor infrastructure in terms of poor roads, energy and communication is known to be one of the non-tariff barriers to trade, preventing women to access international markets. Across value chains infrastructure, access to market is a critical element, especially for transportation of perishable agricultural products, which are mostly traded by women informal cross-border traders.

A few of the research areas that should be further investigated while designing regional transport corridor development schemes are as follows:

- What are the inter-linkages between transport provision (both hard infrastructure and gendered access to the transport resources), value chains and gender inequality?
- What are the main mobility challenges and opportunities for women entrepreneurs in the context of cross-border trade?
- What are the potential ways of improving regional mobilities?
- What are the potential ways of integrating gendered mobilities concerns into value chain of development projects and programs to assist women in maximizing their profitability and competitiveness?

It is to be borne in mind that limited sex-disaggregated data on gender and cross-border trading was found in this review and this inhibits sound empirical evidences. Studies should further delve into existing knowledge gaps on the opportunities for businesses owned by women in the context of procurement schemes, training facilities and access to markets.

Though the relationship between space and informality is beginning to emerge and accepted at the urban levels, this relationship is largely hidden in discussions on cross-border trade. Sadly, in tune with the neatly compartmentalized discussions in different sectors, though informal work or trading has been examined closely, the relationship between space, informality and access lacks similar engagement. Studies documenting the incidence of cross-border trading highlight that international borders often create a space where income earning activities are being practiced, assigning both status and new roles to women [115–117] but due to the lack of public transportation and high transportation costs, women, in particular, faced hardships in directly accessing the market and claiming the revenues. Kusakabe [105] highlights this case through two examples drawn from Lao PDR—(i) cotton-weaving activities in Sayaboury province, and (ii) sticky rice box production in Kammoune province.

### 3.9. Development Planning and Transport

Lack of gender balance in decision-making, ranging from the macro level in parliaments to community and household levels have been routinely noted. Consequently, women's perspectives and priorities are not reflected in budget allocations and development related decisions. Seminal works like Male Bias in the Development Process [118] highlight that male bias is easy to understand, empirically testable and in principle can be rectified [119].

Given that a large share of low-income women is employed in the informal market, governance issues built on employment and related schemes need to strengthen their focus on accessibility. The fact that a large share of daily travel needs in the developing economies are met by informal public transport modes like tuk-tuk and jitneys, such bottom-up initiatives need to be both recognized and bolstered in the government action plans for transport, urban/rural planning and development at large. Even most basic infrastructure like mapping of these informal intra-city routes are largely amiss (most transport plans lean towards the case plotted in Box 5).

Women are typically less likely to find employment in the transport sector at large, but this is especially pronounced for the developing economies. This inadvertently results in rendering women's needs invisible, even in the provision of informal transport, which is largely market responsive and highly adaptable.

**Box 5.** National urban transport policy (NUTP) of India.

Realizing the importance of public transport and cycling and walking in the cities and equity concerns, the national urban transport policy (NUTP) spelt out the following objectives: (i) To bring about a more equitable allocation of road space to various users through building a people-centric rather than vehicle-centric focus. (ii) To encourage greater use of public transport and non-motorized transport (NMT) modes by offering central financial assistance for this purpose and (iii) to enable the establishment of multi-modal public transport systems that are well-integrated, providing seamless travel across modes [120]. Under the Jawaharlal Nehru National Urban Renewal Mission's (JNNURM) Urban Infrastructure and Governance (UIG) component, 24.2 per cent of the total allocations were for transport related projects. However, 13.3 percent of the transport related project was road widening and flyover building and only 8.66 percent was for mass transit development and the rest for parking and other small transport projects [121]. Despite a positive shift in policy objectives and priorities, the largest portion of the transport related funds in JNNURM were spent on road-building instead of allocating them to public transport projects.

The case of Dantewada, India, establishes how women can themselves tackle the deep-rooted inadequacies of the transport system, culminating from the inaction of both 'state' and 'private' actors. In the lack of any form of public transport connectivity, the tribal women of this remote rural area organized themselves through self-help groups. Through government-sponsored subsidy schemes, they have been successfully running E-rickshaws connecting rural markets, schools, etc (Figure 3).

The Kenyan case is also worth mentioning here (Figure 4). Though the Kenyan transport sector remains largely male dominated, women are increasingly participating in the (informal) transport sector as owners, drivers, touts, stage clerks and fleet managers [122]. The female workers employed in the informal transport sector in Kenya comment that they have no contact with the formal social protection schemes as they are categorized as being 'informal'. There are two sets of facts to be considered here for Global South at large. Firstly, there is a predominance of informal transport systems in the developing countries, which might provide employment opportunities to females. Secondly, given the eventuality that women do find a space in the transport sector, as in the case of Kenya, Kamau [122] raises the following questions:

1. Can we design and integrate social protection for women transport workers?
2. How can this be guaranteed?
3. Are their services being valued? What is the evidence?

**Figure 3.** Taking the daily mobility needs head-on. E-rickshaws and tribal women of Dantewada, India (source: Chhattisgarh:E-rickshaws drive a change in the lives of Dantewada's tribal women. (http://www.newindianexpress.com/thesundaystandard/2018/feb/03/chhattisgarh-e-rickshaws-drive-a-change-in-the-lives-of-dantewadas-tribal-women-1767909.html)).

**Figure 4.** Female (informal) transport workers in Kenya (source: [122]).

## 4. Methodologies and Data Needs

Though the presented findings are consistent for developing countries across the globe, they cannot be periodically corroborated as most developing countries do not conduct periodic travel surveys or qualitative evaluation of the situation. This means that the gendered patterns of different and differentiated mobilities do not get highlighted and consequently are not acknowledged in the planning and design processes. The increasing penetration of mobile phones, even in the remote and poverty-stricken areas, in developing countries has to some extent changed the landscape of how women negotiate their daily mobilities. Developing routines and systems for data collection on revealed, preferred and digital mobilities, thus, holds the potential of assisting policy-making. The analyses emerging from these surveys can help the research and policy-making field to comment on:

i. Given the current system, roles, jobs and gendered divisions—what are the current mobility needs of women (both urban and rural)?
ii. Assuming a world, where women have an increasing degree of freedom, what kind of arenas will be available to them? For example, more female related growth of economy. We can look at the case of advanced (and equitable) economies like Norway for the division of labor in different sectors. What are future mobility needs?
iii. What kind of transportation policies and investment will ensure an increased accessibility to women?
iv. Are there systemic and systematic flaws responsible for gender-bias transport policies?
v. How do land use regulations or the lack of it affect the mobilities in urban and peri-urban areas?
vi. Which kinds of mobilities (both urban and rural) are being supplemented/complimented or substituted by mobile phones?
vii. How can women's priorities and needs be included in planning and monitoring transport systems?
viii. How can transport be made affordable for women, particularly the poorest women?

Rather than advocating simple data collection, it is imperative that data collected in surveys is segregated at the level of gender, activities (land use) and time-use, which can essentially inform the transport planning authorities to take a more needs-oriented approach. Mobility-gap analyses should form an inevitable part of routinized data collection (travel behavior surveys) and analyses. An explanation of the gap figures, for example in terms of average daily trips and time used by different types of households, sheds light on the distribution of mobility opportunities among the respective genders.

Further, given the high incidence of no travel to work [123] and limited trips on a daily basis (a recent one-day trip diary from the city of Bengaluru, India revealed that only 20% of female respondents had taken a trip on the interview day (source: Personal correspondence, Tanu Priya Uteng), it is crucial that the popular methodology of the Global North focusing on a one-day travel diary is not adopted in the developing economies. Instead, at least a one-week trip diary approach should be adopted to filter out the trips that women are making over an expanded period of time (Box 6 illustrates a case from Indonesia).

**Box 6.** Time use allocation, transport poverty and social exclusion.

> Through exploring the time use allocation and immobile behavior using a three-week time use diary collected in Bandung metropolitan area (BMA), Indonesia (one of the first multi-day time use studies within developing countries context), Susilo and Liu [124] establish clear differences of weekday and weekend patterns of time use allocations and mobility behaviors across different socio-demographic groups and gender. There exists a strong tendency of social exclusion due to transport poverty. As a typical patriarchal society in developing country, women are still responsible for housework and thus have a shorter contract time, free time and travel time but longer committed time than men on both weekday and weekend. Housewives also have a particularly low travel time, indicating that either due to unaffordable public transport services or lack of time, they may not travel a far distance.

Integrating data collection across sectors also holds the potential for addressing the needs of women in a more robust and target-oriented fashion. For example, maternal mortality indicators have received a lot of attention and are a good indicator for demonstrating the efficiency of an entire health system. However, unpacking this indicator, availability of transport, availability of medical supplies, presence of trained health staff and access to health facility are major factors affecting maternal mortality rates. Thus, programs aiming at decreasing maternal mortality should have a detail analysis of mobility component as well. In areas where this access is problematic, complementary programs to address this issue specifically should be introduced, either through training the nurses to ride bicycles/motorbikes, providing support for owning a community cart, etc. The broad aim of building capacity of the public health authorities to promote equitable access to primary health care services needs to be broken down into workable components, based strictly on the contextual realities. Given such benchmarking, it will become easy to assess the specific kinds of alterations needed in the mobility systems to adapt towards gendered needs (for example, usage of mobile phones to substitute the missed trips and access information). This applies at both rural and urban levels. Community health workers in rural Bangladesh and Indian's ASHA workers are both response to inaccessible hospitals in the rural parts of developing world.

Further, there exists a need to link the 'soft' or qualitative information to the 'hard' data information. This can aid in developing a model that corresponds much more to 'everyday transport functioning' than the much-used classical, techno-economical approach to transport model designing.

Projects can employ both traditional methods like focus groups/questionnaire surveys/measuring actual behavioral response to different measures (for example, concessionary bus cards, channelizing feeder services, a change in the bus frequency, assistance in getting mobile phones as part of the development schemes like self-help groups, etc.), and new methods like mobile app-based data collection, to understand the existing travel behavior and adaptive preferences of different groups. Studies conducted on these lines (e.g., [125–127]) have found that discretionary trips have a greater number of adaptation alternatives available from which to choose than non-discretionary work trips. This is applicable to the case of gendered mobilities in developing countries where a major share of women's trips caters to discretionary purposes of combining various household/social/shopping related purposes.

Analyses of travel behavior, constraints and accessibility need to be triangulated and complement quantitative and qualitative data collection techniques.

### *4.1. Roadmap to Applied Research*

#### 4.1.1. Travel Survey

Travel survey data contains relevant information on all personal daily mobility, including information on multimodal trips, transport modes, distances and times. Such datasets additionally contain an extensive list of respondent and trip background variables, including socio-demographics, occupational status, home and work locations (possibly multiple), weekly working hours, occupational status, education, income, etc. Such datasets are generally not collected in the developing countries or collected as an adhoc exercise rather than being part of a regular data collection regime. We suggest using both the traditional and wherever applicable, the mobile interface to collect this dataset in the developing countries. Given the high penetration of mobile phones in the developing countries, an app-based method can be a viable option for data collection in the urban areas.

A one-day trip diary will invariably fail to capture the nuances of women's movement as there is a strong prevalence of a no-trip phenomenon on single days. Thus, the least time period assigned for trip data collection should be a one-week period to capture the clustered travel phenomenon.

Existing studies demonstrate that the use of location-based technologies reduces missing data on travel days and trips and improves the accuracy of departure/arrival locations and times over traditional travel survey data [128]. Modules are already available to extract the entire travel diary by accessing the smartphone data provided by the telephone operators. This data can be further enriched with local and regional spatial data from external sources, such as Open Street Maps, census, register and cadastral data.

#### 4.1.2. Attitudinal and Preference Survey

Attitudinal and preference travel survey distributed over different types of residential environments (inner-city, outer-urban, suburban, peri-urban and rural) and population categories within these regions will assist in capturing habits, attitudes, subjective well-being, perceived barriers and motivations with regard to the access to health, education, and employment opportunities [129]. Background information on personal and household attributes, preferences and lifestyles should also be collected as part of this exercise.

#### 4.1.3. Multi-Sited Ethnography Combined with In-Depth Traveller Interviews

Quantitative data inquiries should ideally be preceded and paralleled by collection and analysis of qualitative data. This is necessary not only to provide a full understanding of mobility experiences, barriers and motivations underlying the decision-making process of multimodal route, access, egress and transfer practices, but also to ensure that key questions are adequately identified for the quantitative surveys.

#### 4.1.4. Document Analysis and Informant Interviews

Document analysis and informant interviews form an important research tool in mobility mapping through (i) examining documents and records relevant to regional and urban policy-packaging, and sectoral development decisions having spatial dimension and (ii) key-informant interviews. Use of documentary analysis has become quite popular in transport research, especially when we are trying to evaluate the impact of an initiative. For example, a national government led approach to increase access to health facilities. Key informants, including politicians, international funders, policy makers, humanitarian organizations, the local and regional authorities, ministry officials directly involved in decision making on sustainable transport, etc., should be engaged to shed light on the topic of mobility and its inter-sectoral development dimensions. These experts can provide valuable insights when making recommendations for solutions.

#### 4.1.5. Exploring the Links Between Activity Participation and Subjective Well-Being

Concepts, terminologies and the will to explore linkages between the quality-of-life (QOL), subjective wellbeing (SWB) and daily travel have taken roots in the western research world in the past decade. This strand of research is primarily absent in the Global South. This could be partly due to lack of data for conducting robust studies to explore the relationship between QOL, SWB and daily travel. We propose that statistical methods like choice modeling, structural equation models, factor/cluster, etc. should be employed to identify and estimate the links between gender, daily travel times, time use, subjective well-being (SWB) and the extent to which the overall quality of life is affected by lack of access.

Studies based on similar approaches suggest important gender differences in activity participation, travel time, time use outcomes and resultant SWB [130]. They further find that travel times were unassociated with SWB for both genders. Instead, results were consistent with travel times serving as inputs in activity participation and therefore—at least for women—indirectly contributing to higher levels of SWB. These findings suggest that focusing on activity participation as a chief policy objective could yield higher quality of life benefits than a policy focus on travel time savings [130] (p. 10).

Transport planning in the Global South has been dominated by the logic of travel time savings, underpinning decisions on building highways, flyovers and road expansion programs. The field has simultaneously and consistently failed to serve a majority of the population. We therefore propose a shift in the transport planning approach—shifting the focus from travel time reductions as the chief policy objective to a focus on activity participation, enhanced SWB and QOL.

Activity participation should ideally be analyzed at the macro, meso and micro levels and policies designed for these three levels should be interlinked and complement each other (refer Table 1).

**Table 1.** Typology levels.

| Spatial Level | Methodology | Output |
|---|---|---|
| Macro-regional/city level | Mapping of interaction between land use and accessibility; app-based survey plotting the travel behavior, travel impedances, public transport route mapping, time use studies, mapping activity participation, personal interviews, ethnographic studies, document analysis and popular media discourse and history of planning to unearth biases. | Accessibility mapping; route mapping; time use mapping and mobility potential of women (and specifically low-income women) |
| Meso-zonal level | Questionnaire surveys, mapping activity participation, ethnographic studies and focus groups. | Accessibility levels to i. Work; ii. Education; iii. Social opportunities; iv. Health amenities and v. Training facilities. |
| Micro-ward level | Safety audits, walkability and bikeability audits; assessment of walking and bicycling infrastructure and mapping unsafe areas, routes, hotspots. | Micro-level information, which can be collated and scaled up. For ex. design interventions based on local conditions. |

#### 4.1.6. Accessibility Mapping

Continuing the above discussion, there is a dominance of the transport modeling approach based on the logics of travel-time savings underpinning the major transport infrastructure projects. Rarely is comprehensive land use and transport interaction (LUTI) models employed in the Global South to see the linkages between land use/activity participation and transport infrastructure provision.

We proposed that mapping of the interactions between accessibility with different transport modes to different land uses, and facilities like schools, employment and health centers, be undertaken for informing the design and implementation of transport infrastructure like public transport routes, bicycle network, etc.

In the following example, we presented a similar exercise for Bergen, Norway based on InMap (a simplified LUTI model), which links job accessibility by bicycle (and E-bike) and the land use plans (refer Figures 5 and 6). In this case, the supply of land was determined by the local municipalities through land use plans, while the demand for the land was estimated as a function of the accessibility to jobs, trade, general services and health services in the areas (from comparing E-bike accessibility with the land use growth potential, the authors found that it was possible to develop land use strategies to enhance the use of E-bikes. High job-accessibility with E-bikes close to the city centers supports the current general strategy of pursuing high density developments/transformation projects in these areas. The findings that the green field development areas were, in general, not found to provide any substantial accessibility with E-bikes).

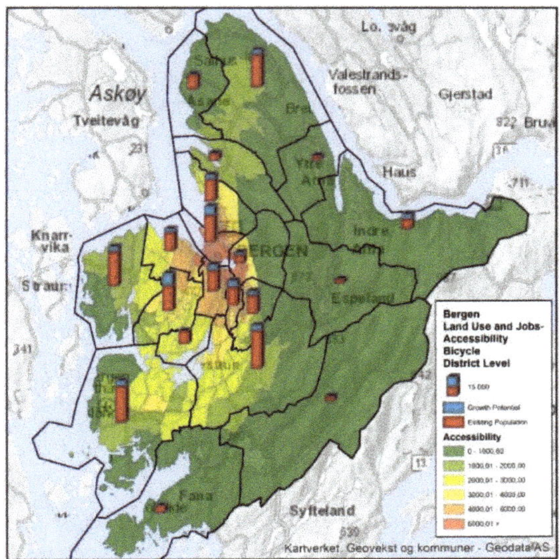


**Figure 5.** Accessibility with bicycle and growth potential, Bergen, Norway. (Source: [131]).

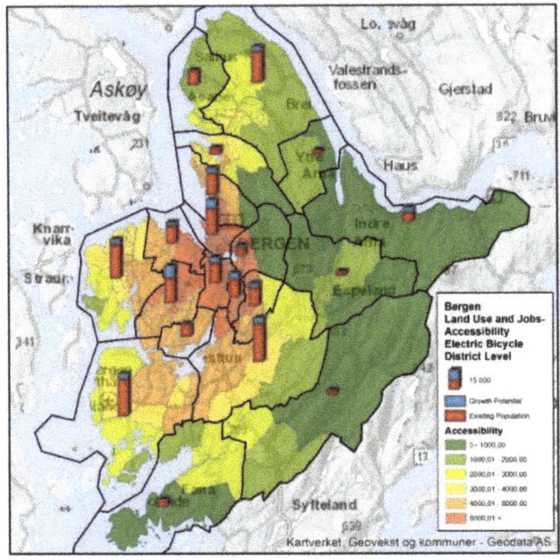


**Figure 6.** Accessibility with E-bike and growth potential, Bergen, Norway. (Source: [131]).

## 5. Research Gaps and Recommended Research Directions

The research gaps identified in this study deal with the following three components and the interface between these components: (i) Organization/governance, (ii) infrastructure provision and planning and (iii) travel (spatial) behavior culture/education. These three components can be further broken down to address the specificities of rural, urban and peri-urban areas. The relationship between gender and transport should underline all developmental policies—social protection programs, welfare policies, rural development, slum resettlement programs and urban planning policies. We comment specifically on the identified research gaps and research directions in the following points.

### 5.1. RESEARCH GAP 1: Inclusion of Gender from the Costs–Benefits Analyses, Which are Typically Used to Justify Investments Promoting Male-Biased Car-Based Urban Mobility

Research direction: How can we insert gender in the equation of costs–benefits assessments and other routinized protocols that are used for decision making?

Accessibility to employment, health and education opportunities have tangible, economic benefits. These numbers can be calculated, and easy to convert into economic indicators. If these indicators are taken into the cost–benefit assessments, then the chances of prioritizing transport infrastructure facilitating public transport, walking and cycling will be substantially increased.

It is of utmost importance that research findings are linked to policy making and program formulation. Currently, we have active research engagement on the topic of gender, transport and spatial planning in the Global South and yet the research findings are almost routinely ignored in policy making and seldom taken forward at the program formulation stage.

### 5.2. RESEARCH GAP 2: Acceptance of the Importance of (Affordable and Connected) Public Transport Provision for Facilitating Women in Transport Policy and Program Formulations

Research direction: What kind of capacity, knowledge and cooperation modalities need to be built to include both formal and informal public transport systems in the different hierarchies of transport plans?

How should we design specific programs on the tax structure for public transport and fund allocation to prioritize PT (Public transport)?

A most basic and simple message is to prioritize public transport over providing for cars. Informal public transport options like shuttle bus, jitneys, boda-boda, etc. should be included in the different hierarchies of transport planning from national, regional to local transport plans. Provide adequate tax benefit and subsidies to promote public transport. Address issues of unsafe running and operation of the informal public transport modes. Design policies to respond to the 'special needs' of women in terms of trip duration, access-egress, length, trip-chaining and trip purposes (with special reference to accompanying trips—traveling with children, elderly, etc.).

Research direction: How can we insert economic support systems in the welfare domain to address transport affordability issues for low-income women of the developing economies?

Affordability (and thereby accessing public transport) is a major impediment for low-income women. This is further compounded if the employment spaces of women living in slum and squatter settlements are hit by resettlement programs. Programs to address affordability issues needs to be sewed in with other welfare programs. Free public transport tickets, bicycles and other innovative solutions can have positive impacts on low-income women's access to education, employment and health opportunities.

### 5.3. RESEARCH GAP 3: Focus on Efficiency of Formal Sector Mobility and Transport System Performance Takes Precedence over Provision for Accessibility to Services and Opportunities for the Informal Economy and Communities

Research direction: How to incorporate the transport needs for informal economy engaged in regional trade into formal regional transport corridor developments?

Research direction: How to fuse spatial development and relocation policies to cater to women's opportunities with respect to education, health and employment?

Conduct accessibility mapping for different transport modes and prioritize areas for future growth that support walking and bicycling. In areas that are already built, create infrastructure that ensures safe walking and bicycling.

Further, it is vital that land use planning and development programs recognize and make space for informality, as informal markets provide the lion share of working space available to low-income women in both rural and urban parts of the developing world.

Route mapping and route planning of both the formal and informal public transport supply can aid in linking low-income areas to the employment and education hubs.

Develop toolkits for the incorporation of informal economy engaged in cross-border flows into regional transport corridor planning

*5.4. RESEARCH GAP 4: Integration with Other Sectors is Missing—Transport is Operated as a Separate Sector without Recognizing its Deep Connections to the Health, Employment and Education Domains*

Research direction: How to frame a multi-sectoral approach—ensure the mobility for women (both urban and rural) to ensure their access to employment/markets, education and health centers?

Link policies of other social development sectors with the transport sector. Welfare and social protection programs can be built around the issue of access to promote access to education, health and employment (refer to the cycling program of Bihar, India).

*5.5. RESEARCH GAP 5: Lack of Clear and Visible Efforts to Map and Address Safety Concerns and Incidences of Women on the Move*

Research direction: What are the exact policies, programs and tools needed to enhance women's personal safety? How to design more gender-sensitive public transport, walking and cycling space?

Safety is a major concern for women. Spatial and transport projects need to prioritize creating safe spaces at both macro and micro levels. At the macro level, smart solutions like Safetipin apps and safety auditing routines can be employed to map unsafe areas. Mapping without following up will be a wasted endeavor. Protocols need to be established on how to transform these unsafe areas, routes, etc. into safe, accessible areas [132].

At the micro level, for example, for spaces within the public transport, bus drivers and bus conductors need to be trained to deal with situations of sexual harassment. When the driver or conductor themselves are found to be culprits, punitive measures need to be in place for dealing with such actions [133].

Further, there exists a strong need for putting more emphasis on non-work related travel issues concerning trips made for care, household and household-based industry works.

*5.6. RESEARCH GAP 6: Labour Issues—Social Protection Programmes are not Linked with Employment in the (Informal) Transport Sector and Lack of Efforts to Address Male-Dominance in the Transport Sector Workforce*

Research direction: How can we encourage more female participation in the spatial planning and transportation sector?

How can the stigma and discrimination associated with women in the transport industry—especially in roles like driver, conductors, etc.—be tackled? How can laws that restrict women's participation in the transportation sector be removed?

Women face high levels of discrimination to both enter and work in the transport sector, which is often supported by restrictive laws in some developing countries (refer Figure 7).

Typically, women are either absent or marginally present on the different levels of transport domain. Actively encouraging and engaging women in the transport field, through targeted programs, can have major impact on the future of this field. The current transport field has blatantly ignored the needs and preferences of women, which may be corrected through inserting women in this sector. A first step could be to collect data on female employment both in the formal and informal transport sector and chart out ways to protect these workers through social protection programs.

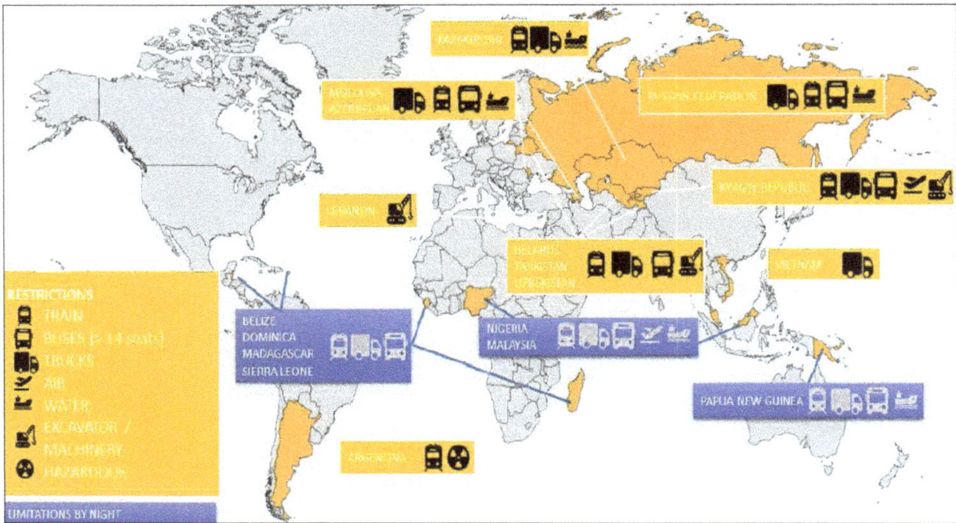


**Figure 7.** Legal restrictions on women's employment in the transport sector (Source: [134]) Note: Ukraine removed restrictions after the data collection period ended.

### *5.7. RESEARCH GAP 7: Lack of Smart City, Smart Solutions Research from an Inclusive Perspective*

Research direction: How can we design smart mobility and smart-city solutions to create inclusive settlements?

Smart cities and smart mobility solutions have changed the ways in which urban areas are being planned, utilized and consumed. It is vital that the element of inclusive settlements is inserted in this development on an immediate and urgent basis to avoid further pitfalls. Questions like who has access to these solutions? Who are the current users? How can we facilitate access to the other groups? What kind of solutions are needed? Are solutions based on bottom-up feedbacks? etc. needs to be routinely asked, monitored and fed into design solutions.

### *5.8. RESEARCH GAP 8: Lack of Simple, Standard Indicators*

Research direction: How can the findings emerging on gender and transport be supported by simple (and non-complex) indicators for the benefit of policy makers?

How can programs like the results-based budgeting technical assistance (TA), which the World Bank has adopted in the field of Health Planning, be designed and developed in the field of transport planning in the developing economies?

The knowledge generated through periodic assessments and estimations of the macroeconomic and welfare effects of creating accessible (education, health and employment) opportunities, could potentially benefit two important groups of policy makers and government stakeholders—(i) officials responsible for infrastructure evaluating road construction programs as economic investments and (ii) welfare sector officials focused on promoting gender equality, social development and poverty reduction. For the first set of officials, core area of interest revolves around the relationship between the provision of road infrastructure and national income growth as measured by GDP metrics while the second group is interested in social development outcomes. These two analytical perspectives are seldom brought together to complement each other. Through creating the simple, standardized and context-informed methodology for the measurement of economic and social impact of road building and other transport intervention programs, both sets of policy concerns can be monitored.

Simple indicators, however, run the risk of oversimplifying situation, yet they hold the potential for providing an overview to the policy-makers. Indicators should be developed and routinely used, but subjected to periodical assessment and upgrading, based on both quantitative and qualitative assessments and contextualization.

### 5.9. RESEARCH GAP 9: *Traffic Accident Data is Under-Reported and not Sex Disaggregated*

Research direction: Collect sex-disaggregated data on traffic accidents, which should include all forms of mobility—walking, cycling, public transport and car-based accidents.

Pedestrians form the biggest group who get either seriously injured or die in traffic accidents. If we analyze the predominance of walking among women in the developing countries, it will be no surprise that a greater number of (pedestrian) women might be dying in traffic accidents and yet a sex disaggregated analysis of traffic deaths remains unavailable. Disaggregated data analyses can assist in answering questions on the spatiality and temporality dimensions of the accidents—where (locations) and when (time) are male and female respectively meeting accidents?

### 5.10. RESEARCH GAP 10: *Unpacking the Linkages Between Feminization of Informal Settlements, Relocation and Livelihood Opportunities*

Research direction: Collect sex-disaggregated data on livelihood issues in light of relocation.

Informal settlements are increasingly becoming feminized with a relatively higher concentration of females than males living in slums in the developing world. Relocation decisions impact livelihoods of women to a much greater extent than men. Access to employment opportunities that suits the livelihood profile of slum women is hardly ever brought forth in relocation decisions. The interconnection between livelihood and spatiality needs to be out in focus.

### 5.11. RESEARCH GAP 11: *Knowledge on Mobility Options and Cross-Border Trading*

Research direction: A few of the research areas that should be further investigated while designing regional transport corridor schemes are as follows:

- What are the inter-linkages between transport provision (both hard infrastructure and gendered access to the transport resources), value chains and gender inequality?
- What are the main mobility challenges and opportunities for women entrepreneurs in the context of cross-border trade?
- What are the potential ways of improving regional mobilities?
- What are the potential ways of integrating gendered mobilities concerns into value chain of development projects and programs to assist women in maximizing their profitability and competitiveness?

## 6. Conclusions

We have looked at the specificities of women's mobility and its implications for transport planning along with urban, regional, rural and social planning. We posit that the multiplicity of the issues involved in understanding a complex topic like gendered mobilities with both its traditional and emerging 'smart' variant necessitates that we borrow and build on a host of theories and analyses. To that end, we have used a combination of socio-psychological and feminist theories, socio-technical transition theory, mobility biographies and social practice theory framework to frame a discussion around the differentiated patterns of women's transport needs, usage and implications for future planning.

Currently, the transport sector in the Global South is becoming aware of the environmental consequences but is grappling with how to balance the development and the environmental agenda. Given this scenario, government agencies would do well to understand how the environmental meaning operates. If environmental meaning is a better retention mechanism than a recruitment mechanism, then appeals to environmental sustainability or reduced carbon footprint are not likely to change user behavior. Determining effective mechanisms for recruitment to sustainable mobility is an important agenda and our analysis reveals that women could be champions here. They exhibit sustainable travel behavior and yet continue to be ignored. Transport systems will benefit greatly by becoming gender-responsive as the sector can then simultaneously cater to the three pillars of sustainable development—environment, social and economic pillars. As an initial step, it would be useful to have a simple and overarching framework that underlines the relevance of transport to women's employment/livelihoods/income security and unpaid domestic care work (i.e., 'production' and 'reproduction'). In other words, mobility/transport are needed to facilitate women's access to markets and jobs and to facilitate their access to services (e.g., health centers, schools, childcare centers, shops, etc.). Additionally, the transport sector itself can be a source of employment for women, contributing to becoming a source of livelihood but more essentially, influencing the norms dictating the field of transport at large to be more mindful of women's needs and preferences. Traditionally, the norms governing the transport sector remains heavily male dominated in terms of both employment and a technical orientation. Issues regarding safety, affordability, accessibility, availability, acceptability and accommodation are vital and need to be taken into transport design and planning.

The transport field is undergoing huge transitions both from a policy perspective with added impetus of Agenda 2030, and practice perspective with insertions of smart solutions. In light of these developments, the policy focus is actively promoting a shift from car-based daily mobility to a more sustainable one. However, given the prevalent 'norms' in the transport field, focus on 'hard' infrastructure persists and social-psychological elements benefitting women's daily mobilities are routinely ignored. Practice theory, for example, could be employed here to highlight and explain user adoption of innovations and provide for a richer understanding of the micro-phenomena that takes place on different life stages by which these stages can be targeted in the right manner. In light of the social (sustainable) practices adopted by women at large, the process of recruitment/defection and retention/reproduction needs to be better understood. In the parlance of practice theory, how a practice such as walking, bicycling and public transport usage can undergo formation and reformation in a manner that it can sustain itself needs to be studied.

The ways in which different groups of women engage with different forms of mobilities vary according to the life-stage, conditions and contextual circumstances. This set of knowledge, on user variations, offers tools for promoters with which the sustainable practices of majority of women can be upscaled to a mainstream or dominant practice of daily mobility. The relationship between user and promoter circumstances and conditions is not static, but co-evolutionary. Promoters respond to how users actually use their products and services, and, in this regard, every transport decision needs to be unpacked and scrutinized as a product delivery with an open discussion on who constitutes the target customers. These analyses, for example, can be used to bolster the agenda of creating the walking- and biking-friendly cities.

Further, the authorities need to seriously engage with the foremost issue of 'How to make data collection and analyses routinized processes?'. Links between transport and housing policies should be inserted in land use, housing and relocation decisions. The research gaps identified so far highlight that very few studies have looked at the interactions between urban housing policies and accessibility wrt subsidized housing, establishment of resettlement colonies and loss of livelihoods of women.

Studies on governance and cross-border trade should discuss ways in which firstly, informality is recognized and secondly, the transport sector is restructured to facilitate women's employment both within and outside the transport sector.

Additionally, how can the transport sector engage with women's organizations to hear their needs and demands? Such analyses should be carried at micro, meso and macro levels to ensure that local, regional and international (wrt cross-border trading) markets are available to women for expanding their capabilities.

From a methodological perspective, a multi-method approach (e.g., [135]) should be adopted to integrate both qualitative and quantitative data in knowledge generation and decision making. Plotting of existing data on travel behavior and survey data on individuals' subjective and self-reported preferences, acceptance of different mobility options, and attitudes related to the adoption of the proposed solutions can be a suitable starting point. An easily accessible way to move forward would be through creating predictive scenarios of travel behavior by interlinking demographics data with travel behavior. Such interlinking should be augmented through identification of major trends and driving forces in the past and projections for future. Along with mapping out the gender effects, these scenarios could be used to discuss and assess the environmental, social and economic impacts of the projected travel behavior and future mobility. For the regulating and implementing agencies, such scenario assessments will map out the opportunities and challenges involved in formulating policies and programs to reach sustainability goals and assess the scale of the required changes. Ways to sew this knowledge into the specifics of designing social protection programs or niches for transition demands further exploration. The case of the Cycle program for rural students in the state of Bihar, India is an excellent example of such a niche.

**Author Contributions:** Both authors were involved in drawing the conceptual plan of this article. T.P.U. framed and wrote the first draft of the article, while J.T. provided valuable inputs on the content, structure and framing of the final draft.

**Funding:** This research was funded by UK AID through the UK Department for International Development under the High-Volume Transport Applied Research Program, managed by IMC Worldwide. Discussions on the 'smart' elements are inspired from a similar study conducted in the Global North by Dr. Priya Uteng for a Nordic Flagship Project SHIFT, funded by The Nordic Energy Research (grant number 77892) http://www.nordicenergy.org/flagship/project-shift/.

**Acknowledgments:** We remain deeply grateful for the constructive comments received by the anonymous reviewers.

**Conflicts of Interest:** The authors declare no conflict of interest.

## References

1. Urry, J. *Sociology beyond Societies. Mobilities for the Twenty-First Century*; Routledge: London, UK, 2000.
2. Sheller, M.; Urry, J. The new mobilities paradigm. *Environ. Plan. A* **2006**, *38*, 207–226. [CrossRef]
3. Schipper, L.; Fabian, H.; Leather, J. *Transport and Carbon Dioxide Emissions: Forecasts, Options Analysis, and Evaluation*; ADB Sustainable Development Working Paper Series; ADB: Mandaluyong, Philippines, 2009.
4. Singh, S.K. Urban transport in India: Issues, challenges, and the way forward. *Eur. Transp.* **2012**, *52*, 1825–3997.
5. Richardson, B.C. Sustainable transport: Analysis frameworks. *J. Transp. Geogr.* **2005**, *13*, 29–39. [CrossRef]
6. Kaijser, A. How to describe large technical systems and their changes over time. In *Urban Transport Development: A Complex Issue*; Jönson, G., Tengström, E., Eds.; Springer: Berlin/Heidelberg, Germany, 2005.
7. Razavi, S. Indicators as Substitute for Policy Contestation and Accountability? Some Reflections on the 2030 Agenda from the Perspective of Gender Equality and Women's Rights. *Glob. Policy* **2019**, *10*, 149–152. [CrossRef]
8. UN Women. *Turning Promises into Action: Gender Equality in the 2030 Agenda for Sustainable Development*; UN Women: New York, NY, USA, 2018.
9. Turner, J.; Fouracre, P. Women and Transport in developing countries. *Transp. Rev.* **1995**, *15*, 77–96. [CrossRef]
10. Priya Uteng, T. *Gender and Mobility in the Developing World*; The World Bank: Washington, DC, USA, 2012.
11. Kronsell, A.; Rosqvist, L.S.; Hiselius, L.W. Achieving climate objectives in transport policy by including women and challenging gender norms—The Swedish case. *Int. J. Sustain. Transp.* **2015**. [CrossRef]
12. Levin, L.; Faith-Ell, C. How to apply Gender Equality Goals in Transport and Infrastructure Planning. In *Workshop 3: Smart Mobilities, Planning and Policy Processes—Gender and Diversity Mainstreaming*; Nos-Sh workshop series; VTI: Stockholm, Sweden, 17–18 September 2018.
13. Ajzen, I. The theory of planned behaviour. *Organ. Behav. Hum. Decis. Process.* **1991**, *50*, 170–211. [CrossRef]
14. Schwartz, S.H. Normative influence on altruism. In *Advances in Experimental Social Psychology*; Berkowitz, L., Ed.; Academic Press: New York, NY, USA, 1977; Volume 10, pp. 221–279.
15. Verplanken, B.; Aarts, H.; Van Knippenberg, A. Habit, information, acquisition, and the process of making travel mode choices. *Eur. J. Soc. Psychol.* **1997**, *27*, 539–560. [CrossRef]
16. Dijst, M.; Farag, S.; Schwanen, T. A comparative study of attitude theory and other theoretical models for understanding travel behaviour. *Environ. Plan. A* **2008**, *40*, 831–847. [CrossRef]

17. Connell, R.W. *Gender*; Polity Press: Cambridge, UK, 2002.
18. Acker, J. Inequality regimes: Gender, class and race in organizations. *Gend. Soc.* **2006**, *20*, 441–464. [CrossRef]
19. Kronsell, A. Gendered practices in institutions of hegemonic masculinity: Reflections from feminist standpoint theory. *Int. Fem. J. Politics* **2005**, *7*, 280–298. [CrossRef]
20. Hirdman, Y. *Genus—Om det Stabilas Föränderliga Former*; Liber: Malmö, Sweden, 2001.
21. Geels, F. Technological transitions as evolutionary reconfiguration processes: A multi-level perspective and a case-study. *Res. Policy* **2002**, *33*, 897–920. [CrossRef]
22. Grin, J.; Rotmans, J.; Schot, J. *Transitions to Sustainable Development. New Directions in the Study of Long-Term Transformative Change*; Routledge: New York, NY, USA, 2010.
23. Kemp, R.; Avelino, F.; Bressers, N. Transition management as a model for sustainable mobility. *Eur. Transp.* **2011**, *47*, 25–46.
24. Rip, A.; Kemp, R. Technological change. In *Human Choice: An Climate Change*; Rayner, S., Malone, E.L., Eds.; Battelle Press: Columbus, OH, USA, 1998; pp. 327–399.
25. Rotmans, J.; Loorbach, D. Towards a Better Understanding of Transitions and Their Governance. In *Transitions to Sustainable Development. New Directions in the Study of Long Term Transformative Change. Part II*; Grin, J., Rotmans, J., Schot, J., Eds.; Routledge: New York, NY, USA, 2010; pp. 105–220.
26. Schwanen, T.; Banister, D.; Anable, J. Scientific research about climate change mitigation in transport: A critical review. *Transp. Res. Part A* **2011**, *45*, 993–1006. [CrossRef]
27. Beige, S.; Axhausen, K.W. Interdependencies between turning points in life and long-term mobility decisions. *Transportation* **2009**, *39*, 857–872. [CrossRef]
28. Scheiner, J. Mobility biographies: Elements of a biographical theory of travel demand. *Erdkunde* **2007**, *61*, 161–171. [CrossRef]
29. Schoenduwe, R.; Mueller, M.G.; Peters, A.; Lanzendorf, M. Analysing mobility biographies with the life course calendar: A retrospective survey methodology for longitudinal data collection. *J. Transp. Geogr.* **2015**, *42*, 99–109. [CrossRef]
30. Chatterjee, K.; Scheiner, J. Understanding changing travel behaviour over the life course: Contributions from biographical research. In Proceedings of the 14th International Conference on Travel Behaviour Research, Windsor, UK, 19–23 July 2015.
31. Chatterjee, K.; Sherwin, H.; Jain, J. Triggers for changes in cycling: The role of life events and modifications to the external environment. *J. Transp. Geogr.* **2013**, *30*, 183–193. [CrossRef]
32. Clark, B.; Lyons, G.; Chatterjee, K. Understanding the process that gives rise to household car ownership level changes. *J. Transp. Geogr.* **2016**, *55*, 110–120. [CrossRef]
33. Priya Uteng, T.; Julsrud, T.E.; George, C. The role of life events and context in type of car share uptake: Comparing users of peer-to-peer and cooperative programs in Oslo, Norway. *Transp. Res. Part D Transp. Environ.* **2019**, *71*, 186–206. [CrossRef]
34. Müggenburg, H.; Busch-Geertsema, A.; Lanzendorf, M. Mobility biographies: A review of achievements and challenges of the mobility biographies approach and a framework for further research. *J. Transp. Geogr.* **2015**, *46*, 151–163. [CrossRef]
35. Sattlegger, L.; Rau, H. Carlessness in a car-centric world: A reconstructive approach to qualitative mobility biographies research. *J. Transp. Geogr.* **2016**, *53*, 22–31. [CrossRef]
36. Reckwitz, A. Toward a Theory of Social Practices: A Development in Culturalist Theorizing. *Eur. J. Soc. Theory* **2002**, *5*, 243–263. [CrossRef]
37. Schatzki, T. *Where the Action Is (On Large Social Phenomena Such as Sociotechnical Regimes)*. SPRG Working Paper 1. 2011. Available online: http://www.sprg.ac.uk/uploads/schatzki-wp1.pdf (accessed on 29 April 2018).
38. Pantzar, M.; Shove, E. Understanding innovation in practice: A discussion of the production and re-production of Nordic Walking. *Technol. Anal. Strateg. Manag.* **2010**, *22*, 447–461. [CrossRef]
39. Priya Uteng, T.; Lucas, K. The Trajectories of Urban Mobilities in the Global South: An introduction. In *Urban Mobilities in the Global South*; Priya Uteng, T., Lucas, K., Eds.; Taylor & Francis: Abingdon, UK; Routledge: New York, NY, USA, 2017.
40. Levy, H.; Voyadzis, C. *Morocco—Socio-Economic Impact of Rural Roads*; Operations Evaluation Department, World Bank: Washington, DC, USA, 1996.

41. Porter, G. Transport planning in sub-Saharan Africa. Progress report 1. Improving access to markets and services. *Prog. Dev. Stud.* **2007**, *7*, 251–257. [CrossRef]
42. Muralidharan, K.; Prakash, N. *Cycling to School: Increasing Secondary School Enrollment for Girls in India*; Working Paper 19305; National Bureau of Economic Research: Cambridge, UK, 2013; Available online: https://www.nber.org/papers/w19305 (accessed on 29 April 2018).
43. Porter, G.; Hampshire, K.; Abane, A.; Munthali, A.; Robson, E.; Mashiri, M.; Tanle, A.; Maponya, G.; Dube, S. Child Porterage and Africa's Transport Gap: Evidence from Ghana, Malawi and South Africa. *World Dev.* **2012**, *40*, 2136–2154. [CrossRef]
44. Carr, M.; Chen, M. *Globalization, Social Exclusion and Work: With Special Reference to Informal Employment and Gender*; Working Paper No. 20; Policy Integration Department World Commission on the Social Dimension of Globalization, ILO: Geneva, Switherland, 2004.
45. Houpin, S. Urban mobility in the Mediterranean CMI. In Proceedings of the Session C: Structuring Transit System and Urban Integration, Urban Mobility Program Damascus Regional Conference, Damascus, Syria, 13 April 2009.
46. Kabeer, N. Globalization, labor standards, and women's rights: Dilemmas of collective (in) action in an interdependent world. *Fem. Econ.* **2004**, *10*, 3–35. [CrossRef]
47. Absar, S.S. Conditions, concerns and needs of garment workers in Bangladesh. *Dev. Bull.* **2000**, *51*, 82–84.
48. Kantor, P. A Sectoral Approach to the Study of Gender Constraints on Economic Opportunities in the Informal Sector in India. *Gend. Soc.* **2002**, *16*, 285–302. [CrossRef]
49. Schuler, S.R.; Hashemi, S.M.; Riley, A.P. The Influence of Women's Changing Roles and Status in Bangladesh's Fertility Transition: Evidence from a Study of Credit Programs and Contraceptive Use. *World Dev.* **1997**, *25*, 563–578. [CrossRef]
50. Mayoux, L. Questioning Virtuous Spirals: Micro-Finance and Women's Empowerment in Africa. *J. Int. Dev.* **1999**, *11*, 957–984. [CrossRef]
51. Ahmed, S.M.; Chowdhury, M.; Bhuiya, A. Micro-Credit and Emotional Well-Being:Experience of Poor Rural Women from Matlab, Bangladesh. *World Dev.* **2001**, *29*, 1957–1966. [CrossRef]
52. Amin, R.; Pierre, M.; Ahmed, H.; Haq, R. Integration of an Essential Services Package (ESP) in Child Reproductive Health and Family Planning with a Micro-Credit Program for Poor Women: Experiences from a Pilot Project in Rural Bangladesh. *World Dev.* **2001**, *19*, 1611–1629. [CrossRef]
53. Omorodion, F.I. Rural Women's Experiences of Micro-Credit Schemes in Nigeria, Case Study of Esan Women. *J. Asian Afr. Stud.* **2007**, *42*, 479–494. [CrossRef]
54. Perry, D. Micro Credit and Women Money-lenders: The Shifting Terrain of Credit in Rural Senegal. *Hum. Organ.* **2002**, *61*, 30–35. [CrossRef]
55. Pitt, M.; Shahidur, R.K.; Cartwright, J. *Does Micro-Credit Empower Women? Evidence from Bangladesh*; The World Bank, Policy Research Working Paper Series, 2998; The World Bank: Washington, DC, USA, 2003.
56. Izugbara, C.O. Gendered Micro-Lending Schemes and Sustainable Women's Empowerment in Nigeria. *Community Dev. J.* **2004**, *39*, 72–84. [CrossRef]
57. Nigam, A. *Give Us Credit: How Access to Loans and Basic Social Services Can Enrich and Empower People*; UNICEF: New York, NY, USA, 2000.
58. Gupta, S.; Sharma, A.K. A Causal analysis of Public Sector Working Women's Mobility in Navi Mumbai: Lessons for Policy Formulation. In Proceedings of the CODATU IX Conference, Mexico City, Mexico, 11–14 April 2000.
59. Desai, R.; Parmar, V.; Mahadevia, D. Resettlement, mobility and women's safety in cities. *IIC Q.* **2017**, *43*, 65–76.
60. World Bank. *Safe, Clean, and Affordable. Transport for Development*; Transport Sector Board, World Bank: Washington, DC, USA, 2008.
61. Thaddeus, S.; Maine, D. Too far to walk: Maternal mortality in context. *Soc. Sci. Med.* **1994**, *38*, 1091–1110. [CrossRef]
62. Mlay, R.; Chuwa, M.; Smith, H. Emergency referral transport needs of pregnant women in rural Tanzania. *Afr. J. Midwifery Womens Health* **2010**, *4*. [CrossRef]
63. Hamlin, C. *Preventing Fistula: Transport's Role in Empowering Communities for Health in Ethiopia*; The World Bank: Washington, DC, USA, 2004.

64. Wagstaff, A. *Unpacking the Causes of Child Survival: The Case of Cebu the Philippines*; World Bank Report; The World Bank: Washington, DC, USA, 2000.
65. Peters, D. *Gender and Transport in Less Developed Countries: A Background Paper in Preparation for CSD-9*; Background Paper for the Expert Workshop "Gender Perspectives for Earth Summit 2002: Energy, Transport, Information for Decision-Making"; UNED Forum: London, UK, 2001.
66. Porter, G.; Hampshire, K.; Dunn, C.; Hall, R.; Levesley, M.; Burton, K.; Robson, S.; Abane, A.; Blell, M.; Panther, J. Health impacts of pedestrian head-loading: A review of the evidence with particular reference to women and children in sub-Saharan Africa. *Soc. Sci. Med.* **2013**, *88*, 90–97. [CrossRef] [PubMed]
67. Mahumud, R.A.; Alamgir, N.I.; Hossain, M.T.; Baruwa, E.; Sultana, M.; Gow, J.; Alam, K.; Ahmed, S.M.; Khan, J.A.M. Women's Preferences for Maternal Healthcare Services in Bangladesh: Evidence from a Discrete Choice Experiment. *J. Clin. Med.* **2019**, *8*, 132. [CrossRef] [PubMed]
68. Murray, S.F.; Davies, S.; Phiri, R.K.; Ahmed, Y. Tools for monitoring the effectiveness of district maternity referral systems. *Health Policy Plan.* **2001**, *16*, 353–361. [CrossRef]
69. Agarwal, S.; Sethi, V.; Srivastava, K.; Jha, P.K.; Baqui, A.H. Birth Preparedness and Complication Readiness among Slum Women in Indore City, India. *J. Health Popul. Nutr.* **2010**, *28*, 383–391. [CrossRef] [PubMed]
70. Mberu, B.; Muindi, K. Ensuring Safer Pregnancies for Kenyan Women in Urban Slums. 2017. Available online: http://aphrc.org/post/7979 (accessed on 29 April 2018).
71. INTALInC. Transport and Mobilities: Pathways to Transformation. 2018. Available online: https://intalinc.leeds.ac.uk/wp-content/uploads/sites/28/2018/06/INTALInC-Uganda-report.pdf (accessed on 29 April 2018).
72. Graham, A. Gender Mainstreaming Guidelines for Disaster Management Programmes, A Principled Socio-Economic and Gender Analysis (SEAGA) Approach. In Proceedings of the Expert Group Meeting on Environmental Management and the Mitigation of Natural Disasters: A Gender Perspective, Ankara, Turkey, 6–9 November 2001.
73. Priya, T. Urban Disaster Management Practises of India: How Sustainable? Case Study: Bhuj, India. Master's Thesis, Norwegian University of Science and Technology, Trondheim, Norway, 2003.
74. International Forum for Rural Transport and Development (IFRTD). Reconstructing Vulnerability or Reconstruction with Change? *Forum News* **2005**, *1*, 1–4.
75. WHO (World Health Organization). *Global Status Report on Road Safety*; WHO: Geneva, Switherland, 2015.
76. GBD 2015 Mortality and Causes of Death Collaborators. Global, regional, and national life expectancy, all-cause mortality, and cause-specific mortality for 249 causes of death, 1980–2015: A systematic analysis for the Global Burden of Disease Study 2015. *Lancet* **2016**, *388*, 1459–1544. [CrossRef]
77. Marquez, P.V.; Banjo, G.A.; Chesheva, E.Y. *Confronting Death on Wheels: Making Roads Safe in Europe and Central Asia—Establishing Multisectoral Partnerships to Address a Silent Epidemic*; World Bank: Washington, DC, USA, 2010.
78. Damsere-Derry, J.; Palk, G.; King, M. Road accident fatality risks for "vulnerable" versus "protected" road users in northern Ghana. *Traffic Inj. Prev.* **2017**, *18*, 736–743. [CrossRef]
79. Williams, S.J.; Kowal, P.; Hestekin, H.; O'Driscoll, T.; Peltzer, K.; Yawson, A.; Biritwum, R.; Maximova, T.; Salinas Rodríguez, A.; Manrique Espinoza, B.; et al. Prevalence, risk factors and disability associated with fall-related injury in older adults in low- and middle-incomecountries: Results from the WHO Study on Global Ageing and Adult Health (SAGE). *BMC Med.* **2015**, *13*. [CrossRef]
80. Moshi, H.I. Physical Trauma and Its Consequences in Rural and Semi-Urban Regions of Low- and Middle-Income Countries. *Curr. Issues Glob. Health IntechOpen* **2018**. Available online: https://www.intechopen.com/books/current-issues-in-global-health/physical-trauma-and-its-consequences-in-rural-and-semi-urban-regions-of-low-and-middle-income-countr (accessed on 29 April 2019).
81. Warr, M. Fear of victimization: Why are women and the elderly more afraid? *Soc. Sci. Q.* **1984**, *65*, 681–702.
82. Mitra-Sarkar Sand Partheeban, P. Abandon All Hope, Ye Who Enter Here: Understanding the Problem of Eve Teasing in Chennai, India. In *Women's Issues in Transportation: Summary of the 4th International Conference, Volume 2: Technical Papers*; The National Academies Press: Washington, DC, USA, 2009; Available online: https://www.nap.edu/read/22887/chapter/9 (accessed on 29 April 2018).
83. Loukaitou-Sideris, A.; Fink, C. Addressing Women's Fear of Victimization in Transportation Settings: A Survey of U.S. Transit Agencies. *Urban Aff. Rev.* **2009**, *44*, 554–587. [CrossRef]

84. Borker, G. Safety First: Perceived Risk of Street Harassment and Educational Choices of Women. 2018. Available online: https://girijaborker.files.wordpress.com/2017/11/borker_jmp.pdf (accessed on 29 April 2018).
85. Verma, M.; Manoj, M.; Rodeja, N.; Verma, A. Service gap analysis of public buses in Bangalore with respect to women safety. *Transp. Res. Procedia* **2017**, *25*, 4326–4333. [CrossRef]
86. Verma, M.; Rodeja, N.; Manoj, M.; Verma, A. Young Women's Perception of Safety in Public Buses: A Study of Two Indian Cities (Ahmedabad and Bangalore). In Proceedings of the WCTR Conference, Mumbai, India, 26–31 May 2019.
87. Meshram, A.; Choudhary, P.; Velaga, N.R. Assessing and Modelling Perceived Safety and Comfort of Women during Ridesharing. In Proceedings of the WCTR Conference, Mumbai, India, 26–31 May 2019.
88. Campbell, M.; Jain, T. Usage Patterns of Public Transport as a Starting Point for Thinking through Gendered Conceptualizations of Delhi. In Proceedings of the WCTR Conference, Mumbai, India, 26–31 May 2019.
89. Abenoza, R.F.; Romero-Torres, J.; Cats, O.; Susilo, Y.O. User experiences and perceptions of women-only transport services in Mexico. In *Gendering Smart Mobilities*; Uteng, T.P., Levin, L., Christensen, H.R., Eds.; Routledge: Abingdon, UK; New York, NY, USA, 2019.
90. Olvera, L.D.; Plat, D.; Pochet, P. Not really a quiet stroll. The perception of insecurity during pedestrian trips in Dakar (Senegal). In Proceedings of the WCTR Conference, Mumbai, India, 26–31 May 2019.
91. Lauwers, D.; Papa, E. *Towards a Smarter Urban Mobility*; Colloquium Vervoersplanologisch Speurwerk: Antwerp, Belgium, 2015.
92. Shaheen, S.A.; Martin, E.W.; Chan, N.D.; Cohen, A.P.; Pogodzinski, M. *Public Bikesharing in North America during a Period of Rapid Expansion: Understanding Business Models, Industry Trends and User Impacts*; Mineta Transportation Institute: San Jose, CA, USA, 2014.
93. Priya Uteng, T.; Singh, Y.J.; Lam, T. Safety and daily mobilities of urban women—Methodologies to confront the policy of "invisibility". In *Measuring Transport Equity*; Lucas, K., Martens, K., Di Ciommo, F., Dupont-Kieffer, A., Eds.; Elsevier: Amsterdam, The Netherlands, 2019.
94. Viswanath, K. Using Data to Improve Women's Safety in Cities, Transport. 2016. Available online: https://blogs.adb.org/blog/using-data-improve-womens-safety-cities-transport (accessed on 29 April 2018).
95. Priya Uteng, T.; Singh, Y.J.; Hagen, O.H. Social Sustainability and Transport. Making 'smart mobility' socially sustainable. In *Urban Social Sustainability: Theory, Practice and Policy*; Shirazi, M.R., Keivani, R., Eds.; Routledge: Abingdon, UK; New York, NY, USA, 2019.
96. Safetipin. Using Data to Build Safer Cities: Delhi, Bogota, Nairobi. 2017. Available online: http://safetipin.com/resources/files/Report%20Cities%20Alliance.pdf (accessed on 29 April 2018).
97. Kalms, N.; Matthewson, G.; Salen, P. Safe in the City? Girls Tell It Like It Is. 2017. Available online: https://theconversation.com/global (accessed on 29 April 2018).
98. Azcona, G.; Bhatt, A.; Duerto Valero, S.; Priya Uteng, T. *Feminization of Slums: The Effects of Rapid Urbanization in a Gender Unequal World*; UN Women: New York, NY, USA, 2019; Forthcoming.
99. Lima, J.H.D.; Andrade, M.O.D.; Maia, M.L.A. Measuring accessibility: Effects of implementing multiple trip generating developments. *J. Transp. Lit.* **2016**, *10*, 25–29. [CrossRef]
100. Maia, M.L.A. Land use policy and rights to the city. *Land Use Policy* **1995**, *12*, 177–180. [CrossRef]
101. Nayak, S.; Gupta, S. Travel Probability Fields of Urban Poor: Case Study of Slum Dwellers in Kolkata. In Proceedings of the CTRG Conference, Mumbai, India, 17–20 December 2017.
102. Carlin, J. Buenos Aires Urban Planning Wants Transport to Transform Slums. 2017. Available online: https://www.ft.com/content/de3eed42-0e3e-11e7-a88c-50ba212dce4d (accessed on 29 April 2018).
103. Salon, D.; Gulyani, S. Mobility, poverty, and gender: Travel 'choices' of slum residents in Nairobi, Kenya. *Transp. Rev.* **2010**, *30*, 641–657. [CrossRef]
104. Roy, A. Urban Informality—Toward an Epistemology of Planning. *J. Am. Plann. Assoc.* **2005**, *71*, 147–158. [CrossRef]
105. Kusakabe, K. Women's Work and Market Hierarchies Along the Border of Lao PDR. *Gender, Place & Culture* **2004**, *11*, 581–594.
106. Cheater, A.P. Transcending the state? Gender and borderline constructions of citizenship in Zimbabwe. In *Border Identities: nation and state at international frontiers*; Cambridge University Press: Cambridge, UK, 1998; pp. 191–214.

107. Vila, P. Gender and the overlapping of region, nation, and ethnicity on the U.S.-Mexico border. In *Ethnography at the Border*; University of Minnesota Press: Minneapolis, MN, USA, 2003; pp. 73–104.
108. Biemann, U. Performing the border: On gender, transnational bodies, and technology. In *Globalisation on the Line: Culture, Capital, and Citizenship at U.S. Borders*; Sadowski-Smith, C., Ed.; Palgrave: New York, NY, USA, 2002; pp. 99–118.
109. Soprano, C. Africa Opens the Gates to Cross-Border Trade. *World Economic Forum*. 22 August 2014. Available online: https://www.weforum.org/agenda/2014/08/africa-cross-border-traders/ (accessed on 29 April 2018).
110. Ibrahim, B.S. Cross Border Trade in West Africa: An Assessment of Nigeria and Niger Republic. *Afr. Rev.* **2015**, *42*, 126–150.
111. Blumberg, R.L.; Malaba, J.; Meyers, L. *Women Cross-Border Traders in Southern Africa. Contributions, Constraints, and Opportunities in Malawi and Botswana*; USAID: Washington, DC, USA, 2016.
112. Koroma, S.; Nimarkoh, J.; You, N.; Ogalo, V.; Owino, B.; FAO Regional Office for Africa. Formalization of Informal Trade in Africa. Trends, Experiences and Socio-Economic Impacts. 2017. Available online: http://www.fao.org/3/a-i7101e.pdf (accessed on 29 April 2018).
113. Smith, N. Homeless/global: Scaling places. In *Mapping the Futures: Local Cultures, Global Change*; Bird, J., Curtis, B., Eds.; Routledge: London, UK, 1998; pp. 87–119.
114. UN Women. Opportunities for women entrepreneurs in the context of African Continental Free Trade Area. In Proceedings of the Eastern Africa Women in Transportation Conference, Nairobi, Kenya, 22–23 November 2018; Available online: https://eastafricawitconference.com/ (accessed on 29 April 2019).
115. Staudt, K. *Free Trade? Informal economies at the U.S.-Mexico border*; Temple University Press: Philadelphia, PA, USA, 1998.
116. Marchand, M.H. Engendering globalisation in an era of transnational capital: New crossborder alliances and strategies of resistance in a post Nafta Mexico. In *Feminist Post-Development Thought: Thinking Modernity, Post-colonialism and Representation*; Saunders, K., Ed.; Zed books: New York, NY, USA, 2002; pp. 105–119.
117. Moss, P. Taking on, thinking about, and doing feminist research in geography. In *Feminist Geography in Practice: Research and methods*; Moss, P., Ed.; Blackwell Publishers: Oxford, UK, 2002; pp. 1–17.
118. Elson, D. *Male Bias in the Development Process*; Manchester University Press: Manchester, UK, 1991.
119. White, S.C. Male Bias in the Development Process. *Public Admin. Dev.* **1993**, *13*, 84–85. [CrossRef]
120. Government of India. *National Urban Transport Policy, Ministry of Urban Development*; GOI: New Delhi, India, 2006.
121. Mahadevia, D. *Ahmedabad's BRT System*; Centre for Urban Equity (CUE): Ahmedabad, India, 2011.
122. Kamau, A. Women Transport Workers: Social Protection and Labour Issues. In Proceedings of the Eastern Africa Women in Transportation Conference, Nairobi, Kenya, 22–23 November 2018.
123. Ghate, A.T.; Tiwari, G. A gendered perspective of the phenomenon of 'no travel to work' in Indian cities: A case study of Jaipur city. In Proceedings of the WCTR Conference, Mumbai, India, 26–31 May 2019.
124. Susilo, Y.O.; Liu, C. Exploring patterns of time-use allocation and immobility behaviors in the Bandung Metropolitan Area, Indonesia. In *Urban Mobilities in the Global South*; Routledge: Abingdon, UK, 2017; pp. 111–133.
125. Gärling, T.; Gärling, A.; Johansson, A. Household choices of car-use reduction measures. *Transp. Res. A* **2000**, *34*, 309–320. [CrossRef]
126. Loukopoulos, P.; Scholz, R.W. Sustainable future urban mobility: Using 'area development negotiations' for scenario assessment and participatory strategic planning. *Environ. Plan. A* **2004**, *36*, 2203–2226. [CrossRef]
127. Loukopoulos, P.; Jakobsson, C.; Gärling, T.; Meland, S.; Fujii, S. Choices of activity- and travel- change options for reduced car use. In *Progress in Activity-Based Analysis*; Timmermans, H., Ed.; Elsevier: Amsterdam, The Netherlands, 2005; pp. 481–501.
128. Cottrill, C.; Pereira, F.; Zhao, F.; Dias, I.; Lim, H.; Ben-Akiva, M.; Zegras, P. Future mobility survey: Experience in developing a smartphone-based travel survey in Singapore. *Transp. Res. Rec.* **2013**, *2354*, 59–67. [CrossRef]
129. Ettema, D.; Gärling, T.; Olsson, L.E.; Friman, M. Out-of-home activities, daily travel, and subjective well-being. *Transp. Res. Part A Policy Pract.* **2010**, *44*, 723–732. [CrossRef]
130. Sweet, M.; Kanaroglou, P. Gender differences: The role of travel and time use in subjective well-being. *Transp. Res. Part F* **2016**, *40*, 23–34. [CrossRef]
131. Priya Uteng, T.; Uteng, A.; Kittilsen, O.J. *Land Use Development Potential and E-bike Analysis*; Institute of Transport Economics: Oslo, Norway, 2019.

132. UNEP Driving Gender in India's Public Transport Policy. 2017. Available online: https://www.unenvironment.org/news-and-stories/story/driving-gender-indias-public-transport-policy (accessed on 29 April 2018).
133. UN Habitat. Improving Public Transport Services for Women: A Story from Cairo. 2018. Available online: https://unhabitat.org/improving-public-transport-services-for-women-a-story-from-cairo/ (accessed on 29 April 2019).
134. Iqbal, S. *Women, Business and the Law*; The World Bank: Washington, DC, USA, 2018.
135. Jeppesen, S.L.; Leleur, S. Sustainable Transport Planning—A Multi-Methodology Approach to Decision Making. Ph.D. Thesis, Technical University of Denmark, Lyngby, Denmark, 2009.

*Article*

# Disability, Mobility and Transport in Low- and Middle-Income Countries: A Thematic Review

**Maria Kett [1], Ellie Cole [2] and Jeff Turner [3,*]**

[1] Honorary Reader in Disability and International Development, UCL Institute of Epidemiology and Healthcare, London WC1E 6BT, UK; m.kett@ucl.ac.uk
[2] Honorary Research Associate, UCL Institute of Epidemiology and Healthcare, London WC1E 6BT, UK; ellie.cole@ucl.ac.uk
[3] Gender & Inclusion Theme Leader, High Volume Transport Research Programme, IMC Worldwide Consultants, 64–68 London Rd, Redhill RH1 1LG, UK
* Correspondence: Jeffreymturner@hotmail.com

Received: 3 October 2019; Accepted: 2 January 2020; Published: 13 January 2020

**Abstract:** This paper discusses issues affecting the transport and mobility needs of people with disabilities in middle- and low-income countries and how disability intersects with a range of other factors to impact on transport needs, use and engagement. The paper is intended to stimulate discussion and identify areas for further research, and identifies a number of key issues that are salient to discussions around equitable and inclusive transport provision, including patterns of transport use, behaviour and experiences, solutions and policy directions, measuring access and inclusion, policies and intersectionality. The paper also identifies gaps in knowledge and provision, barriers to addressing these gaps, and some possible solutions to overcoming these barriers. These include shifting the focus from access to inclusion, reconceptualising how 'special' transport might be provided, and most importantly listening to the voices and experiences of adults and children with disabilities. Despite lack of transport often being cited as a reason for lack of inclusion of people with disabilities, there is surprisingly little evidence which either quantifies this or translates what this lack of access means to people with disabilities in their daily lives in low- and middle-income countries.

**Keywords:** people with disabilities; inclusive transport; high volume transport; accessible transport; low- and middle-income countries

---

## 1. Introduction

This paper aims to capture the current state of knowledge about the transport needs of people with disabilities in low- and middle- income countries (as defined by the Organisation for Economic Cooperation and Development's Development Assistance Committee list); and how disability intersects with a range of other factors to impact on transport needs, use and engagement. It complements other papers in this special issue, which focus on the mobility needs of young people and older adults; and highlights how these identities intersect and impact on choices. In line with the overarching aims of the special issue in focusing on sustainable High-Volume Road and Rail Transport in low-income countries, we have included the experiences of people with disabilities undertaking both urban and long-distance journeys.

While there has been some research that links a lack of transport to barriers to other services for adults and children with disabilities, such as education [1] or healthcare [2], there has been much less focus on the actual mode of transport used, or the journey itself. So, despite a lack of transport often being cited as a reason for lack of inclusion of adults and children with disabilities, there is surprisingly little evidence which either quantifies this lack or translates what this lack of access means to people with disabilities in their daily lives. Without this information, it remains challenging for planners and policymakers to understand where, what and how they should best invest in making transport more inclusive of people with disabilities.

It is worth reiterating at the outset that while there is some literature on patterns of travel behaviour, types of travel and journey experiences in higher income countries, there is surprising little from low-income countries. Previous work has tended to focus on transport exclusion, rather than inclusion, and in turn how transport exclusion can create and perpetuate social exclusion [3]. As the authors note, social exclusion as a term is often poorly defined and measured, with the result that it is operationally and theoretically difficult to assess [3] (p. 2).

The paper starts by placing inclusive transport in the context of existing international agreements and global frameworks relating to disability rights. We then move on to discuss patterns of travel behaviours and journey experiences identified in the literature. These have tended to be from higher-income countries, but it is possible to identify some possible 'solutions' to current transport challenges that could be, or are already being, explored in low- and middle-income countries. These include adaptations of physical transport infrastructure, development of holistic/door-to-door journey approaches, specialised transport services, fare subsidies and technological innovations. The knowledge gaps about evidence of 'what works' in terms of disability inclusion align with existing gaps more broadly. Most reviews, for example around inclusive education or health, tend to argue for a 'twin track' approach [4]—both mainstreaming disability into services, as well as providing specialist targeted services for those that need them. From the limited evidence there is in the transport sector, the indications are that this is also the required approach to inclusive transport. Finally, we also explore the role of national policies and other institutional factors. The paper ends by identifying the gaps in the literature, highlighting the lack of research around transport issues for people with disabilities in low-income countries, and makes a series of recommendations for future research.

### *Transport Issues Affecting People with Disabilities in Middle- and Low-Income Countries*

Transportation issues rate highly as a challenge for people with disabilities globally [5–7]. However, measuring access to transport, or indeed understanding who is the most severely affected, remains challenging [8]. However, the United Nations (UN) Convention on the Rights of Persons with Disabilities (CRPD) [9], has had an impact on both national and international policy and focus around transport. While the CRPD does not have a specific Article on transport, it does acknowledge the centrality of transport for people with disabilities to access a range of services including homes, schools, healthcare facilities, workplace and leisure [CRPD Article 9]. The CRPD enshrines the right of people with disabilities to access transportation on an equal basis with others. However, barriers to the enjoyment of these rights can be broadly divided into three main areas: institutional (legislation, political will, policy, etc.); environmental (infrastructure, vehicles, information); and attitudinal (transport staff, other passengers, lack of accessible information, etc). In reality, these often overlap.

Globally, there have been increasing efforts to address inclusion and ensure 'no one is left behind', culminating in the Sustainable Development Goals (SDGs) [10], with the aim of equity for all by 2030. While all 17 goals are relevant for people with disabilities, SDG 11 'Making cities and human settlements inclusive, safe, resilient and sustainable' has a specific target (11.2) that 'By 2030, provide access to safe, affordable, accessible and sustainable transport systems for all, improving road safety, notably by expanding public transport, with special attention to the needs of those in vulnerable situations, women, children, persons with disabilities and older persons'; as well as an indicator (11.2.1): 'Proportion of population that has convenient access to public transport, by sex, age and persons with disabilities'. However, SDG targets and indicators are adapted to the country context, and it is left to individual countries to set country-specific goals, indicators and targets to monitor their progress. Moreover, in the most recent SDG progress report, there is no mention of progress towards this target. The recent UN Flagship Report on Disability and Development [11] also reviewed progress toward SDG 11 as it relates to people with disabilities, though again, there is very little data or information on specific progress with regards to accessible transport at global level. This omission may be due to a lack of agreed definitions on what accessible transport is, and/or a lack of standardised targets and indicators. It could also mean that few countries have set targets, monitored their progress, or indeed actually focused in making any improvements.

In the Asia Pacific region, the Incheon Strategy (2012) is the benchmark for progress on 'making the rights real' for people with disabilities in the region [12]. Goal 3 of the strategy is to: 'Enhance access to the physical environment, public transportation, knowledge, information and communication', although it is interesting to note that while the overall target is: (3.B) Enhance the accessibility and usability of public transportation, the overall core indicator is (3.2) Proportion of accessible international airports. While this is a laudable target, it could be argued that it does not address the most likely day-to-day transport barriers faced by the majority of people living in the region.

Similarly, in 2013, a global online survey carried out by the Global Alliance on Accessible Technologies and Environments (GAATES) Transport Committee with its members (completed by 257 people from 39 countries) aimed to better understand what mobility issues people with disabilities faced around the world [13] (p. 1). According to the data presented, the two biggest challenges were inaccessible public transport vehicles and attitudes of drivers/staff. In terms of possible solutions, respondents were fairly evenly divided on four main areas: technical guidance on inclusive design solutions aimed at civil engineers, planners, etc.; guidance/information aimed at transport providers/senior management in transport companies; guidance/information aimed at people with disabilities to empower them; and guidance/information aimed at politicians about the UN Convention, etc. [13] (p. 7).

What these initiatives highlight is that even with a range of measures in place to facilitate inclusion, there may still be a gap between availability and use—demonstrating how these barriers often intersect. A second, related challenge is to measure the size of this gap. Estimating how many adults and children with disabilities have access to, and more importantly use transport systems, as well as their safety, affordability and reliability is challenging. Some countries collect transport data as part of a national census or other largescale surveys. However, even if they highlight a lack of access and availability, it is still challenging to attribute this as a sole cause to a lack of access to healthcare, education or employment, though it is certainly a contributing factor [c.f. 14]. However, regardless of the cause, it is unequivocal that barriers to accessing inclusive transport are found across most countries, in a range of contexts and for a spectrum of impairments, and result in a loss of education, employment and overall wellbeing (see for example, on reduced access to education [1,14]; for employment [15,16]; wellbeing more broadly [17]).

The knowledge gaps about evidence of 'what works' in terms of disability inclusion align with existing gaps more broadly. Most reviews, for example around inclusive education or health, tend to argue for a 'twin track' approach–both mainstreaming disability into services, as well as providing specialist targeted services for those that need them [4]. From the limited evidence there is in the transport sector, the indications are that this is also the required approach to inclusive transport. There are some innovative and potentially paradigm-shifting ways to deliver this, including 'Mobility as a Service' (MaaS), which use locally available and adapted structures with technology to provide the necessary 'total journey'. More work needs to be done to test these for a range of adults and children with disabilities in a range of contexts.

## 2. Materials and Methods

Whilst this was not intended to be a systematic review, a search of the literature was done to identify the key themes. There is a large volume of literature on the broad theme of 'accessible transport', and an initial search was done on the Web of Science database in January 2019, in line with the other articles in this series. The search operator was developed, based on similar operators used in other thematic areas of this study [18], to ensure consistency and include as wide a range of types of impairment as possible in selected geographical regions, and return results based on various transport-identifying expressions.

Results that were written in English from the year 2008 to present were chosen for review. This year was chosen as it represented both an approximate 10-year timespan but also marks papers that were written following the coming into force of the CRPD in 2008. Inclusion criteria were reports that focused on people with disabilities' experience and issues with using various forms of transport. Papers that were not primarily focused on transport were excluded; so, for example, the large volume of literature that exists around active transportation or that of road safety. These are of course relevant to wider discussions about policy, given the prevalence of road traffic-related injuries and disabilities [4] (p. 60), but were not the main focus of this review. Road safety—and the relation between road traffic accidents and disabilities in low- and middle-income countries—is one of the other themes of the overall review. It is worth noting that data on the number of people who survive crashes but live with disabilities are almost non-existent [19].

The search of the Web of Science databased returned 295 results. These results were screened by title and abstract, and a total of 23 results were selected for inclusion. A number of articles were excluded on the basis that they conceptualised 'mobility' as it relates to impairment, rather than transport mobility, so were screened and excluded on this basis. During the full-text review a further 12 articles were excluded due to inappropriate content or unavailability. Following the review of the returned results of the literature review, a further manual review was undertaken using Google Scholar using the same inclusion criteria. This review returned an additional 38 articles from both 'grey' and scientific literature. This literature was included to capture national guidelines on transport and disability inclusion, and reports from civil society and other actors which were not included in the database review. This is important as it provides a context as well as potentially identifies policies and practices not yet featured in the published literature. A total of 49 articles were included in this review. We manually coded and grouped together the main themes that emerged from this body of literature, as well as from existing literature and knowledge in the field using theme content analysis.

*Limitations*

This review included results from one database search—Web of Science—in line with other articles from this series. This limited the number of results included in the review, and some may have been omitted. This literature review is not intended to be systematic but rather thematic and comprehensive; the results included here give a helpful set of themes that emerge from the literature.

The search terms and inclusion criteria excluded returns from high-income countries, although some were used in the manual search to allow for comparisons to be made and to highlight important points in the absence of comparable low-income country returns. Transport issues in higher-income countries were not a focus of the review and these may have overwhelmed the search returns and skewed the emergent results. A further limitation to the operator was that if an article was about a specific country (for example, Kenya but it did not mention Africa), it may not have been picked up.

## 3. Results

In this section, we present the set of themes emerging from the literature and aim to highlight the persistent gaps, and some of the measures being developed to close them, in other low- and middle-income countries. These may also enable low-income countries to target this issue, and reach the global goals, more effectively

### 3.1. Patterns of Travel Behaviour and Experiences

Given the paucity of data on transport disadvantage, identifying 'what works' to overcome the disadvantage can be even more challenging, given the array of factors involved. Moreover, there is very little research on the types of journeys adults and children with disabilities make, the modes of transport they use, and their overall experiences of the journey, in particular from a participatory perspective.

To deliver an inclusive transport system requires a joined-up approach that focuses not only on the what, but also on the how. The type of transport provided, and how it is provided, varies according to country and priorities. Many countries—particularly the higher-income ones—link the issue of accessible transport to broader legislation and commitments, including the CRPD (e.g., [20]). Higher-income countries can obviously provide more comprehensive transport options, though as recent analysis of the in the UK demonstrates, even with these in place, people with disabilities still face a number of challenges in using public transport. [21].

### 3.2. *Transport Infrastructure*

Many of the efforts to reduce car use and improve high volume (often public) transport have been in urban areas, in particular, cities. However, these initiatives have not always been accessible or inclusive for adults or children with disabilities, as for some, cars and/or taxis offer freedom, independence and mobility, which public transport may not. Research in a number of higher-income countries has shown that people with disabilities tend to favour cars—their own or others (such as taxi services) [22]—rather than, for example, waiting for a ramp to enable them to get on or off trains [23], or where there is limited provision of services (more so the case in rural areas). Cars can confer a level of independence, autonomy and safety that users do not always feel on public transport. Fear—of how to use the services as well as other passengers' attitudes—is often a key factor, and many users report feeling unsafe on any form of public transport; and parents of children with disabilities cited fear for their child's safety as one of the key barriers in their journeys to school [1,14].

### 3.3. *Long Distance Journeys*

People living in rural areas who need to make long-distance journeys are generally less well-serviced by public transport almost everywhere. This is partly due to limited demand, which in turn leads to reduced services [24]. A complex set of factors need to be taken into account when planning long-distance travel services. In countries with limited or no formal public transport system, cars or taxis provide one of the few means of getting around, particularly in rural or semi-urban areas and especially in emergencies. Use of hired vehicles raises the issue of additional costs for people with disabilities—the 'hidden costs' of disabilities [25,26]. However, there is limited data which quantifies the amount of these additional costs, or costs of lost opportunities, for example, for adults and children with disabilities who cannot afford transport to make (often long-distance) journeys for educational or income-generation purposes, or access other services and activities.

### 3.4. *Affordability and Subsidies for Individual Journeys*

Included in debates around the 'inclusiveness' of sustainable accessible transport are issues of cost and affordability. In part due to inequalities and exclusionary practices, in many countries around the world, people with disabilities experience poverty. One form of poverty is transport poverty, which can be related to cost, affordability as well as accessibility [27,28]. However, transport poverty is hard to measure, and there are no universally agreed definitions. Moreover, some question whether it even exists as a stand-alone phenomenon or whether it is simply an extension of being poor. Put differently, the question remains as to whether transport poverty a 'real' problem for individuals, or is it a systemic problem that has a systemic solution, and if so, what are these solutions? Given this, is it a problem that has a transport solution, or rather is it a broader issue of social welfare? [28] (p. 353).

In many (usually higher-income) countries, a range of measures have been put in place within transport policies, including concessionary fares, subsidised public transport services and free special transport services (STS) for eligible groups such as older people, children, and/or people with disabilities. As such, they are only viable where there are (public) transport systems in place.

South Africa provides a heavily subsidised public transport system for older adults and people with disabilities [22,25]. However, in his study reviewing provision of services, Venter found that it was access, rather than affordability, that was the largest barrier to use, and the solution was therefore to increase access more broadly, rather than to provide targeted subsidies:

> "The overall implication is that the limited funds that are available for improving public transport in cities should go towards improving accessibility for all, rather than towards lowering fares for all disabled persons as a group. This is not to say that subsidisation is not needed: the evidence shows that both disabled and non-disabled commuters benefit substantially from having access to subsidised bus and rail services. But the benefit stems from the subsidies being available to all low-income workers." [25] (p. 138)

Such universal coverage may also help overcome the predictable eligibility challenges, as assessment of disability is already a complex and much debated issue, particularly in low-income countries which have limited mechanisms for assessment [29]. In addition, most of the debates in the literature tend to focus on the delivery structures and mechanisms, rather than eligibility criteria, so there is limited evidence on what are the most effective mechanism for assessment or identification of eligibility.

### 3.5. Measuring Access to Services

In addition to limited evidence about assessment and eligibility criteria, there are also significant gaps in the literature about what specific impact transport restrictions have on people with disabilities' lives (e.g., opportunities lost), or the additional costs that may be associated with this lack of access (e.g., hiring taxis to get to work, or not getting to work at all) [25,26]. This gap exists across high- [20] and low-income countries and is largely due to a lack of agreed measures to estimate access and inclusion. Of the limited literature available that does address this, intersecting issues of age, gender, poverty, ethnicity, disability, etc., come to the fore (e.g., [2,30–32]). These make it difficult to attribute transport—or lack of—as a singular cause for exclusion. In recent research undertaken by authors in Liberia [33], when asking matched household heads about barriers to accessing healthcare services, distance to health facilities was weighted similarly between disabled (11.0%) and non-disabled households (12.8%). Unfortunately, the study did not ask what the most commonly used mechanism of transport was, nor whether this resulted in additional costs, for example if persons with disabilities had to use more expensive means of transport (e.g., taxi cars instead of taxi motorbikes), or spend more on transport overall. Therefore, while the issue of transport—or lack of—is a barrier to access and inclusion in all aspects of life and is highlighted in many papers (e.g., [1,34]), there are few papers that identify mechanisms to quantify this.

One of the few pieces of research available that does attempt to do this is Venter [25] discussing the situation in South Africa. Here the research found that geographic location was a key determinant of the affordability of transport, with transport in urban areas being unaffordable for poor people generally, but not specifically people with disabilities or older people, whereas in rural areas limited travel options constrain everybody as much as affordability, thus the solutions again point to improving the affordability, availability and quality for all users, not just specific groups, as all users would then benefit [25] (p. 129).

### 3.6. Transport Services Available

One area of transport provision that aims to address a lack of mobility are special transport services (STS). These are designed to address services gap and are usually (though not exclusively) for people with mobility difficulties, such as older adults or people with disabilities. There have been some well-reported models in South Africa, Brazil and Russia [4] (pp. 178–184), all high middle-income counties and we found none from low-income countries. STS (also known by a variety of other names, including 'paratransit, or 'dial-a-ride' in other countries) vary not only in name, but also in delivery and funding structures, as well as types of vehicles used. Some supplement existing services, or link to feeder routes (such as mass rapid transport (MRT) services in Cape Town and Brazil), while others offer specific door-to-door services in adapted accessible vehicles (such as Dial-a-Ride in South Africa, which offers users a heavily subsidised service).

There are a number of debates about the provision of STS, particularly around equity, as they tend to offer a segregated service (see for example [16] p. 4), as well as cost. In some countries, provision of STS (e.g., to schools, hospitals, etc.) is a mandatory requirement by law, though provision can be costly, and demand likely to increase with an ageing population. They are often publicly funded, or at least subsidised. Some high-income countries are exploring different models of payment for these services (see for example [35]). Though their findings are from a wealthy country with high levels of service provision (Sweden), they do have some implications for decentralised budgets elsewhere—not the least of which is the need to coordinate service provision (e.g., school buses) and that overall improved public transport services can have a positive impact on access and inclusion more generally. According to one study from eThekwini Municipality (South Africa), the Dial-a-Ride service provided faces an array of challenges, including high costs, difficulty in managing high demand, scheduling of services and lack of flexibility in adapting route planning. In fact, demand was such that it had to be restricted to people going to study or work. Complaints about the tendering process were also made [30] (p. 35).

An excellent practical guidance for setting up STS in low-income countries but drawing on experience from high-income countries, suggests using a variety of state and non-state funding mechanisms [36]. The guide suggests utilising existing locally available transport mechanisms, including motorised auto-rickshaws, cycle-rickshaws, and similar vehicles operated exclusively or partly for mobility-impaired people. The one thing they all have in common is that they are "demand-responsive", which can be interpreted in one of several ways—pre-booked, scheduled or instant access. This approach strengthens the idea that public transport provision needs to move away from traditional delivery approaches, as well as addresses the challenge that the provision of STS does not address equity, inclusion or attitudinal and other barriers to transport access, nor does it confer independence or autonomy on users, as there is still a reliance on others to provide a service. On the other hand, if there is no way to get to a station or bus stop—no matter how accessible the route is, then it could be argued that the provision of special transport enables people with transport restrictions to make necessary journeys. To be genuinely effective, special transport needs to be included as part of a wider package of measures to address barriers to access and inclusion, such as increased and accessible information and financial aids, such as concessionary travel passes [37] (p. 54).

However, if users require a high degree of assistance, there are still limitations in most current transport provisions around driver capacity, attitudes and willingness to provide such services. Another criticism of STS is the extent to which such systems include alternative (as well as sustainable and healthy) modes of transport such as cycling or walking.

### 3.7. Holistic Approaches

People with disabilities in South Africa seem to enjoy more accessible transport opportunities than those in neighbouring countries, including subsidies to address any additional costs they incur. However, even there, when progressive legislation is in place to enhance the rights of people with disabilities, they still face a range of challenges. These include boarding the ostensibly accessible mass rapid transport (MRT) services in Cape Town—anecdotally, not all the MRT buses are fully accessible, so users who need them have to wait for the next accessible vehicles to come along, causing delays to their journey. By far, the most popular method of public transport for most people in South Africa are taxi minibuses. These pose a number of challenges for people with disabilities, ranging from inaccessible vehicles to other passengers and the drivers themselves. Research with people with disabilities and taxi drivers in the Durban area highlights that the actual operational structures of taxis or other private minibus services function, often unintentionally, to actively disincentivise drivers to pick up passengers with disabilities [30]. Given that taxi drivers often lease their cars from owners, time is money and drivers are less keen to stop and spend time trying to get a wheelchair into, or on top of, a vehicle, or wait for somebody with mobility difficulties to board. In their research, Lister and Dhunpath found little understanding of either side's perspective, which they suggest could be ameliorated by incentives, such as cash and/or training to encourage taxi drivers to pick up passengers with disabilities, along with disability awareness training, and subsidised fares for people with disabilities [30].

### 3.8. Mobility as a Service

Taking the idea of shared, often private, services a step further is the idea of 'mobility as a service'. In some respects, such an approach may well be undertaken informally in a range of settings. While the idea of a barrier-free, door-to-door journey is not new, new ways to conceptualise what a 'total' journey, or 'continuity of travel chain' or even 'integrated mobility' might look like are in development. These factor in age, health, mobility status as well as a range of other intersectional aspects and move away from the more traditional 'special transport' models discussed above to more integrated approaches that acknowledge that what works in one location or for one group may not be effective or utilised in another.

One of the most innovative and exciting area that links up discussions about autonomy, choice, continuity of journey, as well as bringing in new technology is that of 'Mobility as a Service' (MaaS). Not specifically designed for people with disabilities, MaaS is a way:

> "... to see transport or mobility not as a physical asset to purchase (e.g., a car) but as a single service available on demand and incorporating all transport services from cars to buses to rail and on-demand services." [38] (p. 583)

Originating in Sweden, it offers users the opportunity for door-to-door integrated services, paying for a package of services 'as they go' via one (often online) payment system. Though most of the research around MaaS has been in higher-income countries, the concept has relevance to low-income countries, making it worth raising in this paper, and some very preliminary research about transposing it to such contexts has begun [39].

MaaS has the potential to cover a range of transport options, from self-drive cars (still under research), through to taxis (similar to Uber, which already operates via an app-based service), bicycles and even walking. However, despite the ideology, such tailored services can be expensive, and inefficient to deliver, so researchers have begun to conceptualise how journeys can be 'bundled', so users can and buy specifically tailored transport packages, in much the same way as they can buy satellite or cable TV packages to suit their specific viewing requirements [38].

The research focused on developing and delivering a broad range of options for both type of journey (work, socialising, etc.) and mode of transport (buses, taxis, etc.), which could be standardised to some extent to reduce costs. These 'service packages' could also be customised, perhaps including add-ons such as household travel-planning, availability of car space in localities, travel training, ICT training, providing a driver for own car and learning to drive [38] (p. 590). These were offered as alternatives to car ownership, increasing car-sharing to reduce individual ownership, and therefore increase sustainability, but at the same time maintaining independence and freedom of movement. As the researchers note, whilst MaaS was not specifically set up for use by people with disabilities, the system could offer opportunities for flexibility and autonomy. Viewing transport as part of a service package has the potential to move discussions away from seeing transport provision for older adults and people with disabilities as a welfare issue, as, they argue, it currently is. Although they caution about the need to take the social benefits of mobility into account when thinking about MaaS too [38].

While MaaS offers an exciting potential for people with disabilities, it was not (necessarily) designed specifically for people with disabilities, but rather it is a transport system that *can* be used by people with disabilities. In this, it is similar to existing 'community' or 'flexible' transport' systems, usually privately-funded or run cooperatively. Community transport can be funded and provided through a variety of mechanisms, including shared transport (such as cars or taxis), and may be a viable option in low-income countries, as they may lessen costs for passengers, as well as reduce the number of vehicles on the road. However, there are debates about the extent to which they are likely to be viable as a long-term solution to reduced mobility in rural or semi-urban areas, in particular for people with disabilities, as they do not solve broader issues around mobility, including social aspects, such as independency and autonomy.

### 3.9. Technology

Technology has enabled significant improvements across a range of domains, including transport-specific services such as Uber and MaaS, as discussed above, as well as facilitate mapping of journeys to better understand patterns and usage and adapt services accordingly. There is already a significant body of literature exploring the benefits (as well as some of the more negative aspects) that technology can bring to the lives of people with disabilities, in particular how it can support and maintain independence [40]. This is reflected in the array of transport-focused literature on technological advances in the transport sector, some of which may have been specifically designed and intended for use by people with disabilities, though not all. However, all have had a major impact on the ability to travel. These fall largely into two categories: technology which provides information; and technology which provides a service. Some, but not all, of these are mainly used in higher-income countries, but all have the potential for transfer and adaptation.

Information-providing technologies include apps which give live updates about planes, trains, buses and other (usually public) transport, as well as live trackers, digital maps, etc. These often use location-tracking devices, such as GPS. Use of these tracking devices has extended to support independent travel for people with disabilities [15,41,42]. Whilst all these examples have been tested in higher-income countries with existing infrastructure and services, what the results demonstrate is that whilst accessible infrastructure is a necessary condition, it is not enough. Many of the challenges faced are related to, but not inherent within, the transport system, such as uncertainly about scheduling or routes. This indicates that not just the transport mode, station or service need to be inclusive, but the whole journey, requiring a joined-up approach to inclusive transport.

Technologies which support service provision include online ticketing systems, as well as automated and integrated payment systems. However, whilst convenient, for some users with disabilities it has been argued that these integrated payment systems can also present challenges. In their work in Durban, Lister and Dhunpath talk about these in relation to the Muvo Card, a single smartcard that can be used across all three of Durban's transport systems. Users in Durban report difficulties with locating the machines, drivers being in a hurry, no signalling facilities for blind or visually impaired users and numerous other problems with the machinery [30] (p. 40).

Similarly, in their findings from the Philippines, Cendana et al. [43] argue that such a single-use smart card could be a mechanism for more equitable urban transport, though their main concern was about implementation and eligibility. To overcome this, they argued for a provider that would enable the implementation of the use of the smart card across multiple transport platforms (hence multiple providers) to enable discounts to be systematic to eligible travellers.

However, in the end, while technology is an enabler, a facilitator for accessible and inclusive transport for people with disabilities, it is not in itself enough. Much more research is needed on the systems within which it is embedded—including the costs (demand and supply) and user needs (especially the voices of adults and children with disabilities who will use the services—see [1]). Moreover, focusing solely on access tends to lead researchers to create solutions to overcome physical and environmental barriers, rather than attitudinal or social barriers [44].

### 3.10. Intersectionalities, Inequalities and the Lifespan

Disability, like gender and age, is a factor that cuts across mobility, access to transport services from operational and employment perspectives [45]. However, much of the existing research around accessible transport has tended to compare transport use between disabled and non-disabled populations, rather than between different groups of people with disabilities. As a result, there is limited evidence of what works for specific groups, but as noted above, it is clear there is not a one-size-fits-all solution to these challenges. Perhaps as a consequence, increasingly in both high- and low-income countries, researchers and advocates have drawn attention to the need for a broader focus on inclusion, making it accessible and inclusive to all, not just people with disabilities, but also those with temporary mobility difficulties, older adults, people using pushchairs or prams, small children, cyclists and many other groups as well. However, while these design-led solutions tend to address the access issues, there are fewer indicators of measures of success around inclusion, or the socio-political changes required more broadly (see [46] for discussion around the application of these principles in South Africa). Moreover, such universal approaches may unintentionally benefit most those who need it the least if underlying issues such as poverty or fear are not addressed.

These issues have been discussed more widely in the fields of gender and to some extent ageing (see for example [47,48]). Ahmad (2015) highlights how in Pakistan, in order to make public transport accessible for women, including women with disabilities, planners and politicians need to consider religious and social issues as much as financial and logistical ones, as it is these that have the most impact on women. He draws on work which focuses on the gendered aspects of public transport access and provision (e.g., [31,47]) to show how dialogue between disability scholars, feminist critiques, and transport planners is needed to address continuing gaps [49].

Overall, less is known about the complex interactions of disability with a range of other factors including age, sex, location, class, caste, etc. One aspect highlighted in the literature is the unavoidable fact that as people age, their ability to drive safely is compromised, so there is also a safety aspect to reducing the number of older drivers on the road—for example, through mandatory vision screening [50]. This raises challenges when much of the research highlights how cars confer a sense of independence and autonomy, but which can be taken away at the very point when it is most needed, often leaving older adults with disabilities without alternatives.

There is a significant body of research that highlights the changing patterns of transport use across the lifespan, as well as according to location (e.g., [31,51]). However, the majority tends to focus on older adults in higher-income countries, with much less focus on children or young adults (for some exceptions, see [52,53]). Other researchers have highlighted the social aspects of public transport for older adults [54]. Older people have also become a focus for advances in travel technology (see for example, [55,56]). However, there is much less evidence on these complex intersections in low- and middle-income countries. One study, from Mexico, highlights the range of factors that mediate access to transport, with subjective 'transport deficiency' being strongly associated with being female, illiterate, having a mobility disability and using assistive walking devices [31]. The researchers also noted the most commonly used mode of transport for older adults in Mexico City is private car, followed by walking, with a range of factors given for this, including fear, geographical location and limited accessibility of transport options for older adults in Mexico City [31]. Similarly, data in South Africa also suggests that travel options are limited by factors other than affordability [25].

Venter suggests that it is spatiality, rather than affordability or even accessibility, that determines use, a finding also found in other higher-income countries [21]. Venter further suggests that initiatives such as road and footpath upgrading in rural areas would improve access and use [25] (p. 138). Moreover, he suggests that current subsidies in South Africa actively disincentivise older adults from using public transport. Sammer et al. [37] go even further in highlighting the intersecting, and accumulating nature of transport inequalities, referring to 'mobility impairment' more broadly:

> "In the past, mobility problems of physically disabled or sensory-disabled people were the focus of attention, whereas, more recently, problems of the elderly have been recognized as well. However, if such problems concern other groups, such as immigrants and people with learning disabilities, they have been more or less neglected" [37] (p.46)

These findings highlight not only how disability can lead to exclusion, which comes about from a complex intersecting of factors, but also the inequalities this exclusion can create. It is clear from the literature that disability can lead to inequalities more broadly within the transport sector, particularly with regards to access and inclusion, but also health inequalities more specifically. As noted above, this paper has not included the vast literature around road traffic accidents, the impact of which can be most severely felt by people in low-income countries, where there is often less regulated transport systems, poor infrastructure and limited availability and access to emergency services.

It is also clear that beyond these negative aspects, adults and children with disabilities may miss out on the mental and physical benefits of travel (see for example Vancampfort et al. [57] for a discussion about the associations between active travel and physical multi-morbidity). As Mindell [58] notes, disability and illness, along with age, is associated with 'non-travel'; and recent research has shown that urban residents in the USA with health conditions that limit travel, particularly driving, are more likely to limit their travel than their rural counterparts (Henning-Smith et al., 2018, cited in [58]). This is problematic from a number of angles, not least of which is social isolation and loneliness, all of which impact on mental and physical health. Transport policies that are good for health and reduce inequalities are low carbon, sustainable approaches, promoting active travel and public transport use, and reducing private car use [58]. Moreover, Mindell notes that in the majority of countries, motor vehicles are owned and used more by the rich while the adverse health effects, such as injuries, air and noise pollution, and community severance are experienced primarily by those with fewer resources [58] (p. 1). This implies that not only are people with disabilities less likely to travel, but they are at higher risk of the consequences of the overall health effects of non-travel.

There is also a cautionary note to this, in that much of the discussions around inclusion in the transport sector have not focused on sustainable alternative modes of transport, such as cycling or walking. It is worth noting that there is almost no literature that focuses on redressing these inequalities, from legislative or other perspectives, for people with disabilities, and there remains a gap in the literature addressing the inequities of the health benefits of active travel and transport for adults and children with disabilities.

## 4. Solutions and Policy Directions

From the evidence above, there remains a gap between global goals toward accessible, sustainable and inclusive transport provision and addressing the specific transport needs of children and adults with disabilities. Nevertheless, there have been attempts to address this gap, starting with policy. What lessons can be learnt from these, and are they transferrable to other contexts?

This growing shift is evident in much of the literature in the transport sector and reflects moves away from a more straightforward focus on accessible transport 'solutions' (e.g., infrastructure, connectivity, adapting environments and policies) to that of 'inclusive transport'—a broader understanding of the wider impact of transport exclusion (see for example [1,14]). Nevertheless, many guidelines still focus on physical access. In the UK, both public companies such as Transport for London, as well as private companies such as Uber promote themselves as accessible and inclusive. Inclusive transport is not only for people with disabilities but reflects a desire for a truly encompassing and integrated system. However, this is the central challenge—creating a system that works for everyone, whilst ensuring that the specific needs of individuals are catered for. Nevertheless, whilst some of these mechanisms have been aimed primarily at people with disabilities, including legislation and access standards, negative attitudinal and other barriers persist [21].

Having political backing may at least enable these issues to be raised in the first place. Many countries have developed context-specific access standards, including the Government of India, which supports a largescale national campaign on accessible public transport and buildings [59], though recent newspaper reports question how successful these have been [60]. However, whilst these access standards and audits can be contextually relevant, another barrier is the absence of agreed (and universally comparable) definitions over what constitutes 'accessible travel' or its opposite, 'transport impaired' travel. How are these measured if, as argued above, there has been a more general shift away from 'access' to 'inclusion'? In the wider development context, there are few tools or indeed markers, of what inclusion, participation or empowerment actually are, let alone how they are measured.

To address this gap, a team of researchers from Australia have developed and tested a set of tools which combined access audits and road safety audits with inputs from people with disabilities in Cambodia to create a 'Journey Access Tool' (JAT). The JAT is used to measure personal and interpersonal experiences on a regular journey taken by a person with a disability, for example when they utilise health, employment or education services, or wider community access. The authors caution that while the tools were overall a success in the trial, the interventions and interpersonal dynamics (e.g., the personal assistants, interpreters and relations of the people with disabilities) were more difficult to address, as there was a tendency for care givers and assistants to speak on behalf of the person with a disability or interpret their views, risking skewing the data. Key messages may also be overlaid or misinterpreted through the interventions of (often well-meaning) others [61]. Though the tool is only at the trial stage, these findings illustrate the crux of the debates about accessible transport provision: the extent to which it is the transport system itself, or wider systemic issues, that create the biggest barriers for people with disabilities. The examples provided here demonstrate that while overall there is an awareness of the need to shift from focusing solely on access—which can be measured and audited by sets of standards and other tools—to broader discussions on inclusion, what actually constitutes meaningful inclusion, as well as what disabled and other 'transport impaired' people want themselves, is still largely missing from the discourse.

Whilst it is clear from the literature that a conducive policy environment is necessary for accessible and inclusive transport, it is also clear that policies alone are not enough. As the examples from South Africa and other countries illustrate, without engaging with local political and other contexts, the best intentions can go awry. Integrating and upgrading locally available transport systems requires more than just accessible vehicles and paved roads. It also requires political will, budgets, monitoring of the process, and recourse if legislation or policy is not upheld. There also needs to be regular engagement with, and training of, public and private transport workers, unions and, crucially, people with disabilities with a range of transport and access needs themselves. Even then, the picture can be mixed. For example, another paper from South Africa highlights the positive changes that, they argue, have come about through constitutional and legislative changes in the country [62]. The paper explores taxi drivers' experiences of, attitudes towards, and beliefs about passengers with communication difficulties in Gauteng Province. In addition to again demonstrating user preference for taxis because of the level of independence that they conferred on people with disabilities, the paper also found that the drivers included in the study viewed people with communication disabilities:

> "... as equals, with no negative stigma to a communication disorder ... Participants regarded individuals with communication disorders as 'good', normal people. This finding, arguably, indicates a positive and embracing culture rather than a negative and discriminatory one, facilitating participation and inclusion." [62] (p. 6)

However, while the researchers credit the positive and enabling policy environment in South Africa for these findings, they acknowledge it is a small study, and the results are not necessarily generalisable to other contexts. They also note that many of the people with communication difficulties did not travel alone, which reduces their autonomy, as well as the strength of the research findings. Another recent piece, again in South Africa, draws attention to the very positive policy context in the country, in particular South Africa's Bus Rapid Transport (BRT) services, and linked upgrading and integration of existing services to promoted accessibility. However, this has provided challenges to realise, due to the limited implementation of the BRT system and the majority of passengers that rely upon a combination of conventional buses, trains and minibus taxis. The authors note that:

> "While we need to continue to improve the access offered in the formal system, the slow rollout means that the number of disabled people benefiting from these changes (i.e., 'horizontal equity') are likely to remain proportionately very small for the foreseeable future." [22] (p. 184)

Behrens and Görgens argue that there are two key reasons for this failure: over-expectation about the role of the state in delivering transport services on the one hand, while on the other, an underplaying of the role—and power—of the private sector (in this case, taxi-minibuses) in South Africa. They argue this has led to a lack of an understanding of the complexity of delivering the promised accessible and integrated transport system in a context where the formal (state) and informal sectors intersect, compete with and occasionally complement each other [22] (p. 185). They conclude that a key focus for the promotion of universal access in South Africa should be the minibus-taxi operations and provide a range of options to encourage this in line with existing policy. These include universally accessible infrastructure and wayfinding information provided at minibus-taxi ranks and at public transport interchanges; state-supported incremental fleet renewal of existing taxis to more accessible vehicles; a (partial) shift to user subsidies, for example, cashless fare collection technologies for designated passengers (eligible for concessions/subsidies) and financial incentives to the minibus-taxi operators to carry passengers who would otherwise not be served [22] (p. 194). It could be argued that the authors could have also included something around disability awareness raising and training—though perhaps, just the use of accessible buses, wayfinding information, etc. would raise awareness of drivers and other users anyway.

Finally, it is worth noting that there is a striking absence of literature from low- and middle-income countries on the extent to which the voices of any people with disabilities are included in arenas where transport policies are decided, monitored, evaluated or governed.

## 5. Discussion

Whilst throughout this paper we have been keen to demonstrate that there is no 'one size fits all' solution to the provision of inclusive transport, it is interesting to note that despite the diverse range of literature included here, the income levels of the countries and types of impairments written about, there are a number of common threads the emerge across the literature. These can offer some guidance about creating sustainable, accessible and inclusive high-volume transport systems in low-income countries that avoid some of the pitfalls outlined above. Addressing legislative, environmental and attitudinal barriers by improving access, delivering specific targeted services and policies and practices, as highlighted in the World Report on Disability [4] remain key. However, whilst transport is widely acknowledged to be a barrier to equity and inclusion, the size of the exclusion gap remains difficult to measure, and what works to close it remains difficult to know, particularly in low-income countries. This may be because of an overall lack of understanding about transport needs, in particular from the perspective of those who need it most and are most affected by these gaps. The voices of adults and children with disabilities themselves are rarely heard in the literature. We found very little evidence of discussions around costs or user satisfaction, particularly around new technologies. It is also apparent that transport is not seen as a right—yet without it, rights will not be attained. Understanding and appreciating the "full social benefit of mobility services" [38] (p. 590) is yet to be realised in many countries. There is then an urgent need for more research in this area.

It is also clear that more work needs to be done to support adults and children with disabilities to use transport services, but this should and could be done by integrating it more fully with other interventions, including education and livelihoods, shifting away from seeing it as a 'transport' issue per se. Overall, the research highlighted the possibility that improving general public transport (including door-to-door access) would enable more people with disabilities to use these services.

A supportive policy environment is necessary, as is a Universal Design framework, but it seems it is not actually the mode of transport that makes the biggest difference to use, rather convenience and autonomy. Even in countries with relatively good, regular public transport systems, and subsidised services (including paratransit services), many people with disabilities prefer to use cars (or taxis, and motorbike taxis—though there is very little on the use of motorbike taxis and people with disabilities). If high volume transport networks are extended, it is worth considering how adults and children with disabilities can and will be included for the whole journey. To do this, there first needs to be a shift by providers about transport access as a right, based on freedom of choice, rather than as a welfare provision (whilst still acknowledging that some people may benefit from additional assistance). Viewing transport provision for people with disabilities as a welfare issue is tied to how current funding mechanisms and delivery structures target and work for people with disabilities (in particular, concessions and STS). Universal concessions remove some of this stigma but can be politically motivated; moreover, the evidence about their usefulness was mixed, with several reviews arguing the money would be better spent on general subsidies [25]. Nor was there much evidence in the literature on the commercial benefits of including adults and children with disabilities into mainstream (existing) transport structures, rather the focus has been on the costs of providing special transport services. Again, this is a lost opportunity.

In the examples from higher-income countries such as South Africa and elsewhere, one common theme that emerged was the need to 'twin track' [63] accessible and inclusive transport—both upgrading and ensuring new buses and mini-buses, etc. are accessible, but also actively incentivise bus and taxi drivers to pick up passengers with disabilities. This could be via a range of methods, including providing additional cash incentives (e.g., subsidising fares), but also providing training and education for drivers and vehicle owners. There should also be public awareness campaigns to promote transport as a safe space for all users, and recourse mechanisms for those who do experience discrimination whilst using public (or where possible private) transport. Such information could be collected through complaints mechanisms, even text-message numbers to report instantly, as well as from passenger surveys. All this information will lead to a rethinking of how to target resources more equitably—and effectively—based on evidence from the users themselves, and in turn help focus delivery of these resources.

However, none of this can be done without listening to the voices of children and adults with disabilities themselves. To date, there is very little research from low-income countries that genuinely includes these perspectives—what the absence of transport means to people with disabilities, why they do not use it, what would enable their access and increase their inclusion. Not only are their voices absent, but there is an almost total absence of any literature from low-income countries on the extent to which people with disabilities are included in governance and decision-making around transportation. Moreover, people with disabilities are not a homogenous group and their identities intersect across age, sex, social class, ethnicity, etc. Listening to these multiple voices and perspectives help us understand the transport and mobility needs (and rights) of other marginalised and excluded groups, including women, older adults and young people. This would seem to be key to moving the inclusive transport agenda forward.

Overall, the focus on inclusion still needs work to shift from a focus on solely access—which can be measured and audited by sets of standards and other tools—to broader discussions on inclusion. However, solutions need to be mindful of ensuring that the onus is put on adapting the environment, not the person, to ensure they are in line with the human rights principles of the CRPD. One issue that has yet to be resolved is how to measure and monitor transport access and inclusion for adults and children with disabilities across a variety of settings, with a range of impairments, incorporating issues of safety and security, independence and autonomy. Crucial to this discussion is what 'inclusive transport' actually means for people with disabilities, and not what other (often well-meaning) people think it means. It is clear from the literature that despite the lack of transport provision to rural areas both within low-income countries and globally, merely providing accessible transport to these areas will not be enough to increase inclusion. Evidence from urban areas shows that provision has not in itself universally increased public transport access in these areas in higher-income countries, and people continue to rely on cars for autonomy and convenience.

There is also a lack of evidence on what inclusion looks like—as noted above, though countries are tasked with developing their own indicators for the SDGs, it is unclear that these are being meaningfully translated regionally or nationally, nor are there any agreed universal targets or indicators for inclusive transport. In order to deliver global commitments, as well as local ones, transport solutions need to move away from putting the onus on adapting the person and focus on adapting the environment and behaviour change. While there are some interesting examples of accessible transport solutions, including from low- and middle-income countries, there are almost no examples from low-income countries of systemic adaptations or accommodations, such as universal travel concessions for eligible passengers, or indeed any examples of prosecutions of companies or people who break the law regarding discrimination in access to transport in low- or middle-income countries. It is clear that significant knowledge gaps remain.

## 6. Conclusions

In order to move this agenda forward, a number of recommendations emerge from this review. The first is the need for much more research on what inclusive transport is—or should be—from the perspective of users in low- and middle-income countries, as noted, there is an overall paucity of research from low-income contexts to review. Secondly, there is a particular need for the voices of adults and children with a range of disabilities to be heard in order to better understand their transport needs. Thirdly, while the review has highlighted some positive examples of favourable policies and joined-up services facilitating access, as the example of STS demonstrates, this has not necessarily led to inclusion. In order to enable a shift to inclusion, planners not only need to consider access from a holistic perspective (Universal Design, whole journey approach, subsidies and adapted/specialised services), but also as a right (for example the right to health, education, etc.). In order to do this, planners should consider a 'twin track' approach to inclusive transport—making improvements to services, training staff, etc., whilst at the same time ensuring that the specific needs of adults and children with disabilities are met. The review has also highlighted how travel and access needs change over time and vary according to a range of factors, including age and location—one size will not fit all. This does not mean accessible approaches cannot be implemented, but planners will need to listen carefully to a range of voices to understand who may be left out, and what alternatives can be developed for them.

In order to facilitate this shift, incentives for existing structures should be developed, taking passengers, transport providers, drivers, unions and other key players into consideration., It is vital that people with a range of impairments are included in planning discussions, on access panels, and in audits of existing services and structures. Linked to this, governments need to develop a locally and contextually appropriate sets of standards, targets and indicators, in collaboration with local disability organisations (e.g., disabled people's organisations) to measure availability, accessibility, affordability, acceptability and quality, as well as safety and security, independence, and autonomy. Complaint mechanisms for users must be developed and adhered to, with appropriate recourse mechanisms, and disciplinary measures for those that do not adhere to them.

Finally, the review has only been able to touch on the range of technologies available to facilitate access and inclusion—much more research on existing and new enabling technologies appropriate for low- and middle-income country use is needed.

**Author Contributions:** Article conceptualisation, M.K. and J.T., methodology, M.K. and E.C.; validation, M.K. and E.C.; formal analysis, M.K. and E.C.; investigation, M.K. and E.C.; resources, J.T.; data curation, M.K.; writing—original draft preparation, M.K.; writing—review and editing, M.K., E.C., and J.T.; visualization, M.K. and J.T.; supervision, M.K and J.T.; project administration, M.K. All authors have read and agreed to the published version of the manuscript.

**Funding:** This research was funded by UK AID through the UK Department for International Development under the High Volume Transport Applied Research Programme, managed by IMC Worldwide.

**Acknowledgments:** The authors would like to thank Gina Porter for her comments on an earlier draft of this paper.

**Conflicts of Interest:** The authors declare no conflict of interest.

## References


1. Kett, M.; Deluca, M. Transport and Access to Inclusive Education in Mashonaland West Province, Zimbabwe. *Soc. Incl.* **2016**, *4*, 61–71. [CrossRef]
2. Mutwali, R.; Ross, E. Disparities in physical access and healthcare utilization among adults with and without disabilities in South Africa. *Disabil. Health* **2019**, *12*, 35–42. [CrossRef] [PubMed]
3. Kamruzzaman, M.; Yigitcanlar, T.; Yang, J.; Mohamed, M.A. Measures of transport-related social exclusion: A critical review of the literature. *Sustainability* **2016**, *8*, 696. [CrossRef]
4. WHO/World Bank. *World Report on Disability Report*; WHO: Geneva, Switzerland, 2011.
5. UNDP. *A Review of International Best Practice in Accessible Public Transportation for Persons with Disabilities*; UNDP: New York, NY, USA, 2010.
6. Frye, A. *Disabled and Older Persons and Sustainable Urban Mobility*; UN-HABITAT: New York, NY, USA, 2013.
7. Frye, A. *Inclusive Public Transport: Meeting the Mobility Needs of Disabled Citizens*; Policy Brief prepared for High Volume Transport Applied Research Programme; IMC: Redhill, UK, 2019.
8. Kuneida, M.; Roberts, P. *Inclusive Access and Mobility in Developing Countries*; World Bank: Washington, DC, USA, 2006.
9. United Nations. *Convention on the Rights of Persons with Disabilities*; UN: New York, NY, USA, 2007.
10. United Nations Development Programme. *Sustainable Development Goals*; UN: New York, NY, USA, 2015.
11. United Nations. *UN Flagship Report on Disability and Development*; UN: New York, NY, USA, 2018.
12. United Nations Economic and Social Commission for Asia and the Pacific. *Incheon Strategy to "Make the Right Real" for Persons with Disabilities in Asia and the Pacific*; UNESCAP: Bangkok, Thailand, 2012.
13. GAATES Survey of Local Transport Needs and Priorities. Analysis of Results. Available online: https://drive.google.com/file/d/1mDzAdxDhKPnqd0tlty4GtqbUvMlUV8da/view (accessed on 13 May 2019).
14. Access Exchange International. *Bridging the Gap: Your Role in Transporting Children with Disabilities to School in Developing Countries*; Access Exchange International: San Francisco, CA, USA, 2017.
15. Davies, D.K.; Stock, S.E.; Holloway, S.; Wehmeyer, M. Evaluating a GPS-Based Transportation Device to Support Independent Bus Travel by People with Intellectual Disability. *J. Intellect. Dev. Disabil.* **2010**, *48*, 454–463. [CrossRef] [PubMed]
16. Grisé, E.; Boisjoly, G.; Maguire, M.; El-Geneidy, A. Elevating access: Comparing accessibility to jobs by public transport for individuals with and without a physical disability. *Transport. Res. A Pol.* **2018**, *125*, 280–293.
17. Porter, G.; Tewodros, A.; Gorman, M. Mobility, transport and older people's well being in sub-Saharan Africa: Review and prospect. In *Geographies of Transport and Ageing*; Curl, A., Musselwhite, C., Eds.; Palgrave Macmillan: Heidelberg, Germany, 2018; pp. 75–100.
18. Sustainability Special Issue "Sustainable High Volume Road and Rail Transport in Low Income Countries". Available online: https://www.mdpi.com/journal/sustainability/special_issues/High_Road_Rail_Transport_Low_Income (accessed on 8 January 2020).
19. Heydari, S.; Hickford1, A.; McIlroy, R.; Turner, J.; Bachani, A.M. Road Safety in Low Income Countries: State of Knowledge and Future Directions. *Sustainability* **2019**, *11*, 6249. [CrossRef]
20. European Disability Forum. *EDF Report on the Situation of Passengers with Disabilities*; EDF: Brussels, Belgium, 2015.
21. Butcher, L. Access to transport for disabled people. *UK Parliament Briefing Paper Number CBP 601*. 30 October. Available online: https://researchbriefings.parliament.uk/ResearchBriefing/Summary/SN00601 (accessed on 13 May 2019).
22. Behrens, R.; Görgens, T. Challenges in Achieving Universal Access to Transport Services in South African Cities. In *The Palgrave Handbook of Disability and Citizenship in the Global South*; Watermeyer, B., McKenzie, J., Swartz, L., Eds.; Palgrave Macmillan: Heidelberg, Germany, 2019.
23. Dejeammes, M. Boarding Aid Devices for Disabled Passengers on Heavy Rail: Evaluation of Accessibility. *TRR J.* **2000**, *1713*, 48–55. [CrossRef]
24. Mattson, J.; Hough, J.; Varma, A. Estimating demand for rural intercity bus services. *Res. Transp. Econ.* **2018**, *71*, 68–75. [CrossRef]
25. Venter, C. Transport expenditure and affordability: The cost of being mobile. *Dev. South. Afr.* **2011**, *28*, 121–140. [CrossRef]

26. Mitra, S.; Palmer, M.; Kim, H.; Mont, D.; Groce, N. Extra Costs of Living with a Disability: A Review and Agenda for Research. *Dis. Health* **2017**, *10*, 475–4828. [CrossRef]
27. Sustrans. Locked Out: Transport Poverty in England. Available online: https://www.sustrans.org.uk/lockedout (accessed on 13 May 2019).
28. Lucas, K.; Mattioli, G.; Verlinghieri, E.; Guzman, A. Transport and Its Adverse Social Consequences. *Proc. Inst. Civ. Eng. -Transp.* **2016**, *169*, 353–365, ISSN 0965-092X. [CrossRef]
29. Mont, D.; Palmer, M.; Mitra, S.; Groce, N. Disability Identification Cards: Issues in Effective Design. *Leonard Ches. Res. Cent. Work. Pap. Ser.* **2016**, *29*, 1–13.
30. Lister, H.; Dhunpath, R. The taxi industry and transportation for people with disabilities: Implications for universal access in a metropolitan municipality. *Transformation* **2016**, *90*, 28–48. [CrossRef]
31. Navarrete-Reyes, A.P.; Medina-Rimoldi, C.T.; Avila-Funes, J.A. Correlates of subjective transportation deficiency among older adults attending outpatient clinics in a tertiary care hospital in Mexico City. *Geriatr. Gerontol. Int.* **2017**, *17*, 1893–1898. [CrossRef]
32. Deka, D.; Lubin, A. Exploration of Poverty, Employment, Earnings, Job Search, and Commuting Behavior of Persons with Disabilities and African-Americans in New Jersey. *Transp. Res. Rec. J. Transp. Res. Board* **2012**, *2320*, 37–45. [CrossRef]
33. Carew, M.T.; Colbourn, T.; Cole, E.; Ngufuan, R.; Groce, N.; Kett, M. Inter- and Intra-Household Relative Inequality among Disabled and Non-Disabled People in Liberia. *PLoS ONE* **2019**, *14*, e0217873. [CrossRef]
34. Eide, A.H.; Mannan, H.; Khogali, M.; van Rooy, G.; Swartz, L.; Munthali, A.; Hem, K.; MacLachlan, M.; Dyrstad, K. Perceived barriers for accessing health services among individuals with disability in four African countries. *PLoS ONE* **2015**, *10*, e0125915. [CrossRef]
35. Hansson, L.; Holmgren, J. Cost effect of reorganising—A study of special transport services. *Res. Transp. Econ.* **2018**, *69*, 453–459. [CrossRef]
36. Rickert, T. *Paratransit for Mobility-Impaired Persons in Developing Regions: Starting Up and Scaling Up*; Access Exchange International: San Francisco, CA, USA, 2012.
37. Sammer, G.; Uhlmann, T.; Unbehaun, W.; Millonig, A.; Mandl, B.; Dangschat, J.; Mayr, R. Identification of Mobility-Impaired Persons and Analysis of Their Travel Behavior and Needs. *Transp. Res. Rec. J. Transp. Res. Board* **2012**, *2320*, 46–54. [CrossRef]
38. Mulley, C.; Nelson, J.D.; Wright, S. Community transport meets mobility as a service: On the road to a new a flexible future. *Res. Transp. Econ.* **2018**, *69*, 583–591. [CrossRef]
39. Mobility as a Service (MaaS) in Developing Countries. Available online: https://citta.fe.up.pt/projects/3-59-mobility-as-a-service-maas-in-developing-countries (accessed on 8 January 2020).
40. Technology and Disability Journal. Available online: https://www.iospress.nl/journal/technology-and-disability/ (accessed on 8 January 2020).
41. Neven, A.; Vanrompay, Y.; Declercq, K.; Janssens, D.; Wets, G.; Dekelver, J.; Daems, J.; Bellemans, T. Viamigo Monitoring Tool to Support Independent Travel by Persons with Intellectual Disabilities. *Transp. Res. Rec. J. Transp. Res. Board* **2017**, *2650*, 25–32. [CrossRef]
42. Schlingensiepen, J.; Naroska, E.; Bolten, T.; Christen, O.; Schmitz, S.; Ressel, C. Empowering People with Disabilities Using Urban Public Transport. *Procedia Manuf.* **2015**, *3*, 2349–2356. [CrossRef]
43. Bustillo, N.V.; Cendana, D.I.; Palaoagm, T.D. E-Purse Transit Pass: The Potential of Public Transport Smart Card System in the Philippines. In Proceedings of the 3rd IEEE International Conference on Computer and Communications, Chengdu, China, 13–16 December 2017.
44. Øksenholt, H.V.; Aarhaug, J. Public transport and people with impairments—Exploring non-use of public transport through the case of Oslo, Norway. *Disabil. Soc.* **2018**, *33*, 1280–1302. [CrossRef]
45. HELPAGE International. *Learning with Older People about their Transport and Mobility Problems in Rural Tanzania*; HAI: Thame, UK, 2015.
46. Thompson, P. Challenges and Successes in the Application of Universal Access Principles in the Development of Bus Rapid Transport Systems in South Africa. *Stud. Health Technol. Inform.* **2016**, *229*, 629–638.
47. Porter, G. Mobilities in Rural Africa: New Connections, New Challenges. *Ann. Assoc. Am. Geogr.* **2016**, *106*, 434–441. [CrossRef]
48. Rama, S. Gendered mobilities: The methodology, theory and practice disjuncture. *Agenda* **2018**, *32*, 113–122. [CrossRef]

49. Ahmad, M. Independent-Mobility Rights and the State of Public Transport Accessibility for Disabled People: Evidence from Southern Punjab in Pakistan. *Adm. Soc.* **2013**, *47*, 197–213. [CrossRef]
50. Desapriya, E.; Harjee, R.; Brubacher, J.; Chan, H.; Hewapathirane, D.S.; Subzwari, S.; Pike, I. Vision screening of older drivers for preventing road traffic injuries and fatalities. *Cochrane Database Syst. Rev.* **2014**, *2*, 1–18. [CrossRef]
51. Li, H.; Raeside, R.; Chen, T.; McQuaid, R.W. Population ageing, gender and the transportation system. *Res. Transp. Econ.* **2012**, *34*, 39–47. [CrossRef]
52. Wheeler, K.; Yang, Y.; Xiang, H. Transport use patterns of US children and teenagers with disabilities. *Dis. Health* **2009**, *2*, 158–164. [CrossRef]
53. Ross, T.; Buliung, R. A systematic review of disability's treatment in the active school travel and children's independent mobility literatures. *Transp. Rev.* **2018**, *38*, 349–371. [CrossRef]
54. Lubin, A.; Alexander, K.; Voorhees, A.M. Achieving Mobility Access for Older Adults Through Group Travel Instruction. *Transp. Res. Rec. J. Transp. Res. Board* **2017**, *2650*, 18–24. [CrossRef]
55. Macagnano, E.V. Intelligent urban environments: Towards e-inclusion of the disabled and the aged in the design of a sustainable city of the future. A South African example. *WIT Trans. Ecol. Environ.* **2008**, *117*, 537–547.
56. Macagnano, E.V. Wireless Portable Computer Systems and Technologies for the Disabled and the Aged towards an Accessible, Inclusive and Intelligent Metropolis of the Future: The South African Context. In Proceedings of the Inclusion Between Past and Future AT from Adapted Equipment to Inclusive Environment Conference, Florence, Italy, 31 August–2 September 2009.
57. Vancampfort, D.; Smith, L.; Stubbs, B.; Swinnen, N.; Firth, J.; Schuch, F.B.; Koyanagi, A. Associations between active travel and physical multi-morbidity in six LMICs among community-dwelling older adults: A cross-sectional study. *PLoS ONE* **2018**, *13*, e0203277. [CrossRef] [PubMed]
58. Mindell, J. Transport and inequalities (editorial). *J. Transp. Health* **2018**, *8*, 1–3. [CrossRef]
59. Accessible India Campaign: Transport Systems. Available online: http://accessibleindia.gov.in/content/makeaccessible/transport-systems.php (accessed on 8 January 2020).
60. Hindustan Times. With just 3% of India's buildings accessible, our disabled are at a huge disadvantage. Available online: https://www.hindustantimes.com/editorials/with-just-3-of-india-s-buildings-accessible-our-disabled-are-at-a-huge-disadvantage/story-Rh2rd4QzNzw9kHpmaTPV1H.html (accessed on 8 January 2020).
61. King, J.A.; King, M.J.; Edwards, N.; Hair, S.A.; Cheang, S.; Pearson, A.; Coelho, S. Addressing transport safety and accessibility for people with a disability in developing countries: A formative evaluation of the Journey Access Tool in Cambodia. *Glob. Health Action* **2018**, *11*, 1. [CrossRef] [PubMed]
62. Green, S.; Mophosho, M.; Khoza-Shangase, K. Commuting and communication: An investigation of taxi drivers' experiences, attitudes and beliefs about passengers with communication disorders. *Afr. J. Disabil.* **2015**, *4*, 1–8. [CrossRef] [PubMed]
63. DFID. *Disability, Poverty and Development*; DFID: London, UK, 2000.

Article

# Older People, Mobility and Transport in Low- and Middle-Income Countries: A Review of the Research

**Mark Gorman [1,*], Sion Jones [1] and Jeffrey Turner [2]**

1 HelpAge International, London WC1H 9NA, UK; sjones@helpage.org
2 Gender, Inclusion & Vulnerable Groups, High Volume Transport Research Programme, IMC Worldwide Consultants, Redhill RH1 1LG, UK; jeffreymturner@hotmail.com
* Correspondence: mgorman@helpage.org

Received: 16 July 2019; Accepted: 22 October 2019; Published: 4 November 2019

**Abstract:** Older populations are rising globally, which in high-income countries has helped to generate a growing literature on the impact of ageing on travel requirements and transport policy. This article aims to provide an initial assessment of the state of knowledge on the impact on transportation policy and usage of the increasing numbers of older people in low- and middle-income countries (LAMICs), through a review of the literature relating to older people and transportation. As both the academic and policy/practice-related literature specifically addressing ageing and transport in LAMICs is limited, the study looks beyond transportation to assess the state of knowledge regarding the ways in which older people's mobility is affected by issues, such as health, well-being, social (dis)engagement and gender. We find significant knowledge gaps, resulting in an evidence base to support the implementation of policy is lacking. Most research in low-income countries (LICs) is either broad quantitative analysis based on national survey data or small-scale qualitative studies. We conclude that, although study of the differing contexts of ageing in LAMICs as they relate to older people's mobilities and transport use has barely begun, institutions which both make and influence policymaking recognise the existence of significant knowledge gaps. This should provide the context in which research agendas can be established.

**Keywords:** ageing; disability; gender; mobility; older people; poverty; transport; urban

---

## 1. Introduction

There is by now a substantial body of literature discussing the impact of an ageing population in developed countries on travel needs and required changes to transport policy. An age-friendly built environment, including safe, affordable, and convenient transportation, has been identified as a critical factor in enhancing the quality of life for increasingly large numbers of older people. Studies across high-income countries have recognised that "Access to transport is regarded as a major determinant to achieving a good quality of life in older age" [1]. A number of reasons have been proposed to explain this growth of interest in ageing and mobility, the most immediate being the global increase in the absolute number and share of older people. This interest has also been facilitated by a substantial growth in the evidence available for high-income countries since 2000. However, very little is known about older people outside high-income countries [2]. This article aims to provide an initial assessment of the state of knowledge on the impact on transportation policy and usage of the increasing numbers of older people in low- and middle-income countries (LAMICs), through a review of the literature relating to older people and transportation. As both the academic and policy/practice-related literature specifically addressing ageing and transport in LAMICs is limited, the study looks beyond transportation to assess the state of knowledge regarding the ways in which older people's mobility is affected by issues such as health, well-being, social (dis)engagement and gender.

While there are many similarities between the experiences of mobility and transport use between high-income countries and LAMICs, there are also important differences. For example, transport infrastructure in many low-income settings is significantly poorer than for the "developed" world and some middle-income countries, and transport choices for older people who have to rely on public or shared transport services are extremely limited in many situations. It has thus been pointed out that context is significant, including, but not limited to, geographical settings: "Mobility is not just about the individual ... but about the individual as embedded in, and interacting with, the household, family, community and larger society" [3].

This consideration of context also highlights the need for a review of research into age-related issues of mobility and transportation to move beyond a narrow biomedical model of ageing. For Schwanen and Paez context comprises four domains: the social relations of household, family, friends, acquaintances and community; the built environment, including transport, communication and other infrastructures; the institutions responsible for the built environment, policies and other forms of regulation; and the cultural norms and expectations that underpin travel practices and the regulation of mobility. In addition to these domains, life-course trajectories provide a history of past events and experiences for individuals, households, families, communities and societies, which impact significantly on issues to mobility in later life. So too do gender and ethnicity. Finally, old age, itself, is often treated as a fixed and stable category in transport studies [2]. This is notwithstanding that as long ago as the 1990s social gerontologists were pointing to the importance of understanding "the causal linkages between aging and social and political structures" rather than seeking inherent cultural meaning in the biological process of ageing [4]. Mobility and utilisation of transport is simply another example of how the meaning and definition of "old age" and "ageing" are culturally and geographically variable and socially constructed.

The article is structured as follows. After a review of the research materials and methods used we address global policy responses to ageing, mobility and transport, and this leads to a consideration of some of the evidence and policy gaps identified. This is followed by sections which address specific dimensions of ageing and its impact on mobility and transport. These are gender, social isolation and social support. Finally, consideration is given to ageing impacts and responses relating to mobility and transport requirements in humanitarian emergencies before a discussion of research gaps and emerging issues.

## 2. Materials and Methods

While this paper is not intended to be a systematic review, a literature search was undertaken utilising the Web of Science database, Google Scholar and other sources (including online material available from major international agencies). The search was undertaken between December 2018 and March 2019, and included material in English from 2000 onwards, just prior to the UN's first major plan of action on ageing relevant to the developing world in 2002. Criteria for inclusion was material focussed both on the experience of older people in utilising transport, and on the barriers due to age related to disability, gender and poverty. Search terms used were Older people AND mobility AND [Africa, Asia, Latin America]; Older people AND transport AND [Africa, Asia, Latin America]; Older people AND travel AND [Africa, Asia, Latin America]; Elderly AND mobility AND [Africa, Asia, Latin America]; Elderly AND transport AND [Africa, Asia, Latin America]; Elderly AND travel AND [Africa, Asia, Latin America]. The search on Web of Science database returned 82 results. After screening by title and abstract, 22 were selected for inclusion. In addition, a manual review was undertaken through Google Scholar and other sources which returned 69 relevant items from scientific and grey literature. Eighty-two items were identified through database searches and 22 items were included; 69 items were included from manual search out of a total of 81 items.

In addition to the Web of Science database, a manual search was also undertaken. The search terms selected here were chosen to extend the inquiry beyond a focus on ageing and older people (which would produce limited results) to include related issues such as gender, disability and poverty.

Countries selected aimed to achieve a representative geographical spread of LAMICs in Africa, Asia and Latin America. As articles and other publications were identified, citations they contained were used to identify other research of potential relevance on the World Bank's Open Knowledge Repository was searched, and the online resources of UN agencies (the World Health Organization, the UN Department of Economic and Social Affairs (which houses the UN's Ageing Programme), the UN Development Programme and UN Habitat. Other "grey" literature, included are the materials of Help Age International, and the publications of other non-governmental organisations working in related fields. (It should be noted that some of this material was developed for advocacy rather than objective research purposes and their conclusions should, thus, be treated with more caution). Our review also draws on the authors' extensive practitioner experience of work with older people, in a major international organisation dedicated to this field, including engagement with international and national policy makers, NGOs and communities in diverse LAMIC contexts. We are confident that we have been able to adequately identify eligibility and quality of the material we present in this paper.

## 3. Results

### 3.1. *Ageing, Transport and Global Policy*

The broadest context for population ageing is at the global level. Demographic change and population ageing are now global trends, not ones confined to high-income countries. The growing numbers and proportions of people living into later life in all societies pose new questions about their transport and mobility requirements and the extent to which transportation policy and practice will be responsive to the mobility needs of older people. The policy response at the global level is an indicator of the extent to which demographic ageing has elicited a policy response, and the degree to which an evidence base to inform policy has been developed.

At a global level, policy parameters are now set by the Sustainable Development Goals (SDGs), whose call to leave no one behind requires that the SDGs are met for all of society, at all ages, with a particular focus on the most vulnerable, including older women and men [5]. Even before the advent of the SDGs various UN agencies addressed the impact of transportation issues on ageing populations in the context of wider agenda-setting debates. Prior to the United Nations' second global assembly on ageing in 2002, Kalache and Keller noted the importance of adequate public transport to enable more independent life well into very old age, noting that "one of the major challenges is to ensure access...to all older persons—including the poor and those who live in remote areas" [6]. The UN's international plan of action on ageing which emerged from the 2002 World Assembly included as a priority "Ensuring enabling and supportive environments—Transportation is problematic in rural areas because older persons rely more on public transport as they age and it is often inadequate in rural areas" and called for investment in local infrastructure, such as transportation [7].

The UN's New Urban Agenda (which lays claim to being "a paradigm shift based on the science of cities; it lays out standards and principles for the planning, construction, development, management, and improvement of urban areas") also makes broad commitments to be age-inclusive, from data collection through consultation to policy-making. Specifically in relation to urban transportation, commitments are made to enable access to safe, efficient and sustainable transport systems for all [8].

In its Global Ageing and Health Report (2015) the World Health Organization addresses the interactions of the different dimensions of an older person's context (social relations, the built environment, policy and regulation, and cultural norms). "Mobility is influenced not only by an older person's intrinsic capacity and the environments they inhabit but also by the choices they make. Decisions about mobility are, in turn, shaped by the built environment, the attitudes of the older person and of others, and having both a motivation and the means to be mobile (such as by using assistive devices or transportation)". The Report goes on, stating: "Specific consideration will need to be given to the needs of older people to ensure that environments are accessible, including homes, public spaces and buildings, workplaces and transportation" [9].

The World Bank has also recognised that "Vulnerable and special-need groups (including women, children, persons with disabilities, and older persons) are underserved by public and private transport systems...because users and providers do not carry the full societal costs of excluding vulnerable groups". The World Bank argues for "equity and inclusivity [to be] at the heart of Universal Access. This objective...places a minimum value on everyone's travel needs, providing all, including the vulnerable, women, young, old, and disabled, in both urban and rural areas, with at least some basic level of access through transport services and leaving 'no one behind'" [10]. Both the Sustainable Development Goals and the New Urban Agenda call for expended, age- and gender-responsive public transportation that responded to the challenges faced by all.

More broadly, the NUA calls for age-and gender-responsive approaches to policy and planning processes, design, budgeting, implementation, evaluation and review, and this indicates the important potential of focussing on universal access, rather than transport-related solutions aimed at older people alone. The World Bank has addressed ways to ensure universal access, but again the research basis is lacking, at least as far as the participation of older people is concerned. The World Bank itself notes that "While there is no widely agreed upon method of measuring universal access, there is a general agreement that sustainable transport should leave no one behind. Data that measure access to transport infrastructure and services for urban areas are not readily available on a global scale. The data that do exist suggest that the accessibility gap is huge, and potentially growing ... " [11].

### 3.2. Evidence and Policy Gaps

Notwithstanding the concern of the UN and other global bodies to establish these broad declaratory frameworks, evidence-building and analysis have not been prioritised. For example, while the World Bank makes significant reference to the evolving requirement for sustainable and accessible transport arising in part from the changing requirements of older populations, no supporting evidence is cited for statements such as "The ageing of the population is likely to have significant effects on mobility..." The evidence base for a research agenda to support the implementation of global policy is therefore lacking [11]. Moreover, these global policy concerns are very rarely reflected at regional and national levels. National policymakers cite resource constraints and the absence of evidence for these policy gaps [12]. Thus, while a number of studies conclude with proposals for policy interventions, these have not been translated into practice [13,14].

More recently greater attention has been paid to the global implications of population ageing for transportation and transport policy [15]. A number of studies in LAMICs have been undertaken, often with the stated intention of influencing transport policies. A study of Hong Kong, for example, took as a starting point the rapid growth of the older population, which was expected to significantly affect the public transport systems, and "to provide suitable policy recommendations that cater to the travel needs of an ageing society" [16]. Other work has relied on the WHO SAGE data for China, India, Mexico, Ghana, Russia and South Africa, which are, thus, studies of small samples of older populations in middle- rather than low-income countries [17]. Datasets comparable to the SAGE data do not exist for low-income countries [18].

Other countrywide studies are rare and those that exist use broad national datasets, such as USAID's Demographic and Health Survey. Although the DHS is conducted in around 90 LAMICs and provides data for approximately 30 indicators in the Sustainable Development Goals, its data on older people are of limited value. While interviewees are women aged 15-49 and men aged 15–49, 15–54, or 15—59, data on older people are collected only through a whole-household questionnaire. Only two countries (South Africa DHS 2016, Haiti DHS 2016) have extended or lifted the upper age cap in individual questionnaires. Thus, in sub-Saharan Africa for example, only one significant national transport study has been undertaken with older people [19]. As we have seen, urban studies are confined to West Africa, particularly Nigeria. For rural areas the literature is very meagre [20]. As we have noted, studies in both south Asia and Latin America also concentrate heavily on urban settings.

These significant research gaps should be seen in the context of the overall lack of both empirical data and theoretical development on ageing and older people in LAMICs [14]. This in turn impacts on planning and policy formation. Commenting on the lack of data sources for an inquiry on active ageing in the Philippines, for example, Pettersson and Schmöcker note that better datasets are a basic requirement for urban planning in developing and newly developed countries [21]. Constraints such as these apply particularly in Sub-Saharan Africa where, despite an intensification of debate on ageing and formal expressions of commitment on the part of national governments, comprehensive policy action is still lacking. "The impasse reflects a lack of political will and an uncertainty about required policy approaches, engendered by wide gaps in understanding...in the region" [21]. A review of research relating to care-seeking behaviours among older people finds that apart from a number of small-scale, qualitative studies, no systematic, country-level evidence exists. Their findings, which point to the negative impacts of physical and logistical access difficulties and financial barriers related to service fees and/or transport costs, thus have no means of translation into policy-making [22].

This highlights a particular problem in relation to population ageing and older people. Parker et al. point out that "Aging in the west is often viewed from a biomedical perspective where the emphasis is on medical treatment and health and social care arrangements. Biomedicine also dominates international health strategies, organizations, and the funding streams for aid..." [12]. From this perspective, the wider implications of ageing as another stage in the life course receive much more scant consideration, and the familial, social and economic relations which structure older people's lives tend to disappear from view. Furthermore, older people are often excluded from data collection mechanisms; for example, much of the data available for LAMICs comes from household surveys, which set age restrictions and, therefore, do not routinely include older respondents. They often provide data and analysis at the household rather than the individual level, and data, where collected, is not disaggregated and analysed by age [23]. Exceptions to this, such as the World Health Organization's "Study on global Ageing and adult health" (a longitudinal study collecting data on adults aged 50 years and older), have very small sample sizes and tend to focus on middle-income countries where datasets may be more reliable than in the least-resourced settings [24].

The inattention to the needs of older populations in transport policy may also reflect the relative lack of political participation by economically and socially disadvantaged older people, and their consequent inability to influence decision-making on mobility services and investment. A low level of political participation has been noted as a key measure of the social exclusion of older people in middle- and low-income countries [25]. While the participation of older people in decision-making processes is restricted, the relative weight given to perceived economic value, social participation and reducing inequalities is likely to remain limited. Porter et al. point out that "Transforming evidence into policy and practice is particularly challenging in the transport sector which is dominated by male, middle-aged, middle-class engineers whose principal focus is road construction rather than transport services and where there is still a common reluctance to engage with users or with qualitative data" [20].

The lack of evidence that the perspectives of older people, amongst others, are considered when planning public transportation has been noted by, for example, the Institute for Transportation and Development Policy (ITDP), which recognises the democratic deficit created as result of local and national government transportation investment and planning decisions which fail to consult with and include urban residents, particularly those who are often the most marginalised [26]. In terms of investment and impact assessment, even in high income contexts, there is little existing guidance for comprehensive transport equity analysis that includes all groups of people [27].

### *3.3. Ageing, Health and Mobility*

The relationship between ageing and health is complex; it is now well recognised that a global epidemiological transition from diseases, which mainly impact children, to non-communicable diseases (NCDs), which are more common in adults, has accompanied the demographic transition with population ageing at its core. The combination of disability and ageing potentially provides a significant limitation on the mobility of older people in LAMICs, though again the evidence base is relatively sparse, since very little research has been conducted on moderate to severe disabilities affecting mobility, communication and mental function in later life [28–30]. For example, a comparative study across urban settings in Latin America and the Caribbean, while addressing gender differences in later life health and functional status, similarly did not establish a connection between limited functional capacity and transport utilisation [31], this notwithstanding that the research utilised data from the SABE study ("Survey on Health, Well-Being, and Aging in Latin America and the Caribbean"), the first health study of the old people in Latin America and the Caribbean to include transportation among the physical environment determinants which it assessed [32].

What does seem to be clear from the evidence is that the outcomes of disease and injuries are increasingly undermining the ability of the world's population to live in full health. A recent review of data from the WHO "Global Burden of Diseases, Injuries, and Risk Factors Study" noted that (after anaemia) the leading cause of impairments (by number of individuals affected) were hearing loss and vision loss. These sensory organ disorders were also the leading causes of impairments in 22 countries in Asia and Africa and one in Central Latin America, while lower back and neck pain was the main cause of disability in most countries [33]. There is some recognition that these disabling conditions can be critical for the wider social participation of older people. There is little published data on the potential health benefits of active travel in low and middle-income countries, although some evidence exists. For example, studies (which again drew on WHO's widely-used SAGE data for six MICs, or the related INDEPTH datasets) found a correlation between increasing age and reduced active travel (walking or cycling), translating into a higher risk of being overweight and raised BMI [34,35]. A smaller-scale study in peri-urban areas of Nepal elicited similar results [36].

Problems over access to health care facilities is the most frequently cited transport-related issue for older people in LAMICs, with physically remote clinics and hospitals necessitating costly and difficult travel a key barrier. A common finding was that poverty and mobility constraints combined to reduced older people's access to healthcare [37]. Data from the WHO's World Health Survey (2002–2004) indicated that, in total, more than 60% of older people in LAMICs did not access health care either because of the cost of the visit, or because they did not have transportation, or they could not pay for transportation. Transportation may be a particularly important issue for older people who live in rural areas because services are often concentrated in large cities far from people's homes and communities [38]. A study in rural Kyrgyzstan, for example, found that one in five older people lived more than a 30-minute travel away from a health facility, with access particularly problematic for those with a limiting longstanding illness or disability [39]. The structure of health care provision is also problematic, noted by a study in South Africa of care and treatment for older people (50+) living with both HIV and other chronic conditions, which found that services were typically provided at different health facilities or by different health providers, necessitating multiple patient journeys [40].

Psychological factors, such as depression, may also play a role in limiting mobility. A study to identify the most burdensome functioning domains in depression and their differential impact on the quality of life using SAGE data from countries in Asia, Africa and Latin America found that affect, domestic life and work and interpersonal activities were the domains most affected by depression, with gender also playing an important differentiating role [41]. A lack of self-confidence in physical capacity, leading for example to a consideration of the risk of falling, as well as concerns regarding traffic, have also been identified as limiting factors. The risk of falls is a prevalent factor in activity restriction by older people in a variety of settings. Even among a physically active older population in the Colombian Andes the risk of falling decreased physical activity with negative effects for self-perceived health and depressive symptoms [42–44].

Again, however, and notwithstanding the high and rising level of death and injuries from road traffic accidents in LAMICs [45], and the significant proportion of older victims in some contexts, research on the impact of traffic accidents on older people remains undeveloped [46]. This may reflect a context in many low- and middle-income countries, where poor quality transport infrastructure (including poorly maintained roads, pavements, crossings, shelters, as well as poorly maintained and inaccessible vehicles) affect people throughout the life course. Accident rates are high, affecting whole populations. In these circumstances it is perhaps unsurprising that the specific impacts on older people receive relatively little attention.

The evidence that does exist indicates that older people, whether walking or using public transport, have a significant exposure to accidents and injury [47]. Even crossing roads in busy urban environments in LAMICs may (rightly) induce feelings of anxiety and fear, with a consequent impact on quality of life [48]. Although data limitations mean that studies have tended to be small-scale, they nevertheless raise important issues. For example, a study of the constraints on the travel of older people in a Nigerian city found that issues such as vehicle design, long access and waiting time as well as poor facilities at bus terminals were serious constraints to the effective mobility of older people [49]. Similar conclusions were drawn from a study in Pakistan, which pointed to safety and security issues, as well as attitudinal behaviours on the part of service providers, as key factors in the mobility and transport utilisation of older people [14]. These are issues to which we return below.

### 3.4. *Ageing, Transport and Gender*

As noted above, a key cross-cutting issue affecting the mobility of older people is that of gender, which has now become a well-recognised issue in transport research, and one that intersects with other factors to increase disadvantage. A study in Bogota, for example, which summarized recent research on unequal access to transport systems, focused on the ways that gender and socioeconomic inequalities may be exacerbated by differences in transport accessibility [50]. Nevertheless, again the intersection of gender with age has attracted far less attention, despite this being a feature that research has shown to be clearly recognised by older people themselves [51]. Moreover, there are no systematic gender and age inclusion procedures for transport, either in terms of training of professionals, participation of users or the design and planning of systems, services and equipment. Again, as international institutions such as the World Bank have pointed out, a lack of evidence limits progress in policymaking, particularly regarding gender issues in transport relating to older people in the developing world [52].

While other international institutions have drawn attention to the intersections of age and gender, again, a clear evidence base substantiating their assertions is lacking. The UN Economic and Social Council (ECOSOC) for example, has pointed out the economic impacts of differential access to transport for women. They noted in 2009 that women in low-income countries were seriously constrained in their access to transport, limiting access to labour markets, increasing production costs and reducing the amount of goods which could be taken to market. The ECOSOC report focussed on issues such as poor access to transport, affecting girls' school attendance, women's use of health and other public services and maternal mortality. While older women were not specifically mentioned, evidence was noted regarding the lack of access to transport services affecting women who spend long hours hauling water and fuel and walking to and from farm plots. Head-loading was cited as a major health hazard to women, as was the potential to suffer higher accident rates through walking on crowded roads with heavy burdens [53]. Other evidence indicates that these are impacts which fall on women who continue to work into later life [20].

A small number of studies do address age and gender as features affecting access to transport and impacting on mobility. Research in urban settings in Pakistan, Iran and Malaysia all found that age and gender, together with other factors, such as car ownership, travel time, travel cost, household size and income, were significant factors in influencing individual choices in transportation. These studies all included higher-income households, and demonstrated the importance of car ownership in travel frequency, though even here gender played an important role, with older men making significantly more journeys than older women [14,54,55]. Other evidence points to the limitations in "choice" that older women in particular may have; a study of a urban setting in Nigeria found that gender, along with increasing age, education and monthly income, were significant in determining walking as the mode choice, in a situation where over 70% of older people lacked access to motorised transport [56].

The impact of gender on transport access in Africa is also examined in a recent field study which assessed whether gender mainstreaming in rural transport programmes in Tanzania has had a transformative effect on women. The study found that despite the attention given to gender issues, women's participation in designing and implementing rural travel and transport programmes was limited by negative views of women's potential to contribute effectively to such programmes. On the other hand, road construction did lead to improvements in transport services and expanded travel options for all women, including those in later life, who had more time both for family and to pursue multiple projects [57].

### 3.5. Mobility, Transport and Social Isolation

Gender thus has clear impacts on mobility and helps to bring into question the role that mobility plays in enhancing or mitigating loneliness and social isolation. The World Health Organization has recognised the importance of attitudinal factors for older people, commenting that "Mobility is influenced not only by an older person's intrinsic capacity and the environments they inhabit but also by the choices they make. Decisions about mobility are, in turn, shaped by the built environment, the attitudes of the older person and of others" [9]. Characteristics of the built environment can function to restrict older people's mobility and participation in urban life [58,59]. Nevertheless, research evidence for the role that transport access plays in influencing older people's social isolation is sparse for low-income countries.

In recent years a significant number of studies have made a direct causal link between transport accessibility and social exclusion [60]. Again, the great majority of these studies have been undertaken in high-income countries where both income poverty and lack of transport are relative rather than absolute states. There are few studies on the relationship between transport and social disadvantage in LAMICs, where income poverty is absolute and where access to transportation is very limited [61]. Studies addressing the social exclusion of older people in relation to transport in low-income countries are still rare [62–64]. While some studies discuss the part played by physical mobility in older people's social isolation, the role that transport access plays in this is not examined [65].

For those living in LICs, social isolation is a significant risk with increasing age, again mediated by factors such as gender and poverty. A primary issue for older people in these situations is the impact of psychological factors. WHO notes the importance of the attitudes of older individuals and of others in decisions about mobility, and the motivation (and the means) to be mobile [9], p.180. A small number of studies has addressed the links between confidence and behaviour, to assess the influences on older people's decision-making relating to mobility. Fear of crime and concerns over the safety of public transport have been identified as a limiting factor in older people's mobility in a number of different settings, albeit with wide national and regional variations [66].

WHO studies highlight a number of issues that influence decision making related to older people's use of transportation. These include availability of services, affordability, reliability and frequency, appropriateness of service destinations and the availability of specialised and priority services. Comfort is also highlighted as a key concern, with respondents to focus groups in studies in Rio de Janeiro and Mexico City citing hard, uncomfortable and bumpy journeys exacerbating existing health concerns and discomfort. Similar concerns were raised by focus groups undertaken in five cities in Argentina, with issues such as struggling to board buses because of the height of the initial step. Multiple elements are necessary to make public transport an attractive alternative for older people, including physical accessibility, the availability of information and ease of way finding [67].

Social exclusion has a number of features which may be identified with the experience of older people in relation to transport in the developing world, and which have been the subject of studies which address older people's mobility and transport needs only obliquely. These features of social exclusion are broader than income poverty alone, and include a lack of participation in social, economic and political life. They are also multidimensional and cumulative: for example, limited financial resources and security are often reciprocally linked to low education and skills, ill-health, and, as noted above, limited or no access to political influence. Social exclusion is also dynamic, and subject to changes over time, as well as directly affecting individuals and households as well as neighbourhoods and local communities [68]. Curl and Musselwhite have pointed out that, despite policy and discourse (at least in high income countries) placing strong emphasis on the maintenance and extension of independence and "ageing in place" as vital requirements for a dignified healthy later life, changes in later life income, lifestyle and ability to use transport present significant challenges for the provision of appropriate transport services [15]. Where and how people live impacts day-to-day life, particularly in the urban built environment. Spatial barriers interact in complex and specific ways with the intersecting identities that individuals carry, creating unique patterns of disadvantage. For example, lower income people (in which older people, together with other disadvantaged groups such as women, children, are also disproportionately represented) are increasingly pushed to the periphery of cities in their search for affordable housing [26].

In some cases, this is as a result of deliberate government policy and this has an impact on access to affordable transport. This was noted by a study of peri-urban areas in South Africa, which found that location along the urban-rural continuum significantly affects both transport expenditure levels and the perceived severity of transport affordability problems for marginalised people, notably those with disabilities and older people [19].

### 3.6. Mobility, Transport and Social Support

One important motivation for mobility and travel for older people is to access social support, particularly from family members, and a factor mitigating against the need for mobility in many low-income settings in Asia remains the fact that normative support for filial obligations to ageing parents is widespread. A recent study of older people in Myanmar for example, found that the majority of rural-dwelling older people had an adult child co-residing or living nearby, facilitating intergenerational exchanges of material support and personal services, and reducing the need for travel [69]. In the same vein, a review of studies on long-term care systems in sub-Saharan Africa noted that "the provision of long-term care rests overwhelmingly with family members, in line with customary sub-Saharan African norms of family solidarity and obligation". However, the review concluded that the evidence also revealed that "a substantial group of older people received no family care whatsoever" [70]. Again, studies which address the implications the care received by older people in terms of their mobility and transport requirements are almost non-existent for low-income countries. This is equally the case for evidence of older people as caregivers, despite the recognition of the important role that they play in many contexts. A study in Kenya, which found that older women AIDS-caregivers reported high disability scores for mobility and low scores in self-care and life activities domains, indicates that this comes at a high cost to older individuals [71]. Similarly, a Ugandan study showed how older women caregivers faced drastic disruptions of living arrangements, including lengthy travel times and absences from their homes to care for PLWAs [72]. Nor should it be assumed that care-giving is ensured simply by the traditional norms of extended family relationships, when families are spatially separated. For example, a recent study of the mobility constraints experienced by married and externally-resident daughters providing end of life care to parents in northern Ghana shows how these younger women had to negotiate conflicting responsibilities to provide parental care [73]. Similar issues are identified for settings in a recent comparative study on Manila and London [74].

Urban locations are by no means the only settings where security issues play a part in older people's travel choices. For example, both the lack of paved main roads in rural Myanmar and security issues in some regions have been found to be a significant barrier to older people's access to health services [69]. A study in Papua New Guinea examining the impact of road development on people with disabilities found that while road development improved service access, inaccessible road and transport infrastructure remained insurmountable barriers to easy and safe travel. Roads were planned for the needs of vehicle users, and planning around road infrastructure did not involve consultation with people with disabilities [75].

### 3.7. Mobility and Transport in Crises and Humanitarian Emergencies

The particular vulnerabilities of older people in times of systems breakdown is an area which has had some attention in relation to humanitarian response. Clearly mobility plays a decisive role in periods of crisis and humanitarian emergency, and is a major potential issue for older people, (as it is for people with disabilities and children). This has elicited some response from humanitarian agencies which have developed guidance on making services accessible to older people with mobility constraints [76]. However, in comparison to the attention paid to people with disabilities, older people have received less attention from the international agencies, and their transport needs have been largely ignored. The United Nations High Commission for Refugees (UNHCR) has produced guidelines on "Working with Older Persons in Forced Displacement", but while the companion guidance on working with people with disabilities in forced displacement has recommendations for accessible transport, that for older persons does not, confining itself to a broad recommendation to "Help older persons to access services by providing transport" [77].

More recently, a consortium of humanitarian agencies and academic institutions working in the fields of disability and ageing have recognised the overlapping and often coterminous vulnerabilities of older people and those with disabilities in emergencies in establishing key standards for achieving the inclusion of both groups in humanitarian action, by, for example, addressing barriers that affect participation and access to services [78]. However, a more specific focus on transport requirements remains lacking, reflecting a lack of studies on the specific mobility and transport requirements of older people in emergency situations.

The work that has been done reveals some familiar issues. For example, a recent study, based on focus group interviews with older displaced persons in Sudan, found that many older people with disabilities faced a number of physical barriers such as having to travel long distances to distribution points, a lack of accessible transport, as well as inaccessible housing, toilets and public buildings. Family and friends were identified as key providers of both physical and financial assistance, including, notably, paying transport costs, but the cost of transport to key points, such as health facilities, was a constant source of stress [79].

## 4. Discussion

This review indicates that, notwithstanding growing recognition (at least at a global institutional level) of the implications of population ageing for transport policymaking, this has yet to be translated into significant investment in research in LAMICs. Such research as has been done in LICs is small-scale and based on qualitative evidence, which may be dismissed by policymakers as "anecdotal". This lack of systematic national-level evidence makes translation of any research findings into policy highly problematic. We have noted the predominant role that transport infrastructure planning plays in many LAMICs, to the detriment of consideration of social value. Such policy biases reflect the fact that policymaking is not a rational, evidence-based process, but is the outcome of numerous interactions between policymakers and other actors (including researchers, but also politicians, lobby groups and advocacy organisations). While evidence plays a role in illuminating policymaking, so too do other factors, such as political voice. As we have seen, the relative lack of political participation by disadvantaged older people limits their ability to influence policy and, by extension, research decisions. This may be both a cause and effect of the exclusion which we have noted of older people from many data collection mechanisms.

There is equally limited consideration of the ways in which age intersects with other factors, particularly gender, differential physical and mental capacity, and poverty. Here we have seen that the lack of age-disaggregated data is problematic. The imposition of upper age limits in data collection is a significant barrier to understanding the characteristics of ageing, whether at the individual or the population level. Furthermore, despite the diversity to be found within older populations, disaggregation by age has been analysed very little. The lack of gender disaggregated data and analysis is also noticeable, notwithstanding the frequent references in policy documents to the particular challenges of public transport for older women.

Access to services is often a strong focus of current work regarding older people's mobility. The difficulty of accessing health services is, as we have seen, frequently cited, but there is a limited perspective on the wider needs of older people beyond basic health. Issues of mental health, social isolation and loneliness are rarely discussed. The roles played by older people as care-givers or recipients, and their mobility implications has become an increasingly significant area of research in Western contexts, and is also an important issue in LICs. This is an area that clearly needs greater attention. Research on mobilities across generations is also a major gap, with little attention paid to relational mobilities, despite the clear importance of intergenerational connections, notably (but not exclusively) related to care-giving.

However, we have also noted an emphasis in research related to population ageing which focusses attention on medical and social care arrangements, to the detriment of considering the social and economic contexts in which people live the later stages of their life course. We have noted that some work has been done in this regard. For example, a number of studies have emphasised the important place that the continuing need to earn a livelihood has for many who enter later life in poverty. Older people's mobilities and the role of transport in relation to livelihoods is, thus, another important area of inquiry but, again, is a significant research gap in LICs (where income earning must continue into old age in the absence of adequate pension provision). In this regard, the affordability of transport, economic needs, subsidy issues and income categories are all areas for research which require further attention [19].

Research undertaken on psychological effects which act to limit the mobility of older people, and the impacts of social isolation, has been reviewed. Issues ranging from concerns over road safety to harassment, personal security, stigma, shame, discrimination and the impact of crime have all been shown to pose significant barriers to older people's mobility. Many transport services, whether public or private, provide physical hazards for older people, as does poorly maintained physical infrastructure. Again, with limited exceptions, these problems are inadequately addressed in the research literature. While the potential value of virtual connectivity (through mobile phones for example) to replace or complement physical mobilities has been examined, both the benefits and costs of virtual connectivity remain to be researched further [20].

In the absence of access to relevant national-level quantitative data, it may be useful to consider other innovative research approaches. For example, action research, involving interventions followed by in-depth monitoring of impacts over a period of time, which involve older people as active research participants, has potential. However, while there is a growing rhetoric around the involvement of older people in research process, and some action has been taken, older people tend to remain respondents. The value of using co-investigation approaches has been demonstrated by work in Tanzania and in Papua New Guinea [19,20]. New methodologies, such as the use of geo-mapping alongside participatory inquiry methodologies to explore the social and spatial barriers to access of urban services, also need further review and analysis [20]. To develop the necessary expertise, in academic institutions, governments and NGOs, to conduct mobility studies with older people there is also a need to build research capacity, both in-country and external expertise.

This review has indicated the importance of taking account of great diversity in the ageing experience, across widely varying contexts. Influences on ageing range from societal and political attitudes to older people to the built environment, population density, climate, topography and land use. As we have aimed to show, study of these differing contexts of ageing in LAMICs, particularly as they relate to older people's mobilities and use of transport, has barely begun. At the same time, as we have seen, policymaking institutions recognise these issues, and assert the importance of prioritising the widest possible inclusion in policies promoting transport and mobilities of people at all ages, so that, in the words of the Sustainable Development Goals, "no one will be left behind" [80]. We have seen that those institutions which both make and influence policymaking recognise the existence of significant knowledge gaps, some of which have been discussed in this article. This should provide the positive context in which research agendas to answer some of these key questions can be established.

**Author Contributions:** The article was conceptualized by M.G., S.J. and J.T.; the methodology was developed by M.G., who also undertook the validation, and formal analysis.; the investigation was undertaken by M.G. and S.J.; the original draft preparation was by M.G.; the review and editing was by M.G., S.J. and J.T.; project supervision was by M.G and J.T. and project administration was undertaken by M.G.

**Funding:** This research was funded by the UK Department for International Development.

**Acknowledgments:** This research was funded by UK AID through the UK Department for International Development under the High Volume Transport Applied Research Programme, managed by IMC Worldwide. Special thanks to Gina Porter for her support in developing the literature search and reviewing the output.

**Conflicts of Interest:** The authors declare no conflict of interest.

## References

1. Manchester Institute for Collaborative Research on Ageing. Ageing, Transport and Mobility: New Approaches from Researchers and Providers, MICRA Seminar. 2013. Available online: https://www.micra.manchester.ac.uk/connect/events/events-archive/2013/ageing-transport-and-mobility (accessed on 3 March 2019).
2. Schwanen, T.; Paez, B. The mobility of older people—An introduction. *J. Transp. Geogr.* **2010**, *8*, 591–595. [CrossRef]
3. Hanson, S. Gender and mobility: New approaches for informing sustainability. *Gend. Place Cult.* **2010**, *17*, 5–23. [CrossRef]
4. Phillipson, C. The social construction of old age: Perspectives from political economy. *Rev. Clin. Gerontol.* **1991**, *1*, 403–441. [CrossRef]
5. United Nations Development Programme. *Ageing, Older Persons and the 2030 Agenda for Sustainable Development*; UNDP: New York, NY, USA, 2017; p. 7.
6. Kalache, A.; Keller, I. The greying world: A challenge for the twenty-first century. *Sci. Prog.* **2000**, *83*, 33–54. [PubMed]
7. United Nations, Madrid Political Declaration and International Plan of Action on Ageing, Madrid. 2002. Available online: https://www.un.org/development/desa/ageing/madridplan-of-action-and-its-implementation.html (accessed on 1 March 2019).
8. United Nations. New Urban Agenda. In Proceedings of the United Nations Habitat 3 Conference, Quito, Ecuador, 17–20 October 2017; Paragraphs 113, 157. Available online: http://habitat3.org/the-new-urban-agenda/ (accessed on 25 February 2019).
9. Beard, J.; Officer, A.; Cassels, A. *World Report on Ageing and Health*; World Health Organization: Geneva, Switzerland, 2015; p. 180.
10. World Bank. *Sustainable Mobility for All, Global Mobility Report 2017: Tracking Sector Performance*; World Bank: Washington, DC, USA, 2017; p. 25.
11. Mason, J.; Turner, P.; Steriu, M. *Universal Access in Urban Areas: Why Universal Access in Urban Areas Matters for Sustainable Mobility Transport and ICT Connections*; World Bank Group: Washington, DC, USA, 2017; Available online: http://documents.worldbank.org/curated/en/221451537381407483/Universal-Access-in-Urban-Areas-Why-Universal-Access-in-Urban-Areas-Matters-for-Sustainable-Mobility (accessed on 10 March 2019).
12. Parker, S.; Khatri, R.; Cook, I.G.; Pant, B. Theorizing Aging in Nepal: Beyond the Bio-Medical Model. *Can. J. Sociol.* **2014**, *39*, 231–253.
13. Olawole, M.O.; Aloba, O. Mobility characteristics of the elderly and their associated level of satisfaction with transport services in Osogbo, southwestern Nigeria. *Transp. Policy* **2014**, *35*, 105–116. [CrossRef]
14. Ahmad, Z.; Batool, Z.; Starkey, P. Understanding mobility characteristics and needs of older persons in urban Pakistan with respect to use of public transport and self-driving. *J. Transp. Geogr.* **2019**, *74*, 181–190. [CrossRef]

15. Curl, A.; Musselwhite, C. Geographical perspectives on transport and ageing. In *Geographies of Transport and Ageing*; Curl, A., Musselwhite, C., Eds.; Palgrave Macmillan: Basingstoke, UK, 2018; pp. 3–24.
16. Szeto, W.Y. Spatio-temporal travel characteristics of the elderly in an ageing Society. *Travel Behav. Soc.* **2017**, *9*, 10–20. [CrossRef]
17. Wilunda, B.; Ng, N.; Williams, J.S. Health and ageing in Nairobi's informal settlements-evidence from the International Network for the Demographic Evaluation of Populations and Their Health. *BMC Public Health* **2015**, *15*, 1231. [CrossRef]
18. HelpAge International. *Global AgeWatch Insights: The Right to Health for Older People, the Right to be Counted*; HelpAge International: London, UK, 2018; p. 37.
19. Venter, C. Transport expenditure and affordability: The cost of being mobile. *Dev. South. Afr.* **2011**, *28*, 121–140. [CrossRef]
20. Porter, G.; Tewodros, A.; Gorman, M. Mobility, transport and older people's well-being in sub-Saharan Africa: Review and prospect. In *Geographies of Transport and Ageing*; Curl, A., Musselwhite, C., Eds.; Palgrave Macmillan: Basingstoke, UK, 2018; pp. 75–100.
21. Pettersson, P.; Schmöcker, J.-D. Active ageing in developing countries?—Trip generation and tour complexity of older people in Metro Manila. *J. Transp. Geogr.* **2010**, *18*, 613–623. [CrossRef]
22. Aboderin, L. Understanding and Advancing the Health of Older Populations in sub-Saharan Africa: Policy Perspectives and Evidence Needs. *Public Health Rev.* **2010**, *32*, 357–376. [CrossRef]
23. HelpAge International. *Global AgeWatch Insights Report*; HelpAge International: London, UK, 2018; p. 33.
24. World Health Organization. Study on Global Ageing and Adult Health. Available online: https://www.who.int/healthinfo/sage/en/ (accessed on 3 March 2019).
25. United Nations Population Fund (UNFPA); HelpAge International. *Ageing in the Twenty-First Century: A Celebration and a Challenge*; UNFPA: New York, NY, USA, 2012; p. 42.
26. Institute for Transportation and Development Policy. Is Your City Really Made for You? *ITDP*. Available online: https://www.itdp.org/2017/02/14/mag-city-made-for-you/ (accessed on 15 March 2019).
27. Litman, T. *Evaluating Transportation Equity: Guidance for Incorporating Distributional Impacts in Transportation Planning*; Victoria Transport Policy Institute: Victoria, BC, Canada, 2014; p. 20.
28. Berthe, A.; Berthe-Sanou, L.; Konate, B.; Hien, H. Functional disabilities in elderly people living at home in Bobo-Dioulasso, Burkina Faso. *Santé Publique* **2012**, *24*, 439–451.
29. Lima-Costa, M.F.; Facchini, L.; Matos, D.; Macinko, J. Changes in ten years of social inequalities in health among elderly Brazilians (1998–2008). *Rev. De Saúde Pública* **2012**, *46*, 100–107.
30. Capistrant, B.D.; Glymour, M.M.; Berkman, L.F. Assessing Mobility Difficulties for Cross-National Comparisons: Results from the World Health Organization Study on Global Ageing and Adult Health. *J. Am. Geriatr. Soc.* **2014**, *62*, 329–335. [CrossRef] [PubMed]
31. Zunzunegui, M.-V.; Alvarado, B.-V.; Beland, F.; Vissandjee, V. Explaining health differences between men and women in later life: A cross-city comparison in Latin America and the Caribbean. *Soc. Sci. Med.* **2009**, *68*, 235–242. [CrossRef] [PubMed]
32. Gómez, J.; Corchuelo-Ojeda, J.; Curcio, C.-L.; Calzada, M.-T.; Mendez, F. SABE Colombia: Survey on Health, Well-Being, and Aging in Colombia—Study Design and Protocol. *Curr. Gerontol. Geriatr. Res.* **2016**, *24*, 1–7. [CrossRef] [PubMed]
33. Vos, T.; Global Burden of Disease 2015 Disease and Injury Incidence and Prevalence Collaborators. Global, regional, and national incidence, prevalence, and years lived with disability for 310 diseases and injuries, 1990–2015: A systematic analysis for the Global Burden of Disease Study 2015. *Lancet* **2016**, *388*, 1545–1602.
34. Laverty, A.A.; Palladino, R.; Lee, J.T.; Millett, C. Associations between active travel and weight, blood pressure and diabetes in six middle-income countries: A cross-sectional study in older adults. *Int. J. Behav. Nutr. Phys. Act.* **2015**, *12*, 65. [CrossRef]
35. Ng, N.; Hakimi, M.; van Minh, H.; Juvekar, S.; Razzaque, A.; Ashraf, A.; Masud, A.S.; Kanungsukkasem, U.; Soonthornthada, K.; HuuBich, T. Prevalence of physical inactivity in nine rural INDEPTH Health and Demographic Surveillance Systems in five Asian countries. *Glob. Health Action* **2009**, *2*, 44–53. [CrossRef]
36. Vaidya, A.; Krettek, A. Physical activity level and its socio-demographic correlates in a peri-urban Nepalese population: A cross-sectional study from the Jhaukhel-Duwakot health demographic surveillance site. *Int. J. Behav. Nutr. Phys. Act.* **2014**, *11*, 39. [CrossRef]

37. Wandera, S.O.; Kwagala, B.; Ntozi, J. Determinants of access to healthcare by older persons in Uganda: Across-sectional study. *Int. J. Equity Health* **2015**, *14*, 26. [CrossRef] [PubMed]
38. World Health Organization. *WHO World Health Survey: 2002–2004*; World Health Organization: Geneva, Switzerland, 2015.
39. Sleap, B. *The Rights of Older People in Kyrgyzstan*; HelpAge International: London, UK, 2013; p. 5.
40. Knight, L.; Schatz, E.; Mukumbang, C. "I attend at Vanguard and I attend here as well": Barriers to accessing healthcare services among older South Africans with HIV and non-communicable diseases. *Int. J. Equity Health* **2019**, *17*, 147. [CrossRef] [PubMed]
41. Kamenov, K.; Caballero, F.F.; Miret, M.; Leonardi, M.; Sainio, P.; Tobiasz-Adamczyk, B.; Chatterji, S.; Ayuso-Mateos, J.L.; Cabello, M. Which Are the Most Burdensome Functioning Areas in Depression? A Cross-National Study. *Front. Psychol.* **2016**, *7*, 1342. [CrossRef] [PubMed]
42. Curcio, C.L.; Gomez, F.; Reyes-Ortiz, C.A. Activity Restriction Related to Fear of Falling Among Older People in the Colombian Andes Mountains Are Functional or Psychosocial Risk Factors More Important? *J. Aging Health* **2009**, *21*, 460–479. [CrossRef]
43. Kalula, S.Z.; Ferreira, M.; Swingler, G.H.; Badri, M. Risk factors for falls in older adults in a South African Urban Community. *BMC Geriatr.* **2016**, *16*, 51. [CrossRef]
44. Romli, A.S.; Tan, M.P.; Mackenzie, L.; Lovarini, M.; Suttanon, P.; Clemson, L. Falls amongst older people in Southeast Asia: A scoping review. *Public Health* **2017**, *145*, 96–112. [CrossRef]
45. Wismans, J.; Skogsmo, I.; Nilsson-Ehle, A.; Lie, A.; Thynell, M.; Lindberg, G. Commentary: Status of road safety in Asia. *Traffic Inj. Prev.* **2016**, *17*, 217–225. [CrossRef]
46. Mabunda, M.M. Magnitude and categories of pedestrian fatalities in South Africa. *Accid. Anal. Prev.* **2008**, *40*, 586–593. [CrossRef]
47. Odofuwa, O. Enhancing mobility of the elderly in Sub-Saharan African cities through improved public transportation. *IATSS Res.* **2006**, *30*, 60–66. [CrossRef]
48. Amosun, S.L.; Burgess, T.; Groeneveldt, L.; Hodgson, T. Are elderly pedestrians allowed enough time at pedestrian crossings in Cape Town, South Africa? *Physiother. Theory Pract.* **2007**, *23*, 325–332. [CrossRef]
49. Ipingbemi, O. Travel characteristics and mobility constraints of the elderly in Ibadan, Nigeria. *J. Transp. Geogr.* **2010**, *18*, 285–291. [CrossRef]
50. Lecompte, M.C.; Bocarejo, J.P. Transport systems and their impact on gender equity. *Transp. Res. Procedia* **2017**, *25*, 424–425. [CrossRef]
51. HelpAge International. *Gender and Transport for Older People*; HelpAge International: London, UK, 2002; p. 1.
52. World Bank. *Making Transport Work for Women and Men: Challenges and Opportunities in the Middle East and North Africa, Lessons from Case Studies*; World Bank: Washington, DC, USA, 2017; Available online: https://openknowledge.worldbank.org/handle/10986/17648 (accessed on 2 March 2019).
53. UN Economic Commission for Europe. *Inland Transport Committee and Gender Issues in Transport*; UN ECE: Geneva, Switzerland, 2009; Available online: https://www.unece.org/trans/theme_gender.html (accessed on 10 March 2019).
54. Soltani, A.; Pojani, D.; Askari, S.; Masoumi, H.E. Socio-demographic and built environment determinants of car use among older adults in Iran. *J. Transp. Geogr.* **2018**, *68*, 109–117. [CrossRef]
55. Nurdden, A.; Rahmat, R.A.; Ismail, A.B. Effect of Transportation Policies on Modal Shift from Private Car to Public Transport in Malaysia. *J. Appl. Sci.* **2017**, *7*, 1013–1018.
56. Olawole, M.O. Analysis of Intra-Urban Mobility of the Elderly in a Medium-Size City in Southwestern Nigeria. *Mediterr. J. Soc. Sci.* **2015**, *6*, 90–104. [CrossRef]
57. Mulongo, G.; Tewodros, A.; Porter, G. *Impacts and Implications of Gender Mainstreaming in the Rural Transport Sector in Tanzania with Particular Reference to Women with Multi-Dimensional Vulnerabilities*; HelpAge International: London, UK, 2017; pp. 43–47.
58. Madani, M.; Sibai, M.A. Assisted Living Facilities' Accessibility Challenge in the Beirut Area. *Int. J. Architecton. Spat. Environ. Des.* **2017**, *11*, 1–14. [CrossRef]
59. Munshi, T. Out-of-Home Mobility of Senior Citizens in Kochi, India. In *Geographies of Transport and Ageing*; Curl, A., Musselwhite, C., Eds.; Palgrave Macmillan: Basingstoke, UK, 2018; pp. 153–170.
60. Schwanen, T.; Lucas, K.; Akyelken, N.; Solsona, D.C.; Carrasco, J.-A.; Neutens, T. Rethinking the links between social exclusion and transport disadvantage through the lens of social capital. *Transp. Res. Part A Policy Plan.* **2015**, *74*, 123–135. [CrossRef]

61. Lucas, K. Making the connections between transport disadvantage and the social exclusion of low income populations in the Tshwane Region of South Africa. *J. Transp. Geogr.* **2011**, *19*, 320–1334. [CrossRef]
62. Porter, G. Living in a walking world: Rural mobility and social equity issues in sub-Saharan Africa. *World Dev.* **2002**, *30*, 285–300. [CrossRef]
63. Porter, G. Transport planning in sub-Saharan Africa. Progress report 2, Putting gender into mobility and transport planning in Africa. *Prog. Dev. Stud.* **2008**, *8*, 281–289. [CrossRef]
64. Porter, G. Mobilities in rural Africa: New connections, new challenges. *Ann. Am. Assoc. Geogr.* **2016**, *106*, 434–441. [CrossRef]
65. Mudege, N.N.; Eze, A.C. Gender, aging, poverty and health: Survival strategies of older men and women in Nairobi slums. *J. Aging Stud.* **2009**, *23*, 245–257. [CrossRef] [PubMed]
66. Lloyd-Sherlock, P.; Agrawal, S.; Minicuci, N. Fear of crime and older people in low-and middle-income countries. *Ageing Soc.* **2016**, *36*, 1083–1106. [CrossRef]
67. Jones, S.E. *Ageing and the City: Making Urban Spaces Work for Older People*; HelpAge International: London, UK, 2016; p. 9.
68. Harriss-White, B.; Olsen, W.; Vera-Sanso, P.; Suresh, V. Multiple shocks and slum household economies in South India. *Econ. Soc.* **2013**, *42*, 398–429. [CrossRef]
69. Knodel, J.; Teerawichitchainan, B.P. Aging in Myanmar. *Gerontologist* **2017**, *57*, 599–605. [CrossRef] [PubMed]
70. Aboderin, I.; Epping-Jordan, J. *Towards Long-Term Care Systems in Sub-Saharan Africa*; World Health Organization: Geneva, Switzerland, 2017; p. 15.
71. Chepngeno-Langlat, G.; Madise, N.; Evandrou, M.; Falkingham, J. Gender differentials on the health consequences of care-giving to people with AIDS-related illness among older informal carers in two slums in Nairobi, Kenya. *Aids Care* **2011**, *23*, 1586–1594. [CrossRef]
72. Ssengonzi, R. The impact of HIV/AIDS on the living arrangements and well-being of elderly caregivers in rural Uganda. *Aids Care* **2009**, *21*, 309–314. [CrossRef] [PubMed]
73. Hanrahan, K.B. Caregiving as mobility constraint and opportunity: Married daughters providing end of life care in northern Ghana. *Soc. Cult. Geogr.* **2018**, *19*, 59–80. [CrossRef]
74. Plyushteva, A.; Schwanen, T. Care-related journeys over the life course: Thinking mobility biographies with gender, care and the household. *Geoforum* **2018**, *97*, 131–141. [CrossRef]
75. Whitzman, C.; James, K.; Powaseu, I. *Travelling Together: Disability Inclusive Road Development in Papua New Guinea*; University of Melbourne/Ausaid: Melbourne, Australia, 2013; Available online: https://msd.unimelb.edu.au/research/projects/completed/travelling-together (accessed on 21 March 2019).
76. Hartog, J. *Disaster Resilience in an Ageing World: How to Make Policies and Programmes Inclusive of Older People*; HelpAge International: London, UK, 2014; pp. 9–11.
77. World Health Organization. *Older Persons in Emergencies: An Active Ageing Perspective*; World Health Organization: Geneva, Switzerland, 2008; pp. 1–41.
78. Age and Disability Capacity Programme. *Humanitarian Inclusion Standards for Older People and People with Disabilities*; ADCAP: London, UK, 2018; pp. 9–13.
79. Sheppard, P.; Polack, S.; McGivern, M. *Missing Millions: How Older People with Disabilities are Excluded from Humanitarian Response*; HelpAge International: London, UK, 2018; pp. 4, 32.
80. United Nations. *2030 Agenda for Sustainable Development*; United Nations: New York, NY, USA, 2015; Available online: https://www.un.org/sustainabledevelopment/development-agenda/ (accessed on 3 march 2019).

*Article*

# Meeting Young People's Mobility and Transport Needs: Review and Prospect

**Gina Porter [1,*] and Jeff Turner [2]**

1 Department of Anthropology, Durham University, Durham DH13LE, UK
2 Gender, Inclusion & Vulnerable Groups, High Volume Transport Research Programme, IMC Worldwide Consultants, London RH1 1LG, UK; jeffreymturner@hotmail.com
* Correspondence: r.e.porter@durham.ac.uk

Received: 18 August 2019; Accepted: 25 October 2019; Published: 6 November 2019

**Abstract:** This paper reviews published and grey literature on young people's daily transport and mobility experiences and potential, with the aim of identifying major research gaps. It draws on literature across a range of disciplines where interest in mobilities has expanded significantly over the last decade (transport studies; social sciences, notably geography and anthropology; health sciences). We focus particularly on young people from poorer households, since poverty and mobility intersect and interact in complex ways and this needs closer attention. Although youth transport issues are set in their global context, the focus on poverty encourages particular attention to low- and middle-income countries (LMICs), especially countries in Africa and Asia. Key themes include education, employment, travel safety and the role of mobile technology. This review demonstrates how young people's travel experiences, needs and risks are embedded in power relations and vary with gender, age and location. It also points to the scale and range of uncertainties that so many young people now face globally as they negotiate daily mobility (or immobility). Significant research gaps are identified, including the need for more in-depth action research involving young people themselves (especially in Asia), and greater attention to the impact of mobile technologies on travel practices.

**Keywords:** children; mobility; transport; Africa; Asia; youth voice; school; work; road safety

---

## 1. Introduction: Young People's Mobility and Transport in a Rapidly Changing World

Young people remain a remarkably neglected constituency in the world of transport planning globally (with the exception of road safety), despite the vital importance of their access to education, health and other services for our future progress and for sustainable development. This omission is of particular concern in the many low- and middle-income countries (LMICs) where well over half the population is under 30 years old and, in some, over half the population is under 18 years old. Two particularly pertinent trends for the discussion that follows are expanding urbanisation and expanding youth populations. It is estimated that by 2030, 60 percent of the world's population will live in cities and that 37 percent of the world's population will be under the age of 20 [1]. There will be 1.3 billion young people aged 15–24 years by 2030 globally [2]. Given the size and importance of the youth constituency, and growing concerns around the potential political risk implied, especially by large concentrations of un/underemployed youth living in cities, young people's transport and mobility needs require urgent attention from policy makers.

Access to services and facilities has massive implications for young people, not only in terms of their current wellbeing but also for their livelihood potential and life chances. The importance of transport for urban young people is specifically recognised in Sustainable Development Goal (SDG) 11.2: provide access to safe, affordable, accessible and sustainable transport systems...with special attention to the needs of those in vulnerable situations (specifically including children). However, better access to mobility/transport will be crucial to achieving almost all the Sustainable Development

Goals: in particular, SDGs 3, Good health and wellbeing, 4 Quality education, 5 Achieve equality and empower women and girls, 8.5, Full and productive employment and decent work for all...including for young people, and 8.6, Substantially reduce the proportion of youth not in employment, education or training. It is unlikely that current inter-generational cycles of poverty can be effectively broken without significant, sustained attention to the mobility of young people and their access to transport worldwide.

This paper reviews the available published and grey literature on young people's daily transport and mobility experiences and potential in order to identify major research gaps. The review draws on the authors' extensive personal knowledge of published and grey literature in this field (in which both have been active participants since the 1990s), searches in the international transport/mobility journals (for instance the Journal of Transport Geography, Transport Policy, Transport Reviews), child and youth-focused journals (for instance Children's Geographies and Children, Youth and Environments) and Google Scholar, plus Web-of-Science searches undertaken for the period 2000–2019 using the following search terms: Young people/Children/Youth AND Transport/Mobility AND Africa/Asia/Latin America. Overall, references in Web-of-Science are far fewer for Asia than for Africa and even fewer still for Latin America: for instance, a search on Young people AND mobility AND Africa produces 63 references, but only 19 for Young people AND mobility AND Asia (and just 8 for Latin America); Children AND mobility produces 176 references for Africa, 62 for Asia and 28 for Latin America. Literature specifically focused on young people with disabilities was reviewed separately in a companion paper by Maria Kett.

There is insufficient space to review wider mobilities associated with migration in this paper: the focus is specifically around the daily mobilities that shape young people's access to education, work and general well-being. Similarly, space precludes consideration of the substantial evidence regarding the importance of transport services for the health of children and young people beyond the traffic accident issues noted in this paper. Consequently, the impacts of transport and mobility on vaccination, maternal health, access to TB, malaria and eye treatment, ARVs and sexual and reproductive health are not included here. Similarly, the effects transport and road infrastructure have on health professionals' decisions regarding work place selection are also omitted. These latter (health-related) topics have been covered in other publications of the first author, who has undertaken field research and published widely in the field of daily mobility, transport and access to services for vulnerable populations in Africa since the late 1970s and with specific reference to children and young people since 2000.

The paper draws on literature from a diversity of disciplines, extending from transport studies and health sciences through to the social sciences (notably geography and anthropology) where interest in mobilities research has expanded significantly over the last decade. Where possible, the spotlight throughout is on young people from poorer households, since poverty and mobility intersect and interact in complex ways and this needs far closer attention. Youth transport issues are set in their global context but with particular reference to LMICs, especially countries in Africa and Asia. Africa is the focus of particular attention in this paper because it is demographically the world's youngest continent: by 2050, the estimated number of young people entering the labour force in Africa will exceed that of the rest of the world combined [3]. This fact may help explain why there is a more extensive published and grey literature pertaining to Africa than to Asia.

The paper addresses seven topics, the first of which briefly explains the specific age focus of interest, while the following six topics represent key aspects of recent research/debate that can be drawn from the literature reviewed here:

- Age categories/definitions: Age groups commonly considered under the 'young people' heading, with particular reference to those age groups commonly incorporated into policy discussions on young people and transport.
- Youth voice: Young people in governance and decision-making and their roles as actors in the wider development arena, with particular reference to the transport sector and the potential to promote their voices in transport planning and policy.

- Education: Transport and the education sector (including affordability, student discounts on urban informal public transport, harassment).
- Employment: Transport as youth employment and associated space for entrepreneurial and innovative practices; transport as employment and entrepreneurial constraint.
- Safety: Road safety and other aspects of young people's personal mobility safety.
- Mobile technology: The interaction between mobile technology and travel among young people (with particular reference to its potential to aid delivery of youth-equitable urban transport services).
- Gender: Mobility is relational—it is embedded in power relations—and gender, like age, is a critical shaper in this respect. Attention is paid throughout to gender as a critical cross-cutting factor shaping the transport and mobility experiences of the focus age groups in this review.

## 2. Age Categories/Definitions: Who Is Included in the Category 'Youth' and 'Young People'?

Various terms are used when identifying young people—'child', 'young person', 'youth'. Formal definition of these terms may vary according to a country's legal code or to a particular institution/agency's decision, but it is also often informally shaped by cultural context: all these categories are, in essence, social constructs. The papers reviewed below adopt varying age definitions when talking about children and youth so it is not possible to work with any one specific definition. We note, however, that childhood is commonly associated with the period prior to puberty, while 'youth'—particularly in papers with an African focus—now often includes reference to people well into their 30s. This accords with the fact that growing levels of unemployment in Africa and the cost of setting up an independent household are putting many young people into an increasingly prolonged period of dependency [4,5].

## 3. Youth Voice

Young people's role in governance and decision-making globally is small but it is beginning to expand rapidly, associated with a growing recognition of the potential political risks associated with their unmet needs. An increasingly vocal mass of youth is drawing on its expanding expertise in social media to help build extensive communities of interest and promote its growing demands for better representation in governance, whether in local, national or international contexts, and this is well characterised by youth usage of social media in the 'Arab spring'. Youth parliaments, mushrooming across the world (for instance, in India from 2002; African Youth Parliament of 50 countries from 2003 onwards, etc.), are a good example of the efforts currently being made to support and enhance youth voice.

In the transport planning sector, across the globe, young voices have played only a very small part to date (perhaps in part because transport planning has been seen largely in terms of motor-mobility and young people are usually not legally allowed to drive until their mid/late teens) [6]. However, demands for social inclusion and mobility justice are growing [7,8]. Even in those rare cases where transport has been planned with attention to youth needs (mostly in developed country contexts), it has very rarely been planned with youth. Participatory planning has mostly been viewed by officialdom as a process of engagement with adults. However, given increasing recognition of the need for policy change towards low-carbon sustainable transport (more emphasis on public transport, cycling and walking), youth transport strategies that directly engage with the youth of today will be of vital importance: new travel behaviours are more likely to be learned and adopted by young people [9].

The (albeit very limited) literature available shows how young people can provide vital understanding of their transport challenges (which may be inaccessible to adults) and have valuable insights into how these can be best addressed [10]. In LMICs, what appears to have been the first participatory research with children and young people (aged 9–18 years) on transport issues started in 2004 with pilots in India, Ghana and South Africa, followed by a major study across 24 sites in Ghana, South Africa and Malawi. This incorporated in-depth qualitative research taking an ethnographic approach plus an extensive survey (N = 3000) [11–16]. The research went beyond adults simply asking young people for their views on transport issues. In the first place, it involved training 70 school children (aged 11–19 years) as co-researchers to identify key issues and questions, mostly through in-depth interviews, photo-diaries and small surveys with their peers. This study, which was followed up by a further phase of academic-led qualitative and quantitative research in each site, drawing on the key issues and questions raised by the young co-investigators, has arguably enabled a much stronger understanding of significant daily mobility issues affecting young lives than might have been achieved through a conventional academic research study alone.

Subsequently, there has been a slow turn towards promoting young people's voices in the transport sector in LMICs through the adoption of participatory approaches, including Simpson and Collard 2019 [17]. However, the constraints which tend to minimise poorer women's voices (not least supposed lack of competence to contribute to what is perceived as the domain of middle-class male engineer experts) continue to play an even stronger role globally in the case of children and young people [14,18,19].

## 4. Education: Making the Journey to School

There is substantial literature around the journey to school in the Global North but much less work in the Global South. In the Global North, focus has moved from road safety to expanded attention to the decline in children's independent mobility associated with increasing car use and growing concern with child obesity and the need to expand young people's active travel (walking and cycling) [20–22]. A similar trend can now be observed in the Global South [23]: among middle-class urban children, obesity issues are also emerging, with clear linkages to the expansion of motor transport usage for school journeys [24–27]. In China, in particular, the one-child policy encouraged many parents to adjust their life patterns to provide better education for their children, including chauffeuring their children to and from school if they could afford to do so [28]. In a Nairobi context, parental perception of positive neighbourhood social cohesion, positive environs and connectivity, all of which reduced their child safety concerns, encouraged positive child physical activity outcomes [29]: these factors are likely to be pertinent in many LMIC urban contexts [30]. While programmes in many developed country cities now promote the idea of 'walking school buses' and 'bicycle trains', in order to encourage more active travel [1,31], in LMICs, these are much rarer, although there have been experiments, notably in South Africa and Dar es Salaam [32].

Travel to school experiences in LMICs, especially among children from lower-income families, tend to differ substantially from the norm prevailing in many countries of the Global North (where parental escorting of younger children to school is common and private motorised transport often dominates) [33]. In many LMICs, escorting is mostly a task assigned to older siblings and the journey is far less likely to involve private transport. For children from poorer households, in particular, challenges around physical access to school are closely aligned with issues of journey time and distance—in Africa, mostly entirely conducted through pedestrian travel, in Asia by either walking or cycling—and its associated dangers [14] (chapter 3) [34]. Information per se—and particularly qualitative information—on school travel experiences is far more extensive for African than Asian contexts where, in recent years, the focus has been principally on analysing large-scale survey data. Consequently, Africa dominates the following discussion.

### 4.1. Travel from Rural Homes to Schools and Training Opportunities in Urban Centres

Globally, secondary schools, further, and tertiary education are likely to be located in urban centres, restricting opportunities for skills development and the take-up of learning and training opportunities in more remote rural areas [35]. Transport and travel are likely to play a crucial 'cause and effect' role in exacerbating poor skills and low productivity, especially in contexts where transport density is low and subsidised transport is unavailable. In rural Africa, where transport availability is much sparser and more costly than in Asia, the impacts are particularly significant [36]. Thus, in Ghana, it has been observed that the closer the secondary school, the more likely that children are sent to primary school, as continuity of the child's education is feasible [37].

Across sub-Saharan Africa, children will be seen trudging each early morning and mid-afternoon along major highways as traffic hurtles past. Such long daily walks to and from school due to lack of or high cost of transport bring attendant problems of lateness and encourage late 'over-age' enrolment (especially of girls), truancy and early drop-out. The common alternative of 'home boarding' in town (renting a room alone in the absence of sufficient cheap school boarding places) is particularly fraught with dangers for young girls, who are often targeted for sex by predatory older men [14]. Data from the 2003 South Africa National Household Travel Survey indicates that just over three-quarters of 'learners' walked to school and just under one fifth of the 16 million total spent over one hour each day walking to and from their schools, which are commonly located in smaller or larger urban centres [38].

In remoter locations of rural Asia, poverty similarly forces many pupils to walk to school, despite the greater availability of regular transport than in most of sub-Saharan Africa. In such circumstances, in Bangladesh, Matin et al. [39] described how it is the norm for such children to walk in groups to school for safety reasons, just as they do in Africa. Unsurprisingly, parents in India, Thailand and China are reportedly more likely to allow girls to carry on with their education where transport services are reliable and safe [40]. In Delhi, India, where sexual harassment on public transport is now a notorious issue for women, a correlation has been found between young women's college choice and their perceptions of route safety, despite the higher transport fares thus involved (though without evidence of direct causal connection between the two [41].

It is also important to note that the performance of children and young people at school or in training programmes may be shaped by transport constraints at home. Such transport failures, particularly (but not only) in rural environments, may require young people to undertake a series of tasks before they leave for school or after they return home: these commonly include carrying water and (less regularly) firewood, together with other domestic work such as cleaning, sweeping, washing clothes and caring for younger siblings [42].

#### 4.1.1. School Travel within LMIC Urban Areas

Within LMIC urban areas, motorised transport is available, but fare costs limit its use by school pupils. This is a major factor shaping the very heavy dependence on pedestrian travel (together with bicycle travel in Asia). In urban Africa, generally well over 90% of pupils travel to school on foot: 96% in a Nairobi slum according to Salon and Gulyani [43], 98% across eight diverse Ghana sites and 99% across eight diverse Malawi sites studied by Porter et al. [14] (chapter 3). Journey times can be far more substantial than in Asia given such high levels of dependence on pedestrian travel and the low availability of cycles to school pupils [14] (chapter 3). In peri-urban neighbourhoods in Ghana, Malawi, South Africa and Kenya, pupils report walking considerable distances, especially where there is a preference for a (usually better) school, or where the nearest schools are full [14,43]. In Soweto-Johannesburg, for instance, over a third of primary-school-aged children were found to travel more than 3 km one way to school, nearly two-thirds attended schools outside of the suburb where they resided, and only 18% attended the school closest to home [44]. School quality considerations are often a significant factor shaping school selection, even among relatively poor households. See also Hunter's research in Durban [45]; he makes the point that this applies to poor—but not the poorest—households.

In urban Asia, the sparse published material available suggests that school pupils make more substantial use of informal public transport and of bicycles. However, as cities expand in Asia, the school commuting distance is inevitably growing, as Li and Xhao and Zhang et al. showed using data from Third and Fifth Travel Surveys of Beijing Inhabitants, respectively (with some children travelling over 5 km) [28,46].

#### 4.1.2. Hazards of School Travel

Walk-along interviews with pupils are particularly powerful in demonstrating the fear of harassment and attack for both boys and girls that are widespread in low income urban and rural neighbourhoods, as Porter et al. showed for Ghana, Malawi and South Africa [47]. See also Phillips and Tossa 2017 [48] for child-led walks in Thailand. Overall, children from particularly deprived neighbourhoods tend to face the most constraints on their movement, as Adams et al. observed in a study across urban and rural neighbourhoods in Western Cape [49]. But even in somewhat less deprived neighbourhoods, children's mobility can be restricted by diverse factors, as Benwell pointed out with reference to baboon troops, domestic 'guard' dogs, traffic and the impact of family composition in suburban Cape Town [50].

Across LMICs, the wider hazards of school travel by motor transport are numerous, whether dedicated school transport or informal public transport is utilised: overcrowded poorly maintained vehicles without seatbelts are driven by poorly trained drivers over poorly maintained roads. The situation is no different today across most of Africa and South Asia than was observed in rural Brazil two decades ago [51]. However, in South Africa, where the Safe Travel to School Programme was recently implemented by a national child safety agency, there have been some indications of improved practice [52]. This has stemmed from a focus on driver road safety awareness, defensive driver training, eye-testing, vehicle roadworthy inspections with selected upgrades, incentives for safe performance, and implementation of a vehicle telematics tracking system with regular, individual driving behaviour information updates.

#### 4.1.3. Subsidised Pupil Travel

Subsidised transport for pupils, as an effective policy measure to increase access to education, is far less widespread in LMICs than in the Global North and seems unlikely to advantage them significantly in practice in many locations. Pupil reports in Cape Coast, Ghana, of being forced off the bus when seats are needed for full-fare passengers are unlikely to be unique. In Dar es Salaam, where, under a government scheme, children pay 33% of the adult fare (but without compensation arrangements to operators), children are often unable to even board the bus in the first place, being barred by the conductors [53]. Similarly, they face exclusion from buses in Karachi because of the requirement there to charge only half-fare [54]. However, across the globe, pupil transport subsidy remains a sensitive, complex issue [55].

#### 4.1.4. Gender Issues in School Travel

Girl pupils face even higher transport constraints and hazards than boys in LMICs and this contributes to girls' lower school enrolment rates. A review in Niger utilizing DHS surveys noted that only 41 girls per 100 boys were at school in rural areas (as opposed to 80:100 in urban areas), and pointed to distance to school as a key factor behind this difference [56]. Improved road access and associated availability of transport appear to have the potential to improve girls' school attendance significantly in some contexts. In Morocco, assessment of a major road programme (National Rural Roads Program (NRRP-1)) entrusted to the National Highways Authority showed that the probability of girls' attendance at primary school increased by 40% with the opening of a paved road [57–59]. Road improvements in Asia appear to have similarly improved girls' access to education, even more than boys, as Mohsin et al. showed for Bangladesh, and Pilgrim and Chanrith for Cambodia, in the latter case with benefits seemingly accruing principally following improvements to a provincial road and a national road as opposed to purely local rural roads [60,61].

The negative impacts of distance, coupled with poor transport, on rural girls' education in Africa and South Asia can be related to a number of factors: girls' heavy household duties (typically heavier than boys' duties); negative cultural perceptions concerning female mobility and girls' education; also perceived dangers for girls who travel a long way to school or alternatively must board far from home in the school neighbourhood (as noted above) [14,62]. The safest travel procedure in African rural and urban contexts, especially for girls, tends to be to walk together, as a group. However, when there is heavy traffic on urban streets and no separation of pedestrians from motorised transport, such group walks have the potential to cause traffic accidents [14] (Chapter 7). The first cause of death among children aged 5–14 years and young adults 16–29 years is road traffic injury [63] (a point further discussed in the Road Safety section below). Data for low income countries are inadequate but the limited data available indicate that the vast majority of young people in such environments are pedestrians rather than passengers in motor vehicles when they are injured [14].

Even where regular transport is available (mostly in urban contexts), its cost and the potential for harassment tends to impact more strongly on (usually less well resourced) girls [14]. To date, there is less research on gendered patterns of school travel in Asian contexts but survey data from Kanpur City, India, suggests that girls are less likely to travel independently than in sub-Saharan Africa: they are either dropped off from personal vehicles by richer families or accompanied by mothers in families without such transport [64].

Cycles play a relatively small role in school travel scenarios across much of Africa, in particular, and there is much evidence across Africa and Asia to show that critical mass is essential to their widespread use, especially among girls [65,66]. Increasing emphasis on low-carbon transport may help future efforts to promote cycling to school (crucially, if accompanied by training on cycle riding and repair). However, if girls are unable to attend school because of the demand for domestic porterage, as is often the case in Africa, broader Intermediate Means of Transport (IMT) interventions such as push carts aimed at the family may be of greater significance. In Pakistan, a World Bank pilot offered a stipend to girls living a long way from the nearest school [67]. Another initiative which could help protect children against harassment and rape is the development of a 'walking bus' (whereby adults chaperone children along a set route, acting as driver and conductors), if this is adapted to local context [32].

## 5. Youth Employment: Transport to Work and Transport as Work

On average, unemployment is likely to be three times higher for youth than for adults globally, but five times as likely among youth as among adults in South Asia and South-East Asia [68]. Even in situations where open unemployment among youth is relatively low, as in most low income countries in sub-Saharan Africa, there is much 'working poverty': temporary, low-paid work in the informal sector with poor working conditions [69,70]. Transport's relationship with youth employment has two significant components which are explored in this section: firstly, its role in getting young people to work places (whether places of employment or places where they are building their own enterprise), secondly, with regard to employment within the transport sector.

### *5.1. Transport to Work*

Firstly, with regard to accessing employment opportunities (whether provided by others, or of their own making, as, for instance, in the case of much petty trade), young people in both rural and urban locations face considerable challenges. Transport availability, reliability and cost are key factors shaping young people's access to employment. In LMIC rural areas, where employment outside agriculture is often seen as key to improved incomes, accessing non-agricultural employment tends to require long daily journeys to the nearest urban centre. In Western Kenya, this has led to the depressing picture of rural youth 'tarmaccing', as young men trudge along pot-holed (tarmac) roads from rural areas to the city in search of work and back again home when they cannot find employment [71].

Even in urban areas where transport availability is much higher, the distance from affordable suburban dwellings to central employment areas (including potentially more lucrative trading locations in the case of young entrepreneurs) can be a particular challenge for young people in poor households. This is a factor that affects poorer young people's access to work opportunities in cities across the globe [72], but in LMIC cities, transport constraints are often particularly limiting. There, the urban poor often live in unplanned settlements and slums on the periphery of cities while employment densities are commonly greatest within the central area. Transport systems (including road infrastructure, formal or informal motorized transport, and non-motorized transport) linking to these opportunities are often extremely congested, unreliable and unsafe [1]. In cities such as Lagos and Manila, journeys from periphery to the city centre can take hours, especially in wet weather when transport demand everywhere increases and potholed roads become impassable. Moreover, the informal para-transit that dominates motorised travel in such contexts is simply unaffordable for the very poor. Unsurprisingly, many young women run small businesses from their own home, at least in part so that they do not have to travel out of the home to work.

In South Africa, poor access to transport is one of a complex mix of factors that results in young people actually stopping actively searching for work. When they leave school, they may have high hopes of finding good employment, but such dreams tend to be quickly dispelled as job applications fail. 'Actively discouraged' youth not seeking work is estimated at 61 per cent in the 20–24 years cohort: a response to feelings of hopelessness and despair [73]. Elsewhere in sub-Saharan Africa, where the informal sector is much larger and there are no social grants, most young people tend to continue to search for work and to take on whatever tasks they can find because, as is so accurately observed by Filmer et al., 'most Africans simply cannot afford to be idle' [69] (p. 3). This comment also applies to many Asian contexts. However, Jeffrey's observations in a north Indian city, where neoliberal economic change had cut back employment opportunities for educated (lower-) middle class young men, suggest conditions somewhat reminiscent of South Africa [74].

In studies of urban transport provision in 1990s Accra, which included specific consideration of its impact on young people, Grieco et al. and Turner and Kwakye, showed how the falling off in transport provision associated with the economic reform measures in place at that time (increased cost of vehicles and spare parts due to devaluation raising the cost of imports, etc.) actually increased dependence on the work of young women and children [75–77]. Children had become increasingly central to the economic organisation of households and would be taken in as foster-children to reduce the transport stress of middle-aged adults faced with transport under-provision: they acted as domestic anchors, compensating for the absence of adult household members delayed in distant markets by transport problems. Additionally, children experienced high levels of local mobility due to domestic tasks required of them, such as refuse removal, water and fuelwood collection and other activities including petty trade. Such conditions still prevail in Accra and probably also in many other African and Asian cities where traffic congestion is high and transport provision poor. Sibling care, especially in AIDS-affected households, adds to the pressures faced by many girls across LMICs [78].

### 5.2. Transport to Work for Young Women

For young women living in the poorer households of LMICs, conditions are often particularly difficult in the absence of appropriate inexpensive and timely transport, as Venter et al. demonstrated for rural South Africa and Esson et al. for Accra, urban Ghana [70,79]. Cultural barriers to mobility vary (see Kjeldsberg et al. regarding variations in rural Nepal [80]) but for some young married women, these barriers can be insurmountable, especially if they cannot find reliable transport for their return journey home. This tends to be particularly important because of consequent delays to evening meal preparations, possibly coupled with male suspicions of the reason behind their delay. The *unreliability* of transport is a common but under-reported factor constraining young women's trade and entrepreneurial aspirations in both rural and urban sub-Saharan Africa [81,82]. This uncertainty/unpredictability may encompass not only concerns about how long the journey may take, but whether the journey can be done at all. Moreover, it can as significantly affect the young urban woman trying to establish a regional trading business in farm produce or fish as the rural woman hoping to sell her perishable plantain or cassava at a city market. Uncertainty with regards to the transport of the perishable foods that so often underpin young women's early efforts to build a trading business can have a particularly stultifying impact on emergent entrepreneurship.

In urban areas, women often must forego potential travel to find and engage in work because males in their households have first priority on sparse funds for transport fares. Young women are more likely to be unemployed and, if they manage to find work (whether working for someone else or in their own enterprise), walk to their place of employment or undertake this work from their own home. The work tends to be predominantly in very low-paid service-related informal sector jobs. Globally, there is a tendency for women to focus on more local (often less well paid) employment opportunities in their neighbourhood because of the financial costs (and also often the time costs, given family caring demands) [83]. This is particularly evident in LMIC cities such as Nairobi [43], Accra [70], Delhi [84] and Tunis (author fieldwork, 2019). As Langevang and Gough emphasised with reference to Accra, it is important to reflect on young people's movements as tactics of social navigation, recognising the importance of spatial mobility to young people's everyday well-being and their processes of social becoming [85].

In some better-provisioned cities in China, both women's and men's journeys for employment are seemingly less arduous. In Shanghai, for instance, only a small percentage of work journeys (13.0%) are longer than 60 min [86]. However, there, the dominantly residential zones are associated with service jobs and it is likely that women's work will predominate in such contexts. Poor access to education among girls and women in many low-income contexts meanwhile limits their ability to read maps and bus information so they unsurprisingly feel safer working close to home, as has been described in both southern Ghana and Buenos Aires [87,88].

### 5.3. Young People's Employment in the Transport Sector

In LMICs, the transport sector often provides an employment niche for the poorest, including young people. In both rural and urban areas, inadequate or costly transport can encourage adults to look to children to fill the 'transport gap'. This dependence on youth porterage, which can contribute substantially to children's time poverty and deficiencies in schooling, is still regularly overlooked in both transport and education research. There is still insufficient detailed information about children's work as load carriers apart from studies of the 'kayayoo' girl porters in Accra, Ghana [76,89] and the research referenced earlier conducted with young people 9–18 years across 24 urban and rural sites in sub-Saharan Africa, which incorporated research on load-carrying and its (negative) impacts on education and well-being [14] (chapter 4) [42].

Employment in the transport sector is highly gendered across LMICs. Beyond their mid-teens, boys rarely carry (nor are they expected to carry) loads in domestic contexts (e.g., water carrying for the household): this is considered work for women, girls and young boys [14,90]. Porterage of goods for commercial purposes is a different matter and many young men also work as push truck operators, as, for instance, in Ghana's urban market centres, usually for very low returns [75]. Other work in the transport sector involving motor vehicles, by contrast, is widely perceived across the LMICs as belonging firmly within the masculine domain. There is occasional publicity around women taxi drivers such as Ghana's MissTaxi [91] and India's motorbike taxi service Bikxie [92] but these are extremely rare, not least due to perceived safety and security issues where women operators are concerned.

For many young men in Africa and Asia, it is the motorcycle or tricycle taxi which has become the most important employment opportunity in the transport sector. Operating a motorcycle-taxi (known as *boda-boda* in much of East Africa, *okada* in much of West Africa) can offer them a relatively lucrative livelihood, whether as independent riders or, more commonly, through a renting arrangement with the motorcycle owner (usually an older, better resourced man or woman). Young men are, in some cases, demonstrating significant entrepreneurship as they move out of less lucrative activities (for instance, charcoal production in Kibaha district, Tanzania) and into motorcycle taxi operations. However, negative impacts are widely reported in some regions: these may include not only reckless driving and increased accident rates, but also violent crime and expanded sexually transmitted diseases (STDs). The latter is, in part, a product of relatively high incomes and consequent high bargaining power for sex, as noted early in their expansion by Nyanzi et al. for south-west Uganda [93] and Waage for Ngaoundere, Cameroon [94]. In rural areas of Lao PDR, Doussantousse et al. similarly found that motorbikes and mobile phones had expanded the sexual territory of indigenous youth at a time when international commerce and a cash economy along improved highways were bringing new people into the region [95]. Among the concerns for their health and safety are at-risk behaviours involving alcohol and sexual practices, especially HIV and sexually transmitted infections. Such issues have led to much government concern, such that, in Ghana, motorcycle-taxis are still banned nation-wide and many countries have city-centre bans in place. However, such bans do not take into account the crucial level of access that motorcycle taxis deliver for people in informal and peripheral urban areas, such as the satellite towns around Abuja in Nigeria. There, the personal mobility they deliver to young people wanting to be independent is widely appreciated not only by young men but also by the many young women passengers who use them extensively (author fieldwork 2019; see also Adamu regarding the impact on northern Nigerian women of *shari'a*-related campaigns to stop them riding commercial motorcycle-taxis [96]).

There is a rapidly growing literature on motorcycle taxi operations by young men (rarely are women involved, except as passengers), for example Burge on young male entrepreneurs in Sierra Leone, Olvera et al. on west and central African cities, and Jenkins and Peters on post-conflict Liberia [97–99]. An extensive review of the recent literature on this theme is now available [100]. It is important to note that motorcycle maintenance and repair is a growing support industry too, both for private and commercial motorcycles, but again mostly employs men.

## 6. Road Safety and Other Aspects of Young People's Personal Mobility Safety

### 6.1. Road Safety

The bare bones of the global road safety issue are clearly presented in a recent World Health Organisation (WHO) road traffic injuries factsheet [101]. This demonstrates that, while globally, people from lower socioeconomic backgrounds are more likely to be involved in road traffic crashes, more than 90% of road traffic deaths occur in LMICs and road traffic injury death rates are highest in the African region (at 26.6/100,000 people), followed by South-East Asia (20.7/100,000 people). Africa has the highest proportion of pedestrian and cyclist mortalities at 44% of deaths: unsurprisingly, pedestrians, cyclists and riders of two- and three-wheeler motorcycles are especially vulnerable as a result of being less protected than car occupants. However, vulnerable road users are still largely ignored in the planning, design and operation of roads. Across Africa and Asia, most roads still lack separate lanes for cyclists or adequate crossings for pedestrians, while motor vehicle speeds are too high [102].

So far as children and young people are concerned, the statistics are particularly sobering: road traffic injuries are the leading cause of death globally for the 5–29 years cohort. The vulnerability of younger children to road traffic relates to their physical, cognitive and social development stage compared to that of adults. Given their small stature, they may find it difficult to see surrounding traffic and for drivers, in particular, to see them. They are likely to have more difficulty judging the proximity, speed and direction of moving vehicles. Impulsivity and a shorter attention span could affect their ability to cope with simultaneous events. In a road traffic crash, their softer heads will make them more susceptible to serious head injury than an adult. Adolescents, meanwhile, are found to be especially prone to take risks that compromise their road safety [63].

In urban areas, most of those injured in public transport accidents are either paratransit passengers or pedestrians (commonly including young people trying to hawk goods between slow moving traffic). In Ilesa, south-west Nigeria, a small study of child accident victims found that the majority (89%) were pedestrians and most were over 5 years old; 60% of them were injured either while hawking by the roadside or when undertaking an errand [103]. Twenty per cent of cases involved motorcycles. This excludes potentially wider damage to young people's health in urban areas associated with vehicle-generated air pollution.

During interviews in the 24-site child-mobility study across Ghana, Malawi and South Africa, teachers were questioned about road safety education. Whatever national programmes exist, their evidence seems to suggest that many children in school obtain little quality road safety training [14] (chapter 8). This mirrors conditions across many LMICs, despite the level of traffic injuries sustained by children. However, efforts have been in progress in a number of countries to promote road safety, funded, for instance, by the UK Department for International Development (DFID) and the Danish International Development Agency (DANIDA) and, since 2012, through the International Automobile Federation (FIA) Road Safety Grants Programme. The FIA projects aim to meet the objectives of the UN Decade of Action for Road Safety (2011) of halving the number of deaths and injuries from road traffic accidents by 2020, with active programmes in diverse LMIC countries including Tanzania, Nigeria, Brazil, Morocco and China [104].

Recent research in the Global North has drawn attention to the value of training parents about road safety [105] but in LMICs, where many children travel to school and other locations without parental accompaniment (see above), this is unlikely to have a significant impact. There, early training of children on pedestrian road safety is crucial. Salmon and Eckersley proposed that to become skilled pedestrians, children need to move 'beyond a view of traffic as rule-bound and develop dynamic adaptable strategies for crossing roads' [106] (p. 729). The reportedly successful local programme they developed in Ethiopia, based on the UK Kerbkraft concept, entailed practical exercises on local streets, enabling children to develop techniques for identifying safe crossing-places. FIA foundation projects such as South Africa's Safe Schools project take a similar approach [107]. A recent project by the Non-Governmental Organisation (NGO) Amend in a Lusaka school that includes addition of a raised platform pedestrian crossing, footpaths, fencing and a school zone warning, removal of vehicle parking which blocked sight lines, and reduced operating speeds of passing vehicles, has reportedly had significant impact [108]. Other examples show how high the returns from such investment can be: in Korea a school zoning scheme, together with improved school bus regulation and road safety training schemes, reportedly reduced traffic accidents among children under 14 years of age by 95% between 1998 and 2012 [101]. However, finding means to bring road safety training to the many children who either never attend school or leave before the year in which road safety training is introduced, is also vital [109]. Here, road safety NGO interventions that support short courses for groups such as young traders (at particular risk as they rush to vehicles to sell, darting across roads and within the path of other vehicles) would be extremely valuable [14] (chapter 8).

Motorised traffic is also a growing danger for those who operate it or travel in it. In Cambodia, Kitamura et al. argued that speeding by young people is promoted by road improvements that occur alongside underdeveloped traffic legislation and limited public awareness and knowledge of road safety [110]. They emphasise the importance of implementing the "three Es", namely Engineering, Enforcement and Education in low income countries such as Cambodia but note that the role of education to increase people's road safety awareness is neglected compared to the other two dimensions. Across LMICs in Africa and Asia, there are widespread issues associated with poorly regulated (privately operated) public transport, limited vehicle maintenance, deficiencies in law enforcement, high traffic mix and little separation of vulnerable road users from high speed motorized traffic. When coupled with the lack of seat belts, overcrowding/standing passengers, poor road infrastructure and, overall, very hazardous road environments, it is clear that the conditions in which young people navigate the city are often potentially lethal. The dangerous practice of transporting passengers in the cargo area of light delivery vehicles (LDVs) also occurs in many countries: one small study in South Africa found that 35% of passengers treated for injury following ejection from the vehicle were children under the age of 18 (and 11% sustained a permanent disability) [111].

Much recent attention has been paid to the high level of traffic injury associated with motorcycles (mostly driven by young men). Problems associated with a lack of adequate body and head protection (given that helmets are uncomfortable in high temperatures and often of sub-standard manufacture), poorly regulated vehicle and driver safety and the preponderance of young male drivers with a taste for speed, are exacerbated when two, three or more passengers are riding pillion. Phone use when operating a vehicle adds to these hazards [112]. Air pollution from motorcycles is also a growing issue in densely populated urban areas, especially in Asia.

Finally, it is important to note that where gender-disaggregated data are available, road injury patterns show a significant gender dimension, with nearly three quarters (73%) of all road traffic deaths occurring among males under the age of 25 years: they are almost three times as likely to be killed in a road traffic crash as young females [101,113,114]. Unfortunately, in LMICs in Asia and Africa, adequate gender-disaggregated data are still regularly missing from road safety studies. This is likely to be partly a factor of overall poor reportage of road injury by accident victims, their families and carers, the police and hospitals [102]. One review of published and grey literature on road traffic injury in urban sub-Saharan Africa among young people (≤19 years) suggested that boys and young men were twice as likely to be involved as girls and young women [115]. A recent study of primary school pupils in low-income neighbourhoods in Cape Town found that older boys (10–15 years) were most at risk of experiencing a severe pedestrian injury [116]. In India, data for 2014–2016 show that females represented only around 15% of road accident victims [117]. Gender imbalance in RTIs in LMICs, as globally, appears to be associated not only with higher male access to and use of road transport and higher male mobility overall, but also to gender variations in attitude to risk.

### 6.2. *Travel Safety and Security*

Beyond road traffic accidents, travel safety and security is often regarded as primarily an issue affecting females. There is certainly substantial evidence regarding high perceptions of travel danger (from verbal harassment to rape and murder) among girls and young women across the globe. There is also ample evidence of actual harassment of women globally and on a daily basis, with recent statistics suggesting, for instance, that over 70% of women in Karachi had experienced harassment on public transport and 90% in Sri Lanka, while 89% of women in Santiago had either seen or experienced it themselves [83] (pp. 15–16). Participants in a Chennai research study reported 14 years as the mean age at which they first encountered harassment in travel contexts; harassment was worst at night [118]. Jeffrey similarly reported so-called 'eve-teasing' in India [74]. This is rarely reported to the police and, as Anand and Tiwari noted for a Delhi slum [84] and Salon and Gulyani for a Nairobi slum [43], results in women travelling far less than they might otherwise do, thus contributing to their economic and social exclusion.

Lack of reportage means that it is difficult to assess the age-distributed incidence (or impact) of sexual harassment. Women worldwide have reported street harassment even in their 80s [119], so this is not purely an issue for youth. The statistics cited above are for women per se, but young girls may well be at particularly high risk of harassment and are even less likely to report such actions so their situation is often particularly dire. Young girls interviewed in the 24-site study in Ghana, Malawi and South Africa confided (especially to the peer researchers) a range of problems from catcalling and jeering by men to being groped when on public transport and actual rape [14] (pp. 184–186), [120]. Especially in locations with high HIV/AIDS prevalence, rape is clearly life-threatening. Girls' fear of travelling alone often leads them to postpone travel until others can accompany them (travel in groups is usually preferred by them and their parents), to take longer journeys to avoid particular trouble spots or to simply not travel, especially during hours of darkness [14,121] (for urban South Africa). However, it is important to note that boys can also face significant harassment, intimidation and, albeit very rarely, rape as they travel [14,122] (pp. 186–187).

## 7. Mobile Technology: Interactions between Mobile Technology and Travel among Young People

Young people tend to be at the forefront in uptake and use of digital technology across the world. Consequently, there is already substantial evidence of their engagement with mobile technology—mobile phones, the internet and other information and communication technologies (ICTs)—in transport contexts, not least as an aid to help address transport poverty. E-learning and mHealth are expanding rapidly, while smart mobility and smart city solutions are now becoming central foci of urban planning research globally. In remote rural areas, the potential for e-connectivity to reduce transport poverty can be particularly powerful [123,124]. In LMICs, where low cost mobile phone handsets and mobile phone networks have expanded dramatically over the last two decades, the implications for travel practices are extremely significant, as a growing literature attests (for example, Porter on the implications for poorer people's mobility [125], Williams et al. 2015 on the Nairobi digital matatu project [126]). When emergencies arise—not least obstetric emergencies among first time and very young mothers (who are at particularly high risk)—mobile-enabled mobility (whereby a phone-call to the nearest health centre brings in an ambulance) can be life-saving [127]. One issue worthy of note, however, is the varying cost of airtime and data. In countries with a highly competitive ICT sector, such as Kenya, airtime and data are relatively cheap, but in others, including South Africa, running a mobile phone, especially a smart phone, is costly. For young people with limited resources, this is an issue of considerable significance, although two studies conducted with African youth indicated that many young people see these costs as a priority over other consumables and often make considerable sacrifices in order to maintain their access [128–130].

The published literature specifically concerned with *youth* use of phones in daily travel contexts (as opposed to migration contexts) is, to date, relatively sparse globally, although there is a large amount of literature on phone practices based around youth culture and social media, adult or partner surveillance, etc. Nonetheless, there is growing evidence in the Global North of young people embedding technology in their everyday lives to better accommodate the uncertainty in activity and travel scheduling, such that it 'lubricates' modern life without fundamentally changing travel behaviours [131]. In the Global South, technology may have more impact on travel behaviour. The extent to which mobile phones can reduce travel is a particularly important question with regard to resource-poor people and environments and to carbon reduction, especially in urban contexts. Qualitative and survey data regarding young people's perception of the extent to which their use of phones had substituted for travel in the previous year, conducted in 24 sites in Africa, suggests that some reductions in travel are occurring, although the precise patterning varies with location [132–134].

In urban South Africa, many short daily journeys conducted by both males and females, especially walking journeys, seem to be being substituted by phone calls with safety as the main reason behind this. There is also evidence of a perceived reduction in longer (more expensive) journeys by motor vehicle as a result of greater mobile phone usage: in this case, the change is probably mostly attributable to the potential for saving money (although safety considerations could come into play too, given the hazards of long distance travel in the continent) [121]. Among urban residents, perceived reductions in long distance, irregular, and short everyday journeys look substantial, but it is important to note that these are only assessments based on respondent reflections about their travel activities over the past year. Whether this a real change, and if so, whether it is one that has reduced pedestrian and motorised traffic flows in the neighbourhoods where the surveys were conducted—and, if so, with what consequences for health and security—would be worthy of further research.

ICT/transport connectivities can be particularly important for women. As noted earlier, many women and girls in LMICs are restricted in their physical travel by male family members who may not only express concern for the vulnerability of womenfolk travelling alone but also distrust the potential that independent female mobility offers for promiscuity. In such contexts, women's access to mobile phones can be seen as a potential (virtual) mobility aid. However, keeping control of a partner's mobile phone communication is now a regular male endeavour in many households (as Burrell observed for rural Uganda [135]). As handset prices drop and phone ownership increases among young women, this is becoming more difficult to maintain, but surveillance of wives' and girlfriends' phone contact lists and calls and use of the phone as a 'digital leash' to check their physical location and travel movements appear to be growing features of many relationships [136,137]. In this context, it is noteworthy that young men in the motor-cycle taxi business and older 'sugar daddy' male taxi drivers have become notorious in some locations for using their relative wealth built through transport operations to buy phones for their girlfriends [121].

For (mostly male) transport operators, owning a working phone is widely considered essential to running a successful business; this is even the case for bicycle-taxi operators in Malawi [133]. Across Africa, people of all ages keep the numbers of local motorcycle-taxi and taxi drivers on their phones (ibid). For young women in particular, this is often seen not merely as a convenience but as a vital informal safety mechanism [121].

Beyond the transport sector per se, informal use of mobile devices appears to be having some impact on youth entrepreneurship through leapfrogging physical distance and promoting social networking. Qualitative and survey data on phone use for business among young people in 24 African research sites indicate that the phone is used extensively and intensively in small informal businesses. In trade, it is used not only to build relationships with customers and suppliers but also to help with pre-arranging meetings, organising travel, finding staff and for mobile money transfers and ensuring that payments have been made [121]. However, it could be unwise to overrate the phone's potential to promote youth entrepreneurship [138]. In Kibera, one of Nairobi's slum areas, although phones are utilised by young people to ease communication and strengthen existent social ties, this does not necessarily allow them to bypass Kenya's hierarchical class-based society [139].

*New Forms of ICT-Enabled Mobility Service*

While mobile phone usage will continue to be interwoven in diverse ways with human corporeal mobility and with physical transport technologies in LMICs, these patterns of interweaving are constantly being re-shaped. A burgeoning array of inventive phone apps, closely tied to growing smartphone use and the development of wireless infrastructure, appears to have particularly significant potential. In Asia, where the ride-hailing app boom is currently massive, smartphones are now available cheaply; advanced fourth-generation services can be accessed for just a few dollars. In Phnom Penh, for instance, at least four services, including one named CamGo, have been launched recently. This is for tuk-tuks, which are cheap but whose popularity was somewhat marred by drivers charging unreasonably high fares or intentionally taking round-about routes to increase the fare. CamGo is reportedly popular with young people because it offers a fixed rate per km and the route takes the shortest distance to the destination, measured by GPS, confirmed before boarding. Other recent examples include Chiang Mai, northern Thailand, where the Indonesian company Grab has launched a ride-hailing service for microbuses, now with 300 registered drivers. Ride-hailing services are also being used to book home deliveries. In India, Jugnoo, a ride-hailing service specializing in motorized tricycles, has partnered with fast-food restaurants such as Kentucky Fried Chicken and Burger King to deliver meals: it reportedly has 15,000 vehicles operating in 35 cities. We can anticipate that young people—both as customers and operators—will lead in the usage of these apps, although the returns from ownership of the vehicles involved may well go mostly to older, more established entrepreneurs [140].

In Africa, Uber and Uber-style apps are now playing a similar role. In South Africa, for instance, Uber has operated in major cities since 2013, although not without considerable hostility from metered taxi companies [141]. Subsequently, in 2017, the South Africa Meter Taxi Association set up their own app, "Yookoo Rider". This benefits customers through the registration of cab drivers, comprehensive driver vetting and criminal checks with fingerprint technology [141].

Uber-style companies now operate in many Africa cities (including for motorcycle taxis in cities like Kampala). With Little Cab, a Kenyan ride-hailing app backed by telecoms operator Safaricom, customers can pay for their ride through Safaricom's mobile money service, M-Pesa, buy discounted airtime during the trip and access free Wi-Fi. It also lets women exclusively request female drivers from 18:00 to 6:00 for safety reasons [142]. Many smaller operations are now attracting young entrepreneurs—for instance, the mobile application, Tag Your Ride, was launched by a young South African university graduate [142]. Young women customers appear to derive very considerable benefits from these apps, as can women drivers, especially if they are able to build a women-only service. Finally, however, it is necessary to refer back to the potential of phones as a causal factor in transport accidents with potential for impact on all sectors of the population.

## 8. Concluding Summary and Reflections

This paper covered diverse aspects of mobility while taking a specific child and youth perspective and drawing on the voices and evidence of young people themselves. It has emphasised how travel experiences, needs and risks are embedded in power relations and vary with gender, age and location (urban versus rural, rich country versus poor country, Asia versus Africa). It has also pointed to the scale and range of uncertainties that so many young people now face as they negotiate daily mobility (or immobility). Neoliberal economic and social changes have been radically transforming young people's experiences of youth and early adulthood across much of the world over the last decade [143], while climate change and growing environmental fragility are beginning to bring further uncertainties to the fore. In this context, it is important to note that while the majority of emphasis in the literature reviewed has been on daily mobility or immobility and travel experiences, the implication of such daily mobility experiences (physical and virtual) for migration decisions (short and long distance, short and long term) needs far stronger attention, particularly in this era of climate change. The linkages between daily mobility experiences and migration decisions will need far closer investigation globally, but especially in the context of conflict, climate change and growing environmental fragility in many LMICs.

The review pointed to other clear research gaps too: in particular, the need for a realignment of research methods and associated practices. Of prime importance is the need for more in-depth research, particularly in Asia. There is a growing body of detailed evidence regarding children and young people's specific transport and mobility needs and experiences in sub-Saharan Africa, often taking an ethnographic approach, but as this review has demonstrated, data remain sparse (and primarily quantitative) in Asian and Middle East and North Africa (MENA) contexts. But whatever the place context, it is important that mixed methods studies and an interdisciplinary approach are adopted in order to capture a fuller understanding of young people's complex transport needs and constraints. To date, there has been a particular sparsity of research using mixed-methods approaches: the 24-site study of daily mobilities among young people 9–18 years in sub-Saharan Africa [14] seems, to date, to be the only extensive study in LMICs that utilised a range of qualitative and quantitative methods. In Asia, mixed methods studies are rare and a majority of research takes a quantitative survey approach that commonly fails to provide adequate understanding of the patterns that emerge. A triangulation of in-depth ethnographic and survey research drawing on a range of disciplinary skills (possibly coupled with action research where interventions are made, and their impact then studied in depth), can be particularly powerful in understanding mobility experiences, behaviours and opportunities for positive change.

Linked to this point, greater engagement with young people themselves in research and planning processes is essential. Community peer-research with young people as a route to more fully understanding their needs and aspirations in the transport field is gaining growing attention. The mixed-method study cited above, which brought together 70 young researchers (11–19 years) in Africa to help build an extensive academic study, demonstrated the value of this approach. Small studies in Asia and Africa further support the importance of directly engaging young people in the research process [12,17]. At the same time, however, there is also need to build stronger recognition among transport professionals of the value of inputs from more vulnerable groups. There is no point in conducting research with young people if the evidence collected is subsequently ignored. Thus, greater effort is required to draw transport professionals more centrally into the research process with vulnerable groups.

Action research incorporating and assessing both transport service and infrastructure interventions could considerably aid exploration of a diversity of issues in both LMIC cities and rural areas. In LMIC cities, this could include working with young people to pilot and monitor interventions associated with road safety and improving potential for active travel: for instance, walking buses and other interventions such as street lighting and dedicated pedestrian and cycle lanes, as well as safe travel skills training for young women. Given the growing obesity problems among middle-class children in LMICs, particular attention also needs to be paid to improving their active travel to school. Meanwhile, for young people who may have never attended school and thus never had access to any road safety curriculum, piloting of non-school based road safety training interventions could bring great benefit—notably for the many involved in dangerous roadside hawking on busy city streets. In LMIC rural areas, there would be similar value in exploring the potential of walking buses, cycling and cycle maintenance/repair training, especially for girls, dedicated contract transport, and other interventions to improve girls' journey to school and to work. In both urban and rural contexts, new approaches to transport subsidy need exploring (and piloting) that do not result in excluding young people from transport (as has occurred when operators on busy routes have to choose between paying and non-paying customers). Subsidy is contentious in the transport services arena but could significantly improve the mobility opportunities and life chances of pupils, unemployed youth and young workers in urban and rural contexts.

Moving more centrally into the employment sphere, the transport/mobility elements that help shape youth employment, job search, entrepreneurship and unemployment experiences have, as yet, been insufficiently researched. This needs urgent attention across LMIC peri-urban sites, in particular where so many poorer households are located—and not least with specific reference to young women. More research is also needed around mobility aspects of out-of-school activities associated with recreation and social network building, which will be important for overall well-being but also may aid youth employment opportunities. Linked to this, more attention will be required to relationality across age groups (especially the linkages between expanded older people and youth cohorts) and the mobility implications that may extend well beyond work and employment.

Another area for further attention is how mobile technology is reshaping travel practices in low income contexts in the Global South (but also with likely relevance to some poor Global North communities). This includes the potential of mobile technology to reduce motorised transport usage and the extent to which young people may experience negative elements of exploitation or surveillance through digital technology. Mobile technology now helps young people to extend their networks across the world and has potential to support distance management in both emergency and everyday travel contexts and in rural and urban places. The potential for apps (including those developed by young people themselves) to reshape the transport arena globally is very exciting and opens up a potentially dramatic new phase of development. However, the extent to which less powerful groups in society, especially young people, are able to benefit in the longer term while evading potential threats of exploitation (for instance in the gig economy) or the wider surveillance and control also posed by increasingly smart technological innovation is uncertain; the evolving scene will merit careful observation.

Finally, it is important to look beyond the transport sector if we are to make significant improvements in young people's travel experiences and opportunities. Regarding LMICs, far greater attention is needed to youth transport issues from development practitioners working in other sectors outside transport, particularly education, youth employment, ICT and energy (although recent moves in the health sector to incorporate both transport and ICT considerations in their analyses are very encouraging). This will require more sustained efforts among transport practitioners and researchers to reach cross-sectorally and engage productively with those sectors if youth opportunities that are so central to achieving progress across the SDGs are to be fully realised.

**Author Contributions:** Writing—original draft, G.P.; writing—review and editing, J.T.

**Funding:** This research was funded by the UK Department for International Development.

**Acknowledgments:** This research was funded by UK AID through the UK Department for International Development under the High Volume Transport Applied Research Programme, managed by IMC Worldwide. However, some of the material referenced in the paper was drawn from a series of original field research studies on youth mobility (and associated publications) led by the first author and funded in whole or in part by the Department of International Development from 2000 in sub-Saharan Africa. (Studies 2006–2010 and 2012–2015, were jointly funded by UK DFID and the UK Economic and Social Research Council).

**Conflicts of Interest:** The authors declare no conflict of interest.

## References

1. McMillan, T. Children and Youth and Sustainable Urban Mobility. Thematic Study Prepared for Global Report on Human Settlements 2013. Available online: https://unhabitat.org/wp-content/uploads/2013/06/GRHS.2013.Thematic.Children.and_.Youth_.pdf (accessed on 28 October 2019).
2. UNDESA (United Nations Department of Economic and Social Affairs). *The World Youth Report: Youth and the 2030 Agenda for Sustainable Development*; UNDESA: New York, NY, USA, 2018. Available online: https://www.un.org/development/desa/youth/world-youth-report/wyr2018.html (accessed on 28 October 2019).
3. Brookings Institution 2018. Harnessing Africa's Youth Dividend. In *Foresight Africa: Top Priorities for the Continent in 2019*; Chapter 3; Brookings Institution: Washington, DC, USA.
4. By youth, with youth, for youth. Available online: https://en.unesco.org/youth#strategy (accessed on 28 October 2019).
5. Honwana, A. *The Time of Youth–Work, Social Change and Politics in Africa*; Kumarian Press: Boulder, CO, USA, 2012.
6. Barker, J. Passengers or political actors? Children's participation in transport policy and the micro-political geographies of the family. *Space Polity* **2003**, *7*, 135–151.
7. Sheller, M. *Mobility Justice: The Politics of Movement in an Age of Extremes*; Verso: London, UK; Verso: New York, NY, USA, 2018.
8. Martens, K.; Lucas, K. Perspectives on transport and social justice. In *Handbook on Global Social Justice*; Craig, G., Ed.; Edward Elgar Publishing: Cheltenham, UK, 2018; pp. 351–370.
9. Grieco, M. Youth and Transport: The Emergence of Youth Transport Strategies. 2007. Available online: https://www.ssatp.org/sites/ssatp/files/publications/HTML/Gender-RG/Source%20%20documents/Issue%20and%20Strategy%20Papers/G&T%20Rationale/ISGT14%20Youth%20and%20transport%20Grieco.pdf (accessed on 28 October 2019).
10. Developing a School Travel Plan That Caters for the Needs of Pupils with SEN. Available online: http://www.brake.org.uk/facts-resources/21-resources/566-developing-a-school-travel-plan-that-caters-for-the-needs-of-pupils-with-senforoneexample (accessed on 28 October 2019).
11. Children's Transport and Mobility: Towards a Child-Centred Methodology/toolkit. Available online: https://assets.publishing.service.gov.uk/media/57a08c6ee5274a27b20011d1/R8373a.pdf (accessed on 28 October 2019).
12. Lolichen, P. Children in the drivers' seat: Children conducting a study of their transport and mobility problems. *Child. Youth Environ.* **2007**, *17*, 238–256.
13. Porter, G.; Abane, A. Increasing children's participation in African transport planning: Reflections on methodological issues in a child-centred research project. *Child. Geogr.* **2008**, *6*, 151–167. [CrossRef]
14. Porter, G.; Hampshire, K.; Abane, A.; Munthali, A.; Robson, E.; Mashiri, M. *Young People's Daily Mobilities in Sub-Saharan Africa. Moving Young Lives*; Palgrave Macmillan: London, UK, 2017.
15. Porter, G.; Hampshire KBourdillon, M.; Robson, E.; Munthali, A.; Abane, A.; Mashiri, M. Children as research collaborators: Issues and reflections from a mobility study in sub-Saharan Africa. *Am. J. Commun. Psychol.* **2010**, *46*, 215–227. [CrossRef] [PubMed]
16. Porter, G. Reflections on co-investigation through peer research with young people and older people in sub-Saharan Africa. *Qual. Res.* **2016**, *16*, 293–304. [CrossRef]
17. Simpson, E.; Collard, N. Thinking with young people: Transport experiences and aspirations in sub-Saharan Africa and South Asia. Unpublished paper. January 2019.
18. Barker, J. A free for all? Scale and young people's participation in UK transport planning. In *Critical Geographies of Childhood and Youth: Contemporary Policy and Practice*; Kraftl, P., Horton, J., Tucker, F., Eds.; Policy Press: Bristol, UK, 2012; pp. 169–184.

19. Mulongo, G.; Porter, G.; Tewodros, A. Gendered politics in rural roads: Gender mainstreaming in Tanzania's transport sector. In *Proceedings of the Institution of Civil Engineers–Transport*; Thomas Telford Ltd.: London, UK, 2019. [CrossRef]
20. Barker, J.; Kraftl, P.; Horton, J.; Tucker, F. The Road Less Travelled? New Directions in Children's Mobility. *Mobilities* **2009**, *4*, 1–10. [CrossRef]
21. Southward, E.; Page, A.; Wheeler, B.; Cooper, A. Contribution of the School Journey to Daily Physical Activity in Children Aged 11–12 Years. *Am. J. Prev. Med.* **2012**, *43*, 201–204. [CrossRef]
22. Chaufan, C.; Yeh, J.; Ross, L.; Fox, P. You can't walk or bike yourself out of the health effects of poverty: Active school transport, child obesity, and blind spots in the public health literature. *Crit. Public Health* **2015**, *25*, 32–47. [CrossRef]
23. Loo, B.P.Y.; Siiba, A. Active transport in Africa and beyond: Towards a strategic framework. *Transp. Rev.* **2019**, *39*, 181–203. [CrossRef]
24. Malone, K.; Rudner, J. Global Perspectives on Children's Independent Mobility: A socio-cultural comparison and theoretical discussion of children's lives in four countries in Asia and Africa. *Glob. Stud. Childhood* **2011**, *1*, 243–259. [CrossRef]
25. Roya, S.; Hanif, N.R.; Dali, M. Influence of the socio-economic factors on children's school travel. *Soc. Behav. Sci.* **2012**, *50*, 135–147.
26. Andegiorgish, A.K.; Wang, J.; Zhang, X.; Liu, X.; Zhu, H. Prevalence of overweight, obesity, and associated risk factors among school children and adolescents in Tianjin, China. *Eur. J. Pediatr.* **2012**, *171*, 697–703. [CrossRef] [PubMed]
27. Yusoff, Z.M.; Shamin, F.; Arif, H.; Adnan, N.A.; Nordin, N.A. School Location and Mobility Effects to Obesity Cases Among Primary School Children. In Proceedings of the International Conference on Architecture and Built Environment (ICABE), Kuala Lumpur, Malaysia, 5–6 October 2016.
28. Li, S.; Zhao, P. The determinants of commuting mode choice among school children in Beijing. *J. Transp. Geogr.* **2015**, *46*, 112–121. [CrossRef]
29. Muthuri, S.K.; Wachira, L.J.M.; Onywera, V.O.; Tremblay, M.S. Associations Between Parental Perceptions of the Neighborhood Environment and Childhood Physical Activity: Results from ISCOLE-Kenya. *J. Phys. Act. Health* **2016**, *13*, 333–343. [CrossRef] [PubMed]
30. Onywera, V.O.; Larouche, R.; Oyeyemi, A.L.; Prista, A.; Akinroye, K.K.; Heyker, S.; Owino, G.E.; Tremblay, M.S. Development and convergent validity of new self-administered questionnaires of active transportation in three African countries: Kenya, Mozambique and Nigeria. *BMC Public Health* **2018**, *18*. [CrossRef]
31. Kingham, S.; Ussher, S. An assessment of the benefits of the walking school bus in Christchurch, New Zealand. *Transp. Res. Part A* **2007**, *41*, 502–510. [CrossRef]
32. Bwire, H.; Muchaka, P.; Behrens, R.; Chacha, P. Implementation and evaluation of walking buses and cycle trains in Cape Town and Dar es Salaam. In *Non-Motorized Transport Integration into Urban Transport Planning in Africa*; Mitullah, W., Vanderschuren, M., Khayesi, M., Eds.; Routledge: Abingdon, UK, 2017; pp. 150–169. Available online: https://www.researchgate.net/publication/321680296_Implementation_and_evaluation_of_walking_buses_and_cycle_trains_in_Cape_Town_and_Dar_es_Salaam (accessed on 28 October 2019).
33. Barker, J. 'Manic Mums' and 'Distant Dads'? Gendered geographies of care and the journey to school. *Health Place* **2011**, *17*, 413–421. [CrossRef]
34. Porter, G.; Hampshire, K.; Abane, A.; Munthali, A.; Robson, E.; Mashiri, M.; Tanle, A. Youth transport, mobility and security in sub-Saharan Africa: The gendered journey to school. *World Transp. Policy Pract.* **2010**, *16*, 51–71.
35. Owen, D.; Hogarth, T.; Green, A.E. Skills, transport and economic development: Evidence from a rural area in England. *J. Transp. Geogr.* **2012**, *21*, 80–92. [CrossRef]
36. Hine, J. *Good Policies and Practices on Rural Transport in Africa Planning Infrastructure & Services*; SSATP Working Paper 100; World Bank: Washington, DC, USA, 2014.
37. Avotri, R.; Owusu-Darko, L.; Eghan, H.; Ocansey, S. *Gender and Primary Schooling in Ghana*; Institute of Development Studies: Sussex, UK, 1999.
38. Department of Transport (Republic of South Africa). *The first South African National Household Travel Survey 2003: Technical Report*; Republic of South Africa: Pretoria, South Africa, 2005.

39. Matin, N.; Mukib, M.; Begum, H.; Khanam, D. Women's empowerment and physical mobility: Implications for developing rural transport, Bangladesh. In *Balancing the Load: Women, Gender and Transport*; Fernando, P., Porter, G., Eds.; Zed: London, UK, 2002.
40. Cook, C.; Duncan, T.; Jitsuchon, S.; Sharma, A.; Guobau, W. *Assessing the Impact of Transport and Energy Infrastructure on Poverty Reduction*; Asian Development Bank: Metro Manila, Philippines, 2005.
41. India's Women Are Choosing to Go to Worse Colleges Than Men. Available online: https://www.weforum.org/agenda/2017/12/indias-women-are-choosing-to-go-to-worse-colleges-than-men (accessed on 28 October 2019).
42. Porter, G.; Hampshire, K.; Abane, A.; Munthali, A.; Robson, E.; Mashiri, M.; Tanle, A.; Maponya, G.; Dube, S. Child porterage and Africa's transport gap: Evidence from Ghana, Malawi and South Africa. *World Dev.* **2012**, *40*, 2136–2154. [CrossRef]
43. Salon, D.; Gulyani, S. Mobility, poverty, and gender: Travel 'choices' of slum residents in Nairobi, Kenya. *Transp. Rev.* **2011**, *30*, 641–657. [CrossRef]
44. De Kadt, J.; Norris, S.A.; Fleisch, B.; Richter, L.; Alvanides, S. Children's daily travel to school in Johannesburg-Soweto, South Africa: Geography and school choice in the Birth to Twenty cohort study. *Child. Geogr.* **2014**, *12*, 170–188. [CrossRef]
45. Hunter, M. Racial desegregation and schooling in South Africa: Contested geographies of class formation. *Environ. Plan. A* **2010**, *42*, 2640–2657. [CrossRef]
46. Zhang, R.; Yao, E.; Liu, Z. School travel mode choice in Beijing, China. *J. Transp. Geogr.* **2017**, *62*, 98–110. [CrossRef]
47. Porter, G.; Hampshire, K.; Abane, A.; Munthali, A.; Robson, E.; Mashiri, M.; Maponya, G. Where dogs, ghosts and lions roam: Learning from mobile ethnographies on the journey from school. *Child. Geogr.* **2010**, *8*, 91–105. [CrossRef]
48. Phillips, L.G.; Tossa, W. Intergenerational and intercultural civic learning through storied child-led walks of Chiang Mai. *Geogr. Res.* **2017**, *55*, 18–28. [CrossRef]
49. Adams, S.; Savahl, S.; Fattore, T. Children's representations of nature using photovoice and community mapping: Perspectives from South Africa. *Int. J. Qual. Stud. Health Well-Being* **2017**, *12*, 1333900. [CrossRef]
50. Benwell, M. Challenging Minority World Privilege: Children's Outdoor Mobilities in Post-apartheid South Africa. *Mobilities* **2009**, *4*, 77–101. [CrossRef]
51. Vasconcellos, E.A. Rural transport and access to education in developing countries: Policy issues. *J. Transp. Geogr.* **1997**, *5*, 127–136. [CrossRef]
52. Van Niekerk, A.; Govender, R.; Jacobs, R.; Van As, A.B. Schoolbus driver performance can be improved with driver training, safety incentivisation, and vehicle roadworthy modifications. *S. Afr. Med. J.* **2017**, *107*, 188–191. [CrossRef]
53. Sohail, M.; Mitlin, D.; Maunder, D.A.C. *Partnerships to Improve Access and Quality of Public Transport*; WEDC: Loughborough, UK, 2003.
54. Sohail, M. *Urban Public Transport and Sustainable Livelihoods for the Poor: A Case Study, Karachi, Pakistan*; WEDC: Loughborough, UK, 2000.
55. Van Ristell, J.; Quddus, M.A.; Enoch, M.P.; Wang, C.; Hardy, P. Quantifying the impacts of subsidy policies on home-to-school pupil travel by bus in England. *Transportation* **2015**, *42*, 45–69. [CrossRef]
56. DFID. UK Department for International Development. In *Children Out of School*; DFID: London, UK, 2001.
57. Khandker, S.R.; Lavy, V.; Filmer, D. *Schooling and Cognitive Achievements of Children in Morocco*; Discussion Paper no. 264; World Bank: Washington, DC, USA, 1994.
58. Levy, H.; Voyadzis, C. *Morocco Impact Evaluation Report: Socio-Economic Influence of Rural Roads*; Operations Evaluation Department, Report no. 15808-MOR; World Bank: Washington, DC, USA, 1996.
59. Levy, H. *Rural Roads and Poverty Alleviation in Morocco*; World Bank: Washington, DC, USA, 2004. Available online: http://web.worldbank.org/archive/website00819C/WEB/PDF/ (accessed on 29 May 2019).
60. Mohsin, S.M.; Mallorie, E.; Roy, M.A. Construction of village roads by villagers: Creating jobs for women and men in Sunamganj, Bangladesh. In *Gender, Roads and Mobility in Asia*; Kusakabe, K., Ed.; Practical Action: Rugby, UK, 2012; pp. 182–191.
61. Pilgrim, J.; Chanrith, N. Road improvement in Cambodia: Livelihood, education, health, and empowerment. In *Gender, Roads and Mobility in Asia*; Kusakabe, K., Ed.; Practical Action: Rugby, UK, 2012; pp. 192–204.

62. Government of Nepal. *Gender Policy and Operational Guidelines for Local Transportation Sector*; Department of Local Infrastructure Development and Agricultural Roads: Khatmandu, Nepal, 2010.
63. WHO 2015: World Health Organization, Global Health Estimates. Available online: https://www.who.int/healthinfo/global_burden_disease/en/ (accessed on 28 October 2019).
64. Singh, N.; Vasudevan, V. Understanding school trip mode choice—The case of Kanpur (India). *J. Transp. Geogr.* **2018**, *66*, 283–290. [CrossRef]
65. Starkey, P. Promoting the Use of Intermediate Means of Transport—Vehicle Choice, Potential Barriers and Criteria for Success. 2001. Available online: https://pdfs.semanticscholar.org/2969/9a6cfc3fbf3c234fe7fcf2a3966c2dad6caf.pdf (accessed on 28 October 2019).
66. Rao, N. Cycling into the future: The Pudukkotttai experience, Tamil Nadu, India. In *Balancing the Load: Women, Gender and Transport*; Fernando, P., Porter, G., Eds.; Zed: London, UK, 2002.
67. Gatnet Communication. Unpublished work. 21 June 2006.
68. ILO. *The Youth Employment Crisis: Time for Action*; International Labour Organisation: Geneva, Switzerland, 2012.
69. Filmer, D.; Fox, L.; Brooks, K.; Goyal, A.; Mengistae, T.; Premand, P.; Ringold, D.; Sharma, S.; Zorya, S. *Youth Employment in Sub-Saharan Africa*; The World Bank: Washington, DC, USA, 2014.
70. Prince, R. Popular music and Luo youth in western Kenya. In *Navigating Youth, Generating Adulthood*; Christiansen, C., Ed.; Nordiska Afrikainstitutet: Uppsala, Switzerland, 2006; pp. 117–152.
71. Garcia-Palomares, J.C. Urban sprawl and travel to work: The case of the metropolitan area of Madrid. *J. Transp. Geogr.* **2010**, *18*, 197–213. [CrossRef]
72. Graham, L.; Mlatsheni, C. Youth unemployment in South Africa: understanding the challenge and working on solutions. In *South African Child Gauge*; De Lannoy, A., Ed.; Children's Institute, University of Cape Town: Cape Town, South Africa, 2015; pp. 51–59.
73. Jeffrey, C. Youth, class and time among unemployed young men in India. *Am. Ethnol.* **2010**, *37*, 465–481. [CrossRef]
74. Grieco, M.; Turner, J.; Kwakye, E. *A Tale of Two Cultures: Ethnicity and Cycling Behaviour in Urban Ghana. Transport Research Record 1441*; Transportation Research Board: Washington, DC, USA, 1995.
75. Grieco, M.; Apt, N.; Turner, J. *At Christmas and on Rainy Days: Transport, Travel and the Female Traders of Accra*; Avebury: Aldershot, UK, 1996.
76. Turner, G.; Kwakye, E. Transport and survival strategies in a developing economy: Case evidence from Accra, Ghana. *J. Transp. Geogr.* **1996**, *4*, 161–168. [CrossRef]
77. Evans, R. Sibling caringscapes: Time-space practices of caring within youth-headed households in Tanzania and Uganda. *Geoforum* **2012**, *43*, 824–835. [CrossRef]
78. Venter, C.; Molomo, M.; Mashiri, M. Supply and pricing strategies of informal rural transport providers. *J. Transp. Geogr.* **2014**, *41*, 239–248. [CrossRef]
79. Esson, J.; Gough, K.V.; Si, D. Livelihoods in motion: Linking transport, mobility and income-generating activities. *J. Transp. Geogr.* **2016**, *55*, 182–188. [CrossRef]
80. Kjeldsberg, C.; Shrestha NPatel, M.; Davis, D.; Mundy, G.; Cunningham, K. Nutrition-sensitive agricultural interventions and gender dynamics: A qualitative study in Nepal. *Matern. Child Nutr.* **2018**, *14*, 3. [CrossRef]
81. Porter, G. 'I think a woman who travels a lot is befriending other men and that's why she travels': Mobility constraints and their implications for rural women and girl children in sub-Saharan Africa. *Gend. Place Cult.* **2011**, *18*, 65–81. [CrossRef]
82. Seedhouse, A.; Johnson, R.; Newbery, R. Potholes and pitfalls: The impact of rural transport on female entrepreneurs in Nigeria. *J. Trans. Geogr.* **2016**, *54*, 140–147. [CrossRef]
83. Allen, H. *Approaches for Gender Responsive Urban Mobility*; GiZ-SUTP: Bonn, Germany, 2018.
84. Anand, A.; Tiwari, G. A gendered perspective of the shelter-transport-livelihood link. the case of poor women in Delhi. *Transp. Rev.* **2006**, *26*, 63–80. [CrossRef]
85. Langevang, T.; Gough, K.V. Surviving through movement: The mobility of urban youth in Ghana. *Soc. Cult. Geogr.* **2009**, *10*, 741–756. [CrossRef]
86. Li, M.; Kwan, M.-P.; Wang, F.; Wang, J. Using points-of-interest data to estimate commuting patterns in central Shanghai, China. *J. Transp. Geogr.* **2018**, *72*, 201–210. [CrossRef]

87. Porter, G.; Blaufuss, K. Children, Transport and Traffic in Southern Ghana. Paper Presented at the Workshop on Children and Traffic, Copenhagen, Denmark, 2–3 May 2002. Available online: www.dur.ac.uk/child.mobility/ (accessed on 28 October 2019).
88. Mark, L. Daily (Im)Mobility in Slums. A Female Perspective from the Villa 20 in Buenos Aires. Master's Thesis, Technisches Universitat, Berlin/ Universidade de Buenos Aires, Berlin, Germany, 2017.
89. Agarwal, S.; Attah, M.; Apt, N.; Grieco, M.; Kwakye, E.A.; Turner, J. Bearing the weight: The kayayoo, Ghana's working girl child. *Int. Soc. Work* **1997**, *40*, 245–256. [CrossRef]
90. Malmberg-Calvo, C. *Case Studies on the Role of Women in Rural Transport: Access of Women to Domestic Facilities*; SSATP Working Paper 11; World Bank: Washington, DC, USA, 1994.
91. Driving Change: The Story of Miss Taxi—One of Ghana's First Female Taxi Drivers. Available online: https://medium.com/beam-blog/driving-change-the-story-of-miss-taxi-one-of-ghana-s-first-female-taxi-drivers-42e98deeff81 (accessed on 28 October 2019).
92. India's Two-Wheel Taxi Service by Women, for Women. Available online: https://www.bbc.co.uk/news/av/world-asia-india-38326781/india-s-two-wheel-taxi-service-by-women-for-women (accessed on 28 October 2019).
93. Nyanzi, S.; Nyanzi, A.; Kalina, B.; Pool, R. Mobility, sexual networks and exchange among bodaboda men in southwest Uganda. *Cult. Health Sex.* **2004**, *6*, 239–254. [CrossRef]
94. Waage, T. Coping with unpredictability: "Preparing for life" in Ngaoundere, Cameroon. In *Navigating Youth, Generating Adulthood*; Christiansen, C., Utas, M., Vigh, H., Eds.; Nordiska Afrikainstitutet: Uppsala, Switzerland, 2006; pp. 61–87.
95. Doussantousse, S.; Sakounnavong, B.; Patterson, I. An expanding sexual economy along National Route 3 in Luang Namtha Province, Lao PDR. *Cult. Health Sex.* **2011**, *13*, 279–291. [CrossRef]
96. Adamu, F. Gender, Hisbah and enforcement of morality in northern Nigeria. *Africa* **2008**, *78*, 136–152. [CrossRef]
97. Burge, M. Riding the Narrow Tracks of Moral Life: Commercial Motorbike Riders in Makeni, Sierra Leone. *Afr. Today* **2011**, *58*, 58–95. [CrossRef]
98. Motorbike taxis in the "transport crisis" of West and Central African cities. Available online: https://journals.openedition.org/echogeo/13080#quotation (accessed on 28 October 2019).
99. Jenkins, J.T.; Peters, K. Rural connectivity in Africa: Motorcycle track construction. *Proc. Instit. Civil Eng. Transp.* **2016**, *169*, 378–386. [CrossRef]
100. Enhancing Understanding on Safe Motorcycle and Three-Wheeler Use for Rural Transport. Available online: http://www.transaid.org/wp-content/uploads/2019/07/RAF2114A_Final-Report_190909_FINAL_Revised.pdf (accessed on 28 October 2019).
101. WHO 2018. Road Traffic Injuries Factsheet. December 2018. Available online: https://www.who.int/news-room/fact-sheets/detail/road-traffic-injuries (accessed on 28 October 2019).
102. Wismans, J.; Skogsmo, I.; Nilsson-Ehle, A.; Lie, A.; Thynell, M.; Lindberg, G. Commentary: Status of road safety in Asia. *Traffic Inj. Prev.* **2016**, *17*, 217–225. [CrossRef] [PubMed]
103. Adesunkanmi, A.R.K.; Oginni, L.M.; Oyelami, O.A.; Badru, O.S. Road traffic accidents to African children. *Injury* **2000**, *31*, 225–228. [CrossRef]
104. ABOUT THE FIA ROAD SAFETY GRANTS PROGRAMME. Available online: https://www.roadsafety.fia-grants.com/about-us.php (accessed on 28 October 2019).
105. O'Toole, S.; Christie, N. Educating parents to support children's road safety: A review of the literature. *Transp. Rev.* **2019**, *39*, 392–406. [CrossRef]
106. Salmon, R.; Eckersley, W. Where there's no green man: Child road-safety education in Ethiopia. *Dev. Pract.* **2010**, *20*, 726–733. [CrossRef]
107. Safe Journeys to School Are a Child's Right. Available online: http://www.fiafoundation.org/media/45780/safe-to-learn-report.pdf (accessed on 28 October 2109).
108. A World Free of High Risk Roads: Cases Studies of Success. Available online: https://www.irap.org/media-centre/case-studies/ (accessed on 28 October 2019).
109. Bradbury, A.; Quimby, A. Community road safety education: An international perspective. *Proc. Instit. Civil Eng.-Munic. Eng.* **2008**, *161*, 137–143. [CrossRef]
110. Kitamura, Y.; Hayashi, M.; Yagi, E. Traffic problems in Southeast Asia featuring the case of Cambodia's traffic accidents involving motorcycles. *IATSS Res.* **2018**, *42*, 163–170. [CrossRef]

111. Howlett, J.B.; Aldous, C.; Clarke, D.L.; Howlett, J.B.; Aldous, C.; Clarke, D.L. Injuries sustained by passengers travelling in the cargo area of light delivery vehicles. *S. Afr. J. Surg.* **2014**, *52*, 49–52. [CrossRef]
112. Oyedemi, T.; Kgasago, T.J. Always-available communication and technological distractions: Technology use, texting and driving. *Commun.-S. Afr. J. Commun. Theory Res.* **2018**, *43*, 36–53. [CrossRef]
113. Cordellieri, P.; Baralla, F.; Ferlazzo, F.; Sgalla, R.; Piccardi, L.; Giannini, A.M. Gender Effects in Young Road Users on Road Safety Attitudes, Behaviors and Risk Perception. *Front. Psychol.* **2016**, *7*, 1412. [CrossRef]
114. Al-Aamri, A.K.; Padmadas, S.S.; Zhang, L.C.; Al-Maniri, A.A. Disentangling age–gender interactions associated with risks of fatal and non-fatal road traffic injuries in the Sultanate of Oman. *BMJ Glob. Health* **2017**, *2*, e000394. [CrossRef]
115. Hyder, A.A.; Labinjo, M.; Muzaffar, S. A new challenge to child and adolescent survival in urban Africa: An increasing burden of road traffic injuries. *Traffic Inj. Prev.* **2006**, *7*, 381–388. [CrossRef] [PubMed]
116. Koekemoer, K.; Van Gesselleen, M.; Van Niekerk, A.; Govender, R.; Van As, A.B. Child pedestrian safety knowledge, behaviour and road injury in Cape Town, South Africa. *Accid. Anal. Prev.* **2017**, *99*, 202–209. [CrossRef] [PubMed]
117. Elango, S.; Ramya, A.B.; Renita, A.; Ramana, M.; Revathy, S.; Manivel, M. An Analysis of Road Traffic Injuries in India from 2013 to 2016: A Review Article. *J. Commun. Med. Health Educ.* **2018**, *8*, 601. [CrossRef]
118. Mitra-Sarkar, S.; Partheeban, P. Abandon all hope, you who enter here: Understanding the problem of 'Eve Teasing' in Chennai, India. In *Proceedings of the Women's Issues in Transportation, Summary of the 4th International Conference*; Technical Papers; Transportation Research Board: Washington, DC, USA, 2011; Volume 2, pp. 74–84.
119. Logan, L.S. Street Harassment: Current and Promising Avenues for Researchers and Activists. *Sociol. Compass* **2015**, *9*, 196–2011. [CrossRef]
120. Hampshire, K.; Porter, G.; Mashiri, M.; Maponya, M.; Dube, S. Proposing love on the way to school: Mobility, sexuality and youth transitions in South Africa. *Cult. Health Sex.* **2011**, *13*, 217–231. [CrossRef]
121. Porter, G.; Hampshire, K.; De Lannoy, A.; Gunguluza, N.; Mashiri, M.; Bango, A. Exploring the intersection between physical and virtual mobilities in urban South Africa: Reflections from two youth-centred studies. In *Urban Mobilities in the Global South*; Uteng, T.P., Lucas, K., Eds.; Routledge: London, UK, 2018; pp. 59–75.
122. Banks, N. Youth poverty, employment and livelihoods: Social and economic implications of living with insecurity in Arusha, Tanzania. *Environ. Urban.* **2016**, *28*, 437–454. [CrossRef]
123. Velaga, N.R.; Beecroft, M.; Nelson, J.D.; Corsar, D.; Edwards, P. Transport poverty meets the digital divide: Accessibility and connectivity in rural communities. *J. Transp. Geogr.* **2012**, *21*, 102–112. [CrossRef]
124. Aguilera, A.; Guillot, C.; Rallet, A. Mobile ICTs and physical mobility: Review and research agenda. *Transp. Res. Part A* **2012**, *46*, 664–672. [CrossRef]
125. Porter, G. Mobile phones, livelihoods and the poor in sub-Saharan: Review and prospect. *Geogr. Compass* **2012**, *6*, 241–259. [CrossRef]
126. Williams, S.; White, A.; Waiganjo, P.; Orwa, D.; Klopp, J. The digital matatu project: Using cell phones to create an open source data for Nairobi's semi-formal bus system. *J. Transp. Geogr.* **2015**, *49*, 39–51. [CrossRef]
127. Green, C.; Adamu, F.; Rahman, I.A. The Role of a Transport Union in Increasing Rural Women's Access to Emergency Maternal Care in Northern Nigeria. *World Transp. Policy Pract.* **2013**, *19*, 29–45.
128. Porter, G.; Hampshire, K.; Abane, A.; Robson, E.; Mashiri, M.; Tanle, A. Youth, mobility and mobile phones in Africa: Findings from a three-country study. *J. Inf. Technol. Dev.* **2012**, *18*, 145–162. [CrossRef]
129. Porter, G.; Hampshire, K.; Milner, J.; Munthali, A.; Robson, E.; De Lannoy, A.; Bango, A.; Gunguluza, N.; Mashiri, M.; Tanle, A.; et al. Mobile phones and education in sub-Saharan Africa: From youth practice to public policy. *J. Int. Dev.* **2015**, *28*, 22–39. [CrossRef]
130. Porter, G.; Hampshire, K.; Abane, A.; Munthali, A.; Robson, E.; Bango, A.; De Lannoy, A.; Gunguluza, N.; Tanle, A.; Owusu, S.; et al. Intergenerational relations and the power of the cell phone: Perspectives on young people's phone usage in sub-Saharan Africa. *Geoforum* **2015**, *64*, 37–46. [CrossRef]
131. Line, T.; Jain, J.; Lyons, G. The Role of ICTs in Everyday Mobile Lives. *J. Transp. Geogr.* **2011**, *19*, 1490–1499. [CrossRef]
132. Porter, G. Mobile phones, mobility practices and transport organisation in sub-Saharan Africa. *Mobil. Hist.* **2015**, *6*, 81–88.
133. Porter, G. Mobilities in rural Africa: New connections, new challenges. *Ann. Am. Assoc. Geogr.* **2016**, *106*, 434–441. [CrossRef]

134. Rural Transport News December 2015. Available online: http://www.ifrtd.org/index.php/component/k2/item/23-rural-transport-news-december-2015 (accessed on 28 October 2019).
135. Burrell, J. Evaluating shared access: Social equality and the circulation of mobile phones in rural Uganda. *J. Comput.-Mediat. Commun.* **2010**, *15*, 230–250. [CrossRef]
136. Archambault, J.S. 'Travelling while sitting down': Mobile phones, mobility and the communication landscape in Inhambane, Mozambique. *Africa* **2012**, *82*, 393–412. [CrossRef]
137. Stark, L. Transactional sex and cellphones in a Tanzanian slum. *Suom. Antropol.* **2013**, *38*, 12–36.
138. Duncombe, R.A. Understanding the impact of mobile phones on livelihoods in developing countries. *Dev. Policy Rev.* **2014**, *32*, 567–588. [CrossRef]
139. Kibere, F.N. The paradox of mobility in the Kenyan ICT ecosystem: An ethnographic case of how the youth in Kibera slum use and appropriate the mobile phone and the mobile internet. *Inf. Technol. Dev.* **2016**, *22*, 47–67. [CrossRef]
140. Ride-Hailing Apps a Boon for Asia's Motorcycle Taxis. Available online: https://asia.nikkei.com/Business/Ride-hailing-apps-a-boon-for-Asia-s-motorcycle-taxis (accessed on 28 October 2019).
141. Uber: A Game-Changer in Passenger Transport in South Africa? Available online: https://static1.squarespace.com/static/52246331e4b0a46e5f1b8ce5/t/56521e01e4b0e332af41071b/1448222209348/CCRED+Review7.4-6.pdf (accessed on 28 October 2019).
142. Top 7 African Taxi-Hailing Apps Giving Uber a Run for Its Money. Available online: https://www.itnewsafrica.com/2018/06/top-7-african-taxi-hailing-apps-giving-uber-a-run-for-its-money/ (accessed on 28 October 2019).
143. Porter, G.; Hampshire, K.; Abane, A.; Robson, E.; Munthali, A.; Mashiri, M. Moving young lives: Mobility, immobility and inter-generational tensions in urban Africa. *Geoforum* **2010**, *41*, 796–804. [CrossRef]

*Review*

# Road Safety in Low-Income Countries: State of Knowledge and Future Directions

**Shahram Heydari [1,*], Adrian Hickford [1], Rich McIlroy [1], Jeff Turner [2] and Abdulgafoor M. Bachani [3]**

1 Department of Civil, Maritime and Environmental Engineering, University of Southampton, Southampton SO16 7QF, UK; A.J.Hickford@soton.ac.uk (A.H.); R.McIlroy@Southampton.ac.uk (R.M.)
2 High Volume Transport Applied Research Programme, IMC Worldwide Consultants, London RH1 1LG, UK; jeffreymturner@hotmail.com
3 Johns Hopkins International Injury Research Unit, Health Systems Program, Department of International Health, Johns Hopkins Bloomberg School of Public Health, 615 N. Wolfe Street, Baltimore, MD 21205, USA; abachani@jhu.edu
* Correspondence: S.Heydari@soton.ac.uk

Received: 26 September 2019; Accepted: 4 November 2019; Published: 7 November 2019

**Abstract:** Road safety in low-income countries (LICs) remains a major concern. Given the expected increase in traffic exposure due to the relatively rapid motorisation of transport in LICs, it is imperative to better understand the underlying mechanisms of road safety. This in turn will allow for planning cost-effective road safety improvement programs in a timely manner. With the general aim of improving road safety in LICs, this paper discusses the state of knowledge and proposes a number of future research directions developed from literature reviews and expert elicitation. Our study takes a holistic approach based on the Safe Systems framework and the framework for the UN Decade of Action for Road Safety. We focused mostly on examining the problem from traffic engineering and safety policy standpoints, but also touched upon other sectors, including public health and social sciences. We identified ten focus areas relating to (i) under-reporting; (ii) global best practices; (iii) vulnerable groups; (iv) disabilities; (v) road crash costing; (vi) vehicle safety; (vii) proactive approaches; (viii) data challenges; (ix) social/behavioural aspects; and (x) capacity building. Based on our findings, future research ought to focus on improvement of data systems, understanding the impact of and addressing non-fatal injuries, improving estimates on the economic burden, implementation research to scale up programs and transfer learnings, as well as capacity development. Our recommendations, which relate to both empirical and methodological frontiers, would lead to noteworthy improvements in the way road safety data collection and research is conducted in the context of LICs.

**Keywords:** road safety; low-income countries; under-reporting; best practices; vulnerable groups; injury severity; road crash costing; crash data; capacity building

---

## 1. Introduction

Road safety is a major global health issue since large proportions of unintentional injuries are caused by traffic-related crashes. According to the Global Health Observatory, 1.35 million fatalities occur on the world's roads each year [1]. In general, although traffic-related injuries and fatalities have seen a decreasing trend during the past two decades, this reduction has not been as significant as expected [2]. This is despite several improvements in motor vehicle safety standards and features, road safety policies, and road design [2]. In fact, road transport still poses a substantial risk to human health in many regions around the world.

The problem is especially critical in low-income countries (LICs), due to several persisting shortcomings in road safety standards, vehicle safety and maintenance, and in the design and implementation of policies and safe transportation infrastructure. Figure 1 displays national wealth versus road death rate based on data provided by the WHO [1]. Research is thus needed to better understand the underlying mechanisms of road safety in LICs. This will help guide road safety policies and strategies, with the aim of reducing traffic-related injuries and fatalities. Note that we refer to LMICs to indicate low- and lower-middle income countries, while LICs refer specifically to low-income countries, as shown in Figure 1.

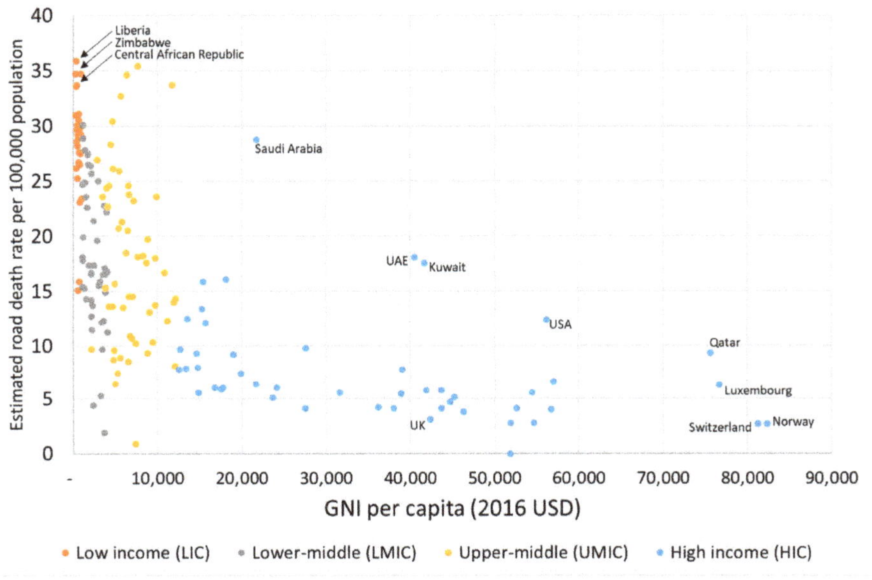


**Figure 1.** Comparing road traffic death rates with national wealth [1].

There are global initiatives that have sought to address such issues, many of which are coordinated through the Global Plan for the Decade of Action for Road Safety 2011–2020 [3], developed by the United Nations Road Safety Collaboration (UNRSC). Examples include the Global Road Safety Partnership (GRSP) [4], the World Bank's Global Road Safety Facility [5], the Bloomberg Philanthropies Initiative for Global Road Safety [6], and the Road Safety in 10 Countries Project [7]. These global efforts have generally focused on the practical implementation of policies and standards to improve road safety in the developing world [8].

Given these developments in the field, and the fact that we are nearing the end of the UN Decade of Action for Road Safety [3], it is important to take stock of where we are, what the state of the field is, and determine what research will be important in the future to make further progress. In line with this approach, our study aims to take a holistic approach investigating different lines of research under the broad topic of road safety in LICs with the aim of (i) analysing and summarising the current state of knowledge; and (ii) proposing a number of future research directions. We have used the Safe Systems framework and the framework for the UN Decade of Action on Road Safety to identify the domains of focus for this work [9,10]. There is a focus on the understanding of road safety engineering issues while also reaching out to other sectors.

The paper is structured as follows: Section 2 discusses our methodology. Section 3 discusses our findings and is centred on an overview of current (available) research focusing on the main road safety issues in LICs. In Section 4, we report on the results of a survey of experts in road safety in

LICs. Section 5 discusses implications and our proposed future directions, where there are several opportunities for improvements, both empirically and methodologically. We conclude with a summary in Section 6.

## 2. Materials and Methods

### *2.1. Review Framework*

This article explores the state of knowledge of road safety in LICs based on a review of literature and consultation with experts in the field. The review panel was formed by an international collaboration of the Transportation Research Group at the University of Southampton (United Kingdom) and Johns Hopkins International Injury Research Unit at Johns Hopkins University (USA). With respect to quantitative results, the literature search was carried out according to the PRISMA (Preferred Reporting Items for Systematic Reviews and Meta-Analyses) guidelines [11] as explained in Section 2.3. Non-quantitative results were identified through online searches as described in Section 2.3. Expert consultation is described in Section 2.4.

### *2.2. Research Questions*

Our research questions were as follows:

i "what are some of the most important topics relating to road safety issues in LICs?"

ii "what is known about these topics?"

iii "what would be some of the most critical future research directions to be considered by researchers?"

### *2.3. Search Strategy, Selection of Studies, and Eligibility Criteria*

We extracted the search terms from published studies (journal/conference articles and reports published by international organisations) in a preliminary search. For the review, while there was no search restriction on the publication date, articles published before 1990 were only considered for inclusion if the number of relevant results was particularly low. The search strategy aimed to limit the search to LICs by combining search results from a list of names of LICs with other relevant keywords applicable to other search topics such as vehicles and crashes, road safety measures, vehicle safety measures or vulnerable road user groups. Articles had to be peer-reviewed and published in English. Articles were screened by title and abstract for relevance, and appropriate bibliographies were also scanned for further relevant articles or reports. The search resulted in over 7000 potential articles to be screened across all the topics under consideration.

Quantitative results were specifically sought for under-reporting, and the literature search for that topic was carried out according to PRISMA guidelines, using electronic databases such as Web of Science, Scopus, PUBMED, TRID, BMC, and EMBASE. Search strings were generated by combining the list of LIC named countries and other relevant search terms such as underreport, under-report, and capture-recapture (a methodology used to address under-reporting). To be included, articles had to be related to a formal assessment of under-reporting. As the number of eligible papers in LICs was relatively few, there was no restriction on whether the assessment was for injury/fatality counts, nor any restriction on date. This in-depth search yielded many papers related to under-reporting as well as the other topics of road safety, both as part of the search results, and from bibliographies within relevant articles and reports. In this regard, further details are discussed in Section 3.1.

With respect to other topics, further relevant articles or reports, published by international organisations, were identified through online searches using appropriate keywords for each of the topics, and from discussions with experts and stakeholders. Search strings were generated by combining the list of LIC named countries, and other relevant search terms; e.g., low-income OR road accident OR traffic accident OR road crash OR traffic crash). Table 1 gives examples of the search strategy developed for this study (the example displayed having been used for searching Scopus). The strategy aims to limit the search to LICs (see Scopus #1), and then limit the search according to the desired topics of vehicles and crashes (see Scopus #2), road safety measures (see Scopus #3), vehicle safety measures (see Scopus #4), or vulnerable road user groups (see Scopus #5). For example, from a total of 69,448 studies relating to LICs, 454 studies were found to be relevant to road crashes in LICs (see Table 1).

**Table 1.** Example Scopus search strategy.

| Set # | Search String | Results |
|---|---|---|
| #1 Low-income countries | TITLE-ABS-KEY("developing country" OR "developing countries" OR Afghanistan OR Benin OR "Burkina Faso" OR Burundi OR "Central African Republic" OR Chad OR Comoros OR Congo OR Eritrea OR Ethiopia OR Gambia OR Guinea OR "Guinea-Bissau" OR Haiti OR "North Korea" OR Liberia Madagascar OR Malawi OR Mali OR Mozambique OR Nepal OR Niger OR Rwanda OR Senegal OR "Sierra Leone" OR Somalia OR Sudan OR Tanzania OR Togo OR Uganda OR Zimbabwe OR "low resource" OR "under-resourced" OR "resource poor" OR "under-developed" OR "underdeveloped" OR "developing world" OR "third world" OR LIC OR (low AND income)) | 69,448 |
| #2 Vehicles and accidents or crashes | TITLE-ABS-KEY((("Motor Vehicles" OR Automobiles OR Motorcycles OR traffic OR vehicle OR vehicular OR car OR cars OR automobile OR motorcycle OR taxi OR cab OR road OR pedestrian OR pedestrians) AND (accident OR accidents OR crash OR crashes OR injury OR injuries))) | 151,221 |
| #3 Road safety measures | TITLE-ABS-KEY("road safety" OR "road safety interventions" OR speeding OR "drink driving" OR helmet OR "motorcycle helmet" OR "seat belt" OR "seatbelt" OR "child restraint" OR "distracted driving" OR "drug driving" OR "traffic calming") | 49,812 |
| #4 Vehicle safety measures | TITLE-ABS-KEY("vehicle safety" OR "vehicle safety standards" OR "advanced braking" OR "anti-lock braking" OR "electronic stability control" OR NCAP) | 6,559 |
| #5 Vulnerable road user (VRU) groups | TITLE-ABS-KEY("vulnerable road user" OR "disability" OR "disabled" OR "gender disaggregation" OR "gender") | 956,942 |
| #6 Crashes in LICs | #1 AND #2 | 454 |
| #7 Road safety measures in LICs | #1 AND #3 | 147 |
| #8 Vehicle safety measures in LICs | #1 AND #4 | 5 |
| #9 VRU crashes in LICs | #1 AND #2 AND #5 | 103 |

### 2.4. *Expert and Stakeholder Consultations*

In the process of synthesizing findings from the literature review and identifying areas for further inquiry, we informally consulted with ten road safety experts from academia and international road safety organizations, including local experts from Bangladesh, Nepal, Uganda, Ghana, and Kenya. We asked a smaller number of experts (n = 6), who have had extensive experience working on road safety in LICs and are well versed with the global road safety landscape, to respond to a survey to provide further input on findings from the literature review. These consultations and surveys aided in the synthesis and prioritization of important areas of focus for future road safety in LICs. This questionnaire was based on three main themes regarding road safety in LICs: (i) main topics of concern; (ii) policy-making and implementation; and (iii) capacity development for road safety. Basically, with respect to each theme, respondents were asked to answer "what needs to be done, and how it can be achieved?" Before undertaking this part of our research, we obtained an approval from the University of Southampton Faculty of Engineering and Physical Sciences Ethics Committee on 03/05/2019 (ERGO II 48744). The outcomes from the survey and other discussions also provided context for topics for which there was a gap in the literature. These consultations contributed to the future research agenda set out in Section 5. Our consultation results are discussed in Section 4.

## 3. State of Knowledge

This section discusses our findings with respect to pertinent previous studies.

### 3.1. Under-Reporting of Road Crashes in LICs

Accurate knowledge of road crashes and their causes can help provide robust motives for the investment of appropriate and effective road safety interventions, and is especially important where such funds are limited [12]. Police road traffic crash data has been the traditional source of such information, although the accuracy of such data is questionable since all countries suffer from some level of under-reporting. The WHO provide estimates of the numbers of fatalities in each country, using negative binomial modelling based on the actual number of reported fatalities [1]. According to their estimates, the average number of road fatalities correctly reported to official sources is likely to be higher for higher income countries, with an average of 88% of road fatalities correctly reported in high-income countries (HICs) and 77% in middle-income countries (MICs). However, this reporting accuracy is significantly lower in LMICs (52%) and LICs (17%).

There are several underlying reasons why under-reporting may occur. A poor understanding regarding the benefits of complete and accurate road crash records means those involved may avoid contacting the police [13], preferring to negotiate with the driver [14,15]. A poorly resourced, capacity-limited police force is likely to focus on those crashes that involve more injuries, fatalities or property damage [13,16,17]. Also, the legal requirement to report road crashes to the police varies from country to country [18]. In built-up areas, victims are likely to be taken to a nearby hospital by relatives or bystanders before any police officers can attend the crash scene [13,14,17]. In rural areas, relatives and neighbours of crash victims may be deterred from travelling to hospitals due to cost implications [14]. There are further limitations due to the reliance on paper records, rather than on the types of electronic recording systems in place in higher income countries [18], especially if such original paperwork is required for subsequent judicial hearings [19].

Numerous studies have investigated how to improve the quality and accuracy of road crash data, particularly through combining police records with hospital records of patients admitted as a result of road crashes. These studies have mainly taken place in the developed world, for example in France [20–23], elsewhere in Europe [24–30], the United States [31,32], Australia and New Zealand [33–38], and China and Japan [39,40].

With respect to quantitative results relating to under-reporting, we found only four studies in LICs (from an initial list of 983 potential articles) that provide a quantitative assessment of the level of under-reporting, comparing police data with those of hospital records (see the PRISMA flow diagram in Figure 2). The four studies were all from LICs in Africa (Ethiopia, Uganda, Malawi, and Mali). Table 2 shows the location, study period, methodologies and metrics used, highlighting the inconsistencies in study designs. The study in Ethiopia [14] focuses on a 264 km stretch of two-way, two-lane road, on which traffic volumes vary from an average of 17,000–20,000 vehicles per day on the 64 km stretch near Addis Ababa, to around 3000–3500 vehicles per day on the remaining 200 km as the road approaches Hawassa. The other studies focus on regional areas (adopting a zone- or macro-level approach), considering the records at police offices and hospitals within a certain region.

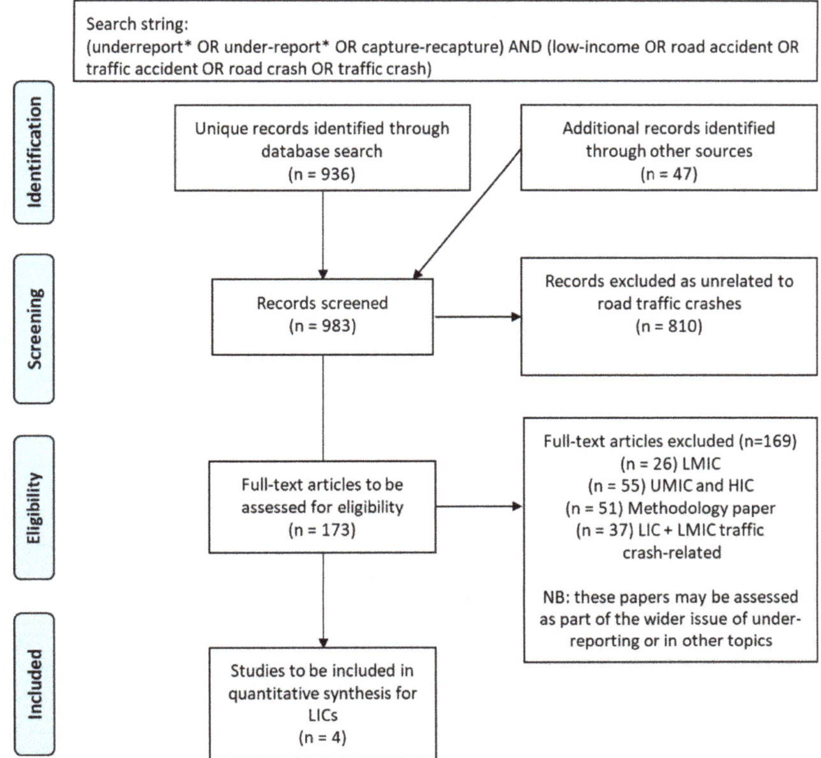


**Figure 2.** PRISMA flow diagram for under-reporting literature search.

The only consistent element to these four studies is the analysis methodology used: the capture-recapture method. The 'Capture-recapture' method is perhaps the most widely used method to rectify inaccuracies in road crash reporting [41]. Originally developed for use in estimating the biometrics of animal populations and subsequently applied to human populations and the injury field as discussed for example in (as discussed for example in [42]), the method involves estimating the number of cases in a defined population using multiple sources of information (e.g., linked databases), assuming that each source alone may under-count the population. While it is a useful (and relatively low-cost) tool for road crash investigators, especially in LICs where under-reporting is frequent, there are certain caveats to understand regarding its use due to population movements and dependence of data sources [43]. However, movement of population has been identified as a minor issue, for example, due to the relatively short timescales between police attending a crash scene and casualties visiting hospital [14,20], or the relative lack of migration within a study region [44]. The four studies identified here all aimed to address any bias introduced by a lack of independence by using a stratified capture-recapture technique to identify those factors that were associated with dependency of the data sources. Yet, whether stratification is satisfactory remains uncertain and the problem may not be fully addressed.

Given the limitations associated with capture-recapture methods, other techniques have been used to estimate the levels of under-reporting. Regression analysis is useful if the two-source approach used in capture-recapture is not appropriate or feasible [1,33]. A modified Poisson regression has been used in France [21], and similar approaches have been used in India [15], the United States [32], and Australia [38]. The impact of under-reporting on crash severity models has also been investigated [45–47]. However, as previously noted, all four quantitative papers in LICs use capture-recapture.

**Table 2.** Summary characteristics of selected studies.

| Reference | Location | Study Period | Method Used | Metric Used |
|---|---|---|---|---|
| Abegaz et al. [14] | Ethiopia (264 km highway) | Jun 2013–May 2014 (12 months) | Capture-recapture, two-source | Fatalities/injuries per billion vehicle km |
| Magoola et al. [13] | Uganda (Jinja) | March 2014–April 2014 (2 months) | Capture-recapture, two-source | Number of road traffic injuries (RTIs) |
| Samuel et al. [16] | Malawi (Lilongwe district) | July 2008–June 2009 (12 months) | Capture-recapture, two-source | Number of fatalities Mortality incidence (fatalities/1000 person-years) |
| Sango et al. [48] | Mali (Bamako district) | January 2012–April 2012 (4 months) | Capture-recapture, two-source | Number of fatalities; Number of records; Incidence per 100,000 people |

The key findings of the studies reported in Table 2 are summarised in Table 3 according to the metric used and sources of data (police and/or hospital). In order to compare results of the four African studies while accounting for uncertainty, the relative accuracy of each data source was calculated in terms of confidence intervals (see Table 3). The accuracy rate is obtained by considering the main metric used in a particular study (e.g., number of fatalities, or injuries per 100,000 vehicle km), and comparing police and hospital results with those generated by the capture-recapture method. These results reveal some consistent patterns. Police records tend to account for more fatalities than hospital records, while hospitals tend to have more complete records of road crashes and injuries. However, neither set of records has a complete record of the numbers of fatalities or injuries. In general, the uncertainties around estimates relating to fatalities are larger than those obtained for injuries and crashes. The study conducted by Abegaz et al. [14], which adopts a micro-level approach, focusing on a specific roadway rather than a region, has the smallest confidence intervals. The results from Ethiopia and Mali indicate that the police records contain around 60% of the total fatalities, and hospital records seem to account for between 40% and 60% of injuries or road crash numbers.

**Table 3.** Accuracy rates of police and hospital records for four LICs (with associated confidence intervals).

| Reference | Fatalities | | Injuries | | Crashes | |
|---|---|---|---|---|---|---|
| | Police | Hospital | Police | Hospital | Police | Hospital |
| Abegaz et al. [14] | 59.2% (57.4–60.9%) | 32.5% (31.5–33.4%) | 23.7% (23.5–23.9%) | 55.6% (55.2–56.0%) | n/a | n/a |
| Magoola et al. [13] | n/a | n/a | 14.4% (13.1–16.5%) | 60.4% (55.1–65.2%) | n/a | n/a |
| Samuel et al. [16] | 37.6% (30.9–48.0%) | 25.5% (21.0–32.6%) | n/a | n/a | n/a | n/a |
| Sango et al. [48] | 57.6% (50.4–67.9%) | 54.5% (47.8–64.2%) | n/a | n/a | 16.8% (15.9–17.8%) | 42.1% (39.9–44.6%) |

### 3.2. Lessons Learned from Global 'Best Practice' and Its Applicability to LICs

We believe it is important to draw on the experience of successful road safety campaigns around the world. For example, Wegman [49] identifies Spain in particular as an exemplar of success in road safety practice, as the number of fatalities there has decreased by over 70% between 2000 and 2013, with neighbours Portugal close behind.

According to the Spanish Government's Traffic Directorate report on their road safety strategy [50], Spain's improved safety level in the 2000's stemmed from their adoption of the European Road Safety Strategy in 2000, followed by the increased use of in-car safety systems, the increased use of helmets (from 73% to 99%), the increased use of seat belts (from 70% to 91%), and reduced risks from slower average speeds (reduced by 2 km/h) and a downward trend in drink-driving (the percentage of drivers who died when over the limit of 0.3 g/L fell from 35% to 29%).

The Spanish experience is one example of how a cohesive national road safety strategy may result in significant reductions in road crashes. However, as a developed country, the challenges and issues involved in improving road safety are likely to be different from those in LICs [51]. While infrastructure improvements and appropriate legislation are, to a relatively large extent, common themes appropriate to all road networks, road user behaviour may vary from one country to another. This is why one should take into account that cultural differences may lead to variation in road user behaviour in different jurisdictions. For example, the same engineering intervention may result in unequal road safety improvements even if all other factors such as traffic exposure and roadway characteristics are similar. However, we speculate that, while such variation in road user behaviour would affect the magnitude of effectiveness, it would not deter the effect of such engineering interventions entirely. Therefore, lessons can be learnt from successful interventions in developed countries. For example, the Safer Africa project [52] aims to utilise knowledge and expertise from successful European road safety projects to implement effective road safety and traffic management interventions in African countries.

In order to attempt to affect road safety positively in LICs and LMICs, 2011–2020 has been designated the World Health Organization's (WHO) 'Decade of Action for Road Safety', seeking to save millions of lives by building road safety management capacity, improving the safety of road infrastructure, further developing the safety of vehicles, enhancing the behaviour of road users and improving the post-crash response. A number of previous studies have carried out reviews of road safety interventions in LICs and LMICs, some of which are shown in Table 4 and discussed in more detail below.

Gupta et al. [53] focused on regulatory and road engineering interventions for preventing road traffic injuries and fatalities among non-motorised and motorised two-wheel (i.e., vulnerable) road users. Of the twenty-five studies in their review, only two were LIC-based. Esperato et al. [54] aimed to evaluate the cost and health impacts of road safety interventions in LMICs, identifying thirteen studies which met their criteria, none of which were based in LICs. Staton et al.'s focus [55] was to determine quantitative impacts of road safety interventions including legislation, enforcement and education campaigns. Of the eighteen studies identified, three were in LICs. Bonnet et al. [56] identified twenty-three articles relating to road safety interventions in Africa; eight were set in LICs. These studies highlight the relative dearth in quantity and quality of research output that relate specifically to the impact of road safety interventions in LICs.

With respect to best practices in LICs, the experience of Ethiopia could be mentioned, where media campaigns coupled with targeted research studies have provided a more complete picture of the impact of road safety legislation [6,57–59]. A summary of these and other research into policy interventions in LMICs is given below under the broad themes of education, enforcement and engineering.

**Table 4.** A sample of papers related to road safety initiatives and interventions in LICs and LMICs.

| Main Themes | Location | | Key Finding |
|---|---|---|---|
| General review of interventions | 12 countries in Eastern and Southern Africa | Bonnet et al. [56] What interventions are required to reduce road traffic injuries in Africa? A scoping review of the literature | There is a lack of road safety interventions and some shortcomings in the associated assessments. |
| | LICs &LMICs | Gupta et al. [53] Regulatory and road engineering interventions for preventing road traffic injuries and fatalities among vulnerable (non-motorised and motorised two-wheel) road users in low- and middle-income countries | There is a lack of studies specifically assessing the impact of engineering interventions in LMICs. Where studies exist, there are mixed results (e.g., while fatalities declined, the number of casualties more than doubled) |
| | LICs & LMICs | Staton et al. [55] Road traffic injury prevention initiatives: a systematic review and meta-summary of effectiveness in low and middle income countries | A multi-faceted approach (involving education, legislative and enforcement measures) is more effective |
| | LMICs | Esperato et al. [54] Projecting the health and economic impact of road safety initiatives: A case study of a multi-country project | Road safety interventions in the Road Safety in 10 Countries project are likely to be cost-effective, but there is a reliance on potentially inaccurate global data sets for evidence of impacts |
| Legislation change | Ethiopia | Abegaz et al. [59] Effectiveness of an improved road safety policy in Ethiopia: an interrupted time series study | Revised road safety policy helped reduce motor vehicle crashes and associated fatalities. However, the overall incidence rate is still very high |
| Road safety education | Pakistan | Ahmad et al. [60] Teaching children road safety through storybooks: An approach to child health literacy in Pakistan | Discussions using bilingual pictorial storybooks helped primary school children better understand crash prevention |
| | LICs & LMICs | Li et al. [61] Children and road traffic injuries: Can't the world do better? | Validation and implementation of road safety interventions are lacking and urgently needed in LMIC |
| | Tanzania | Zimmerman et al. [62] Road traffic injury on rural roads in Tanzania: Measuring the effectiveness of a road safety program | The incidence of road crashes in low-volume rural settings is unacceptably high and most commonly associated with motorcycles. More research is needed to quantify the impact of various prevention strategies |
| | Ethiopia | Salmon and Eckersley [63] Where there's no green man: child road-safety education in Ethiopia | Road-safety education in Ethiopia is often locally inappropriate and impractical, frequently based on dominant but ineffective educational models imported from other contexts |
| | LICs & LMICs | Bradbury and Quimby [64] Community road safety education: an international perspective | Transferring road safety education practices from developed to developing countries is arguably unfeasible because of variations in education systems, teaching methods, traffic regulations and exposure to risk. Road safety practitioners should research and develop bespoke teaching methods and materials in the country in which they will be used |
| | Ghana | Blantari et al. [65] An evaluation of the effectiveness of televised road safety messages in Ghana | Televised advertisements concerning speeding and alcohol-impaired driving, targeted towards commercial drivers, were largely understood, but road safety activities would be strengthened by increasing accompanying law enforcement activities |

**Table 4.** *Cont.*

| Main Themes | Location | | Key Finding |
|---|---|---|---|
| Enforcement | Kenya | Bachani et al. [66] Helmet wearing in Kenya: prevalence, knowledge, attitude, practice and implications | A traffic amendment bill resulted in negligible impact on helmet use, highlighting the need for a multi-faceted strategy that includes media campaigns and widespread enforcement in addition to legislative change for improving helmet use |
| | Cambodia | Bachani et al. [67] Knowledge, attitudes, and practices around drinking and driving in Cambodia: 2010–2012 | A multi-pronged, coordinated approach is needed to effectively address drink-driving, including social marketing and public education campaigns, and enhanced enforcement |
| | Nigeria & Vietnam | Stewart et al. [68] Drink driving prevention initiatives in low- and middle-income countries: challenges and progress in Nigeria and Vietnam | Interventions tailored to local conditions and needs can help reducing drinking and driving in LMICs. Effective partnerships with local governmental and nongovernmental organizations is an important aspect of future progress |
| | Kenya | Bachani et al. [69] Prevalence, knowledge, attitude and practice of speeding in two districts in Kenya: Thika and Naivasha | While this study demonstrates an improvement in the prevalence of speeding in two Kenyan districts over 2010-2012, it also highlights the need for further action to be taken to address the problem |
| | Malawi | Kraemer et al. [70] Cyclists' helmet usage and characteristics in central and southern Malawi: a cross-sectional study | Cycle helmet usage in Malawi is virtually non-existent. Future interventions should be targeted to adult and young adult males who made up 95% of observed cyclists |
| | LICs & LMICs | Johnson [71] Assessing a country's drink driving situation: An overview of the method used in 6 low- and middle-income countries | Situation assessments of drink-driving in six countries are now being used to develop appropriate and relevant pilot projects, taking into consideration the country's culture with respect to transportation, enforcement, health care, and alcohol consumption |
| Roundabout design | Ethiopia | Tulu et al. [58] Investigating pedestrian injury crashes on modern roundabouts in Addis Ababa, Ethiopia | The presence of a public transport terminal beside a roundabout is associated with increased pedestrian crashes. While the maximum gradient of an approach road is negatively associated with pedestrian safety, the provision of a raised median along an approach appears to increase pedestrian safety at roundabouts |

### 3.2.1. Education

Educating the populace, and children in particular, about road safety and good road user behaviour can help to reduce the number and severity of crashes, particularly when used as part of a wider package of interventions [56,61,63]. For example, storybook narratives have been used in Pakistan to improve children's knowledge and attitudes towards road safety, and the relatively low-cost storybooks provided an effective and early strategy towards promoting behavioural change, particularly how to behave at the road side and when crossing. For instance, results indicated that while questions regarding where it is safe to cross were answered correctly by just over 50% of the children, this increased to over 90% correctly answering two months after the initial intervention. However, it was acknowledged that there may be a need for regular road safety education to further ensure that students retain traffic safety messages in the longer term [60], and for any impact to be effective, such measures ought to be coupled with legislative and enforcement measures [55].

Communities and families also have considerable potential to influence the young and road safety programmes need to be developed in support of formal school programmes [64]. In the wider community, research has focused on the impact of televised road safety messaging [65], which was found to be largely effective at promoting lower alcohol consumption when driving, although the language used in the messages was considered a potential barrier in a country with multiple languages spoken. In Tanzania, an education programme was developed based on an assessment of local road crashes, including targeting motorcyclists [62].

The transferral of road safety education practices from developed to developing countries can be difficult due to variations in education systems, teaching methods, traffic regulations and exposure to risk. Road safety education and awareness have been identified as the interventions which are most adapted to LICs [56]. Thus, road safety practitioners should aim to research and develop bespoke teaching methods and materials in the country in which they will be used [64].

### 3.2.2. Enforcement

Improving road safety through enforcement of revised legislation can also help promote road safety, particularly in the five key risk factors identified in the Global Plan for the Decade of Action for Road Safety (speed, drink driving, not wearing motorcycle helmets, not wearing seat belts, and not using child restraints in cars) [3], manuals for each of which have been published by WHO [72]. The drink-driving manual has been further developed into an assessment framework to help understand how LICs and LMICs can better adopt the guidelines [71]. Further examples of research in this area include a study in Addis Ababa in Ethiopia, where joint media and enforcement campaigns have reduced drink driving by 50% [57]. In terms of legislation, Addis Ababa has developed its first ever road safety strategy and implementation plan, and established an inter-agency road safety council chaired by the Deputy Mayor and is considering setting up a road safety fund [6]. Wider road safety policy interventions in Ethiopia were investigated using interrupted time series [59]; the revised road safety policy banned the use of mobile phones while driving, made helmet and seat belt use mandatory, and increased levels of enforcement of excessive speeding, drunk driving and carrying dangerous loads. This was found to have helped reduce motor vehicle crashes by around 19% and associated fatalities by 12% in the year following the intervention; however, the overall crash rate was still very high [57].

Helmet use, for both bicycles and motorcycles, is low throughout LICs. A study in Malawi observed zero bicyclists wearing a helmet over a four-day study period [70]. In Kenya, the low prevalence of motorcycle helmet use remained unchanged, with around 30% of motorcycle riders correctly using helmets following the introduction of a traffic amendment bill in 2012, highlighting the need for a multi-faceted strategy that includes media campaigns and widespread enforcement in addition to legislative change for improving helmet use [66]. The authors reached a similar conclusion after studying the prevalence and attitudes towards drink driving in Cambodia, as Bachani et al. [67] concluded that a multi-pronged and coordinated approach would be needed to effectively address this issue, including social marketing and public education campaigns, and enhanced enforcement measures.

Wismans et al. [73] provide a summary of the WHO report [74] focusing on Asian countries, highlighting that while the interventions suggested have been adopted in many countries, LICs (notably Afghanistan and Nepal) tend to be lacking in such interventions.

#### 3.2.3. Engineering

There is a lack of studies specifically assessing the impact of engineering interventions in LMICs [53], and those that are available have mixed results. Gupta et al. [53] report that the results of the three before-and-after engineering-based studies included in their review show that while fatalities declined, the number of casualties more than doubled. Other more recent research in Ethiopia has focused on reducing traffic-related pedestrian injuries at roundabouts by modelling the effects of different features of the approaches, such as the presence of guard railing and location of pedestrian crossings. The results which emerged from a crash prediction model and development of safety performance functions suggested that the number of crashes involving pedestrians was 50% higher near public transport terminals, where the spatial intensity of pedestrian-vehicle conflicts is high. A change in gradient of the approach of 1% could result in 12% increase in pedestrian crashes, as visibility is reduced and speeds are affected. However, there is less risk of pedestrian crashes where appropriate crossing facilities are provided. For instance, roundabout approaches with central refuges had 44% fewer pedestrian crashes than those without such facilities [58]. However, provision of pedestrian facilities does not imply compliant use. In Ghana, 65% of pedestrians using a zebra crossing displayed some aspect of risky behaviour such as talking, eating or drinking, using telephones, and wearing headphones [75].

The WHO's annual 'Global Status Report on Road Safety' provides an overview of progress that has been achieved in important areas such as legislation, vehicle standards and access to post-crash care. The 2018 report [1] notes that progress has not, however, occurred at a pace fast enough to compensate for the rising population and the rapid motorisation of transport taking place in many parts of the world. At this rate, they note, the Sustainable Development Goals (SDG) target 3.6 to halve road traffic deaths by 2020 will not be met.

### *3.3. Vulnerable Groups and Gender Disaggregation*

The travel behaviour of vulnerable groups, such as the disabled, women and children, can be adversely affected by road safety issues, particularly in LICs. The injury profile of road traffic crashes in LICs differs in important ways from the profile seen in developed countries. Pedestrians, cyclists and passengers on multi-passenger transport (buses, trucks and minibuses) are at particular risk of injury. For instance, Nantulya and Reich [76,77] note that pedestrians, cyclists and passengers in buses and trucks account for around 90% of the casualties in countries in low- and middle-income regions, as opposed to high-income regions where drivers constitute the majority of victims. This large proportion of vulnerable road users in road crash statistics in LICs may be explained by a traffic mix of incompatible users, where pedestrians, cyclists and motorcyclist are forced to share road space with cars and trucks, especially where communities live within the vicinity of roads, where there is a lack of pavement along large urban streets [78], and where children are particularly vulnerable to increasing levels of motorisation [79].

Despite the prominence of vulnerable road users in LICs, they are still largely ignored in the planning, design and operation of roads [1]. Around three-quarters of the casualties and fatalities in LICs are men [80,81], and this may reflect gender disparities in access to economic opportunities and in exposure to road traffic injury risks as drivers and passengers [77]. Unfortunately, the published summary data of injury numbers is seldom disaggregated to look at patterns in factors such as gender [82]. However, as more countries are conducting household surveys, in addition to regular population censuses, more countries can now produce data disaggregated by sex for basic indicators on population, families, health, education and work [83]. Nevertheless, even when such information is collected, it is often not tabulated and disseminated to allow for meaningful gender analysis [83].

Men and women typically adopt different journey patterns, which will differentially change their exposure to risk of involvement in road crashes [82,84]. Uteng [85] suggests that there are a number of factors influencing women's mobility in the developing world, including social and cultural norms in a patriarchal system and transport infrastructure planning and design. Typically, women are less likely to make long journeys or use slower modes, and may not use busy roads as frequently as men, thus crash statistics for men and women will tend to show different patterns [85]. The predominant mode of transport for women in rural Africa, for instance, is walking [86], while men get access and priority for the use of private vehicles [85]. When women use motorised transport, it is likely to be public transport and they are often subject to other risks while travelling [86]. While these gender differences in the use of transport is greatest in rural settings [87], in urban areas, women have less access to either individual or public means of transport than men, due to both economic and social reasons, and are hence dependent on walking or using undesirable and potentially unsafe forms of public transport [88].

In their review of 73 studies of road traffic injuries in developing countries, Odero et al. [80] note that "no study in this review attempted to investigate specific potential factors that would explain the observed gender differences. Such a study is desirable and would need to assess and correct for levels of exposure by gender." Thus, the available literature on gender disaggregation in road crashes specifically relating to LICs is sparse, indicating a clear gap in the knowledge base, and this is a topic which could merit the use of social research mechanisms such as behavioural and attitudinal surveys of these vulnerable groups, especially focussing on countries which have recently improved frequency of and access to household survey data. Car-centric transport policies coupled with increased urbanisation could lead to greater inequity in mobility [85], and further study needs to be made on how such policies affect mobility from a gender-based perspective.

### 3.4. Disabilities Due to Road Crashes

Disability due to road traffic crashes and injuries is a significant proportion of the burden in low-income settings, where appropriate and timely medical care is not usually available for victims. Rehabilitation care systems are sorely lacking in LICs [89,90]. It is estimated that for every road traffic death, there are an additional 20–50 more people who are injured, and often face disability [91]. According to the World Health Organization [91], around 85% of all global road deaths, 90% of the disability-adjusted life years (DALYs) lost due to crashes, and 96% of all children killed worldwide as a result of road traffic injuries occur in low-income and middle-income countries. However, definite data on the number of people who survive road crashes but live with disabilities are almost non-existent [92].

Some indicative data is available as highlighted in the WHO's report on Global Road Safety 2018 [1]. There are data available for 29 of the 175 countries listed in that report, giving details of post-crash response, which provides an estimated value for the percentage of road crash victims with permanent disability. Of these 29 countries, four are classified as LICs and four as LMICs. The range of estimates for the four LICs is quite wide, with an estimated 47% of road crash victims having permanent disabilities in Togo and 40% in the Democratic Republic of Congo, while estimates for Zimbabwe (7%) and Uganda (3%) are much lower. The range is less stark for the LMICs, with an estimate of 19% of road crash victims in Sudan, and 15% in Cambodia suffering from permanent disabilities, compared with 2.4% in Bangladesh and 1% in Palestine (West Bank and Gaza). Of the upper-middle and high-income countries for which data is provided, this metric is highest in Brazil (24%) and Romania (21%), and lowest in Qatar and France (both 1%). There are no reasons offered within that report on why these data should be so disparate, and there is no obvious pattern linking the health care availability and these disability rates.

Zafar et al. [93] carried out secondary analysis of the results of four nationally representative cluster randomized surveys in LICs (Nepal and Uganda in 2014, Rwanda in 2011 and Sierra Leone in 2012) as part of the Surgeons Overseas Assessment of Surgical Needs, which collected information regarding demographics, injury characteristics, anatomic location of injury, healthcare seeking behaviour, and disability from injury. The authors found that among the four LICs, involvement in a road crash was reported by 1.8%–2.6% of the population. These accounted for about 12.9% of all injuries. 'Major disability' was reported by an average of 38.5% of those suffering an injury as a result of a road crash. Respondents from Sierra Leone (49.3%) and Uganda (46%) were most likely to report a disability, whereas those from Rwanda (32.8%) and Nepal (21.1%) were less likely. Patterns of injury varied between countries; however, head and extremity injuries remained the most common.

One potential problem comparing the studies above is that of inconsistent definitions. For instance, the international road safety community has not yet settled on a precise definition of 'serious' injury resulting from a road traffic crash [94]. Injury severity in some developed countries has been assessed by means of the maximum abbreviated injury scale (MAIS), i.e., the maximum score of a six-point scale ranging from 1 (minor injury) to 6 (fatal injury) [95]. However, there is no agreement on which of the central MAIS levels should be used to define serious road injuries as a policy indicator. In the Netherlands, for instance, MAIS 2+ is used to indicate a serious road crash [94], while the International Road Traffic and Accident Database (IRTAD), proposes an injury score of MAIS 3+ to define a seriously injured road casualty [12].

A number of studies (see for example [96–101]) provide means to monitor the prevalence of disabilities among populations, but it should be noted that improved health care and vehicle safety does not necessarily imply that there will be fewer disabled people as a result of road traffic crashes. If healthcare and acute care services improve in LICs and in-vehicle protection devices (e.g., airbags) become more prevalent, there are likely to be fewer fatalities resulting from crashes. However, the implication is that more people will survive with non-fatal but extensive injuries, resulting in higher numbers of disabled people [102], implying a greater need in the future for long-term care and rehabilitation facilities for those who survive road crashes, but are permanently disabled. There are obvious cost implications associated with this [96].

### 3.5. Economic Burden of Road Crashes

Since road crash cost estimates are difficult to obtain, there are few studies that specifically focus on road crash costing or cost-effectiveness of interventions in LICs [103]. A recent summary of the economic costs of RTIs included in the Disease Control Priorities project suggests that these costs could range between 1%–2% of a country's GNP [92]. Delays in implementing road safety measures could also impact severely on a nation's wealth and its population's wellbeing [104]. While estimates on the total costs of RTIs vary based on the methodological approaches used, one large 21-country study estimated the global cost of RTIs at US$518 billion [92,105]. Another recent analysis of the economic impact of road traffic injuries led by the World Bank's Global Road Safety Facility found that if countries were able to halve mortality and morbidity due to road traffic injuries and sustain that over 24 years, they could realize significant increases in their GDP—between 7% and 22% [106].

In addition to the above costs, it is important to note that RTIs result in a significant societal burden as well, information about which is important for evidence-based policy making. However, there is a paucity in information in the global literature about the societal costs of RTIs, especially in LMICs, and more studies are needed. This information would provide insight into the consequences of crashes for the economy and social welfare. In fact, road crash costs can be used as a comparator with other policy areas, to help decision-makers prioritise investments. While epidemiological data for crash-related disability from LICs is scarce, the costs of prolonged care, loss of income and consequences for injured parents and their dependents impose financial pressures on families, threatening sustainable livelihoods [107]. Also, the burden of care for long-term illness and disability may fall disproportionately on women and girls [108]. That said, the tangible costs such as lost productivity (indirect cost) and medical costs (direct costs) can be estimated in economic terms more easily compared to the intangible costs such as pain and suffering [109]. However, regardless of difficulties in estimating road crash costs, accounting for economic costs of crashes is necessary to inform policy makers in prioritising and choosing the most effective countermeasures. Three main approaches used in estimating the cost of road crashes [109] are (i) the human capital approach, in which mainly tangible injury costs to individuals are aggregated at societal, regional and national levels; (ii) the willingness-to-pay approach, which derives a value of pain and suffering based on the preferred amounts that people would be prepared to pay to live in a world where risks are reduced; and (iii) the general equilibrium approach, which uses simulation models to estimate costs from a broader macroeconomic perspective, although this is, as yet, untested for injury cost modelling [92]. The human capital approach has been the most common approach used in LMICs, due to the relative simplicity and structured nature of the approach [109].

### 3.6. Vehicle Safety Standards and Dumping of Old Vehicles

An emerging problem for road safety in LICs is the issue of vehicle safety and technologies. While there is a dearth of literature focusing on this issue in LICs, as part of this project, we have been working to liaise with stakeholders in considering lessons learnt from the developed world in this area to understand this issue further. These stakeholders include experts from the Global New Car Assessment Programme (NCAP), and regional car assessment programs.

With respect to emerging vehicle safety features in LICs (e.g., air bags, crash avoidance systems, etc.), one should take into account risk compensation issues. A number of studies have investigated risk compensation issues mostly in the developed world [110–113]. As discussed by Winston et al. [110], road users may become less vigilant about road safety due to innovations that are designed to improve safety. For example, drivers may trade off improved safety for speedier trips [110]. With the expected change in vehicle fleet (and their safety features) in LICs, it is important to understand how the experience of developed countries in this regard could lend itself to LICs. This also highlights the importance of publicity campaigns to raise awareness among road users in LICs with respect to vehicle safety features. This is particularly necessary to educate and train drivers at the very beginning steps of moving towards advanced vehicle safety features in LICs.

With respect to dumping old vehicles in LICs, the issue has been raised by a number of organisations and research studies, which mostly focus on the environmental impact [114]. The mainly old vehicle fleet in LICs does not meet some of the basic safety standards set in developed countries, increasing the propensity of crashes. This may be exacerbated by vehicle modification, poor maintenance standards, inappropriate use (e.g., overloading) and lack of safety enforcement. In addition, when a crash occurs, drivers and passengers would sustain more severe injuries, for example, due to lack of airbags. To our knowledge, scientific studies that investigate safety implications of exporting old vehicles to LICs are non-existent. Policies may be needed to prevent developed countries from dumping vehicles of a certain age or category in LICs and to encourage scrappage in LICs (e.g., cash for clunkers).

### 3.7. Proactive Approaches to Road Safety

Traditional methods that help detect, prioritise and treat high crash-risk sites have been based solely on prior crash history [115]. However, crash data quality in LICs tends to be poor, at limited numbers of sites, and with high rates of under-reporting. Studies of observed unsafe road user behaviour in LMICs do exist, e.g., [116,117], but manual data collection is time consuming and costly. To overcome these constraints, it may be possible to use new forms of data collection techniques such as video data or remote sensors, and storage and manipulation techniques involving 'big data' to allow for a proactive road safety approach that can address safety deficiencies before crashes occur [118]. For instance, assessing global databases could help identify country-specific determinants of road safety [119]. More locally, it is possible to monitor and analyse road users' trajectories and identify conflicts and near misses [120]. A proactive approach to road safety should ideally complement traditional, reactive methods [121], allowing us to design improvement programs (publicity campaigns, engineering interventions, etc.) before crashes happen. However, applying such approaches in LICs may not be straightforward, due to cost and resource constraints.

### 3.8. Limited Data Conditions, Omitted Variables Problem, and Unobserved Heterogeneity

A series of well-known issues often encountered in road safety analysis are related to data limitations of various types. Crash data may be limited in terms of sample size or the number of risk (contributing) factors available in the data resulting in limited data conditions, the omitted variables problem, and unobserved heterogeneity [122–124]. Note that risk factors are needed to explain the safety of a site (highway segment, jurisdiction, etc.).

When a crash data set is not large enough, the maximum likelihood estimation is prone to bias; therefore, the model estimates are biased. This problem can be addressed by employing Bayesian methods in which prior knowledge can be included in the analysis in the form of the prior distribution, leading to enhanced statistical inferences. In this regard, for instance, Heydari et al. [124] showed how it is possible to obtain reliable statistical models for crash data characterised by a small sample size. With respect to the omission of risk factors, road safety literature discusses that, when important variables that have significant explanatory power are missing from the data, road safety inferences could be misleading [122].

Unobserved heterogeneity is related to the omitted variables problem and leads to spurious road safety inferences as indicated in the crash literature [2]. Several risk factors that affect road safety at a site (intersection, road segment, neighbourhood, etc.) are often missing (being unknown or unmeasured) in crash databases, causing the unobserved heterogeneity problem. A large body of literature discusses how to overcome this problem in order to obtain reliable estimates [2,125]. However, most studies that address the abovementioned issues are conducted in the developed world. Advanced statistical methods can mitigate unobserved heterogeneity and omitted variables problems [123], and such methods should be applicable to data sets in LICs, but their use in such contexts has so far been limited.

### 3.9. Reaching out to Other Sectors—Social and Behavioural Aspects

Road safety education and awareness campaigns in high-income settings are often based on changing people's attitudes. Change someone's attitudes, and they will change their behaviour. This may not, however, work across all cultural contexts. For example, previous research has suggested the link between attitudes and behaviours (in a road safety context) to be weaker in Sub-Saharan African countries than in high-income European countries [126,127]. In a study across six countries, McIlroy et al. [128] found that in Kenya, road safety attitudes predicted self-reported pedestrian behaviours to a significantly lesser extent than in the UK. This cannot, however, be explained by national income or road safety statistics. Across the six countries included in their investigation, McIlroy et al. found no patterns in this respect. This strongly points to the need to conduct preliminary research in the country of interest before applying road safety interventions. Every setting has unique characteristics, and interventions should be designed with this in mind.

Culture is a significant factor in this regard. This has been explored by a variety of researchers, with mixed results. In a study of West African taxi drivers, Kouabenan [129] examined religious, mystical, and fatalistic beliefs. Road crashes were attributed more to external factors (such as poor road maintenance and the absence of pedestrian infrastructure) than to driver behaviours (such as the breaking of road laws or carelessness). Additionally, such beliefs were linked with a disregard for safety measures. A strong belief in luck, fate, or destiny, was linked with a perceived lack of need for things such as helmets or seatbelts. Similar findings have since been reported by Dixey [130], Peltzer [131], Omari and Baron-Epel [132], and Maghsoudi et al. [133]. Additionally, not only are stronger beliefs in fate or destiny related to lower engagement in self-protective behaviours, but also with active engagement in risky behaviours. Results to this effect have been found in a variety of settings, including South Africa [134], Turkey and Iran [135,136], and Cameroon [137].

Results are not, however, clear-cut. In a study in Turkey, Yıldırım [138] found religiosity to have a positive effect on self-reported traffic behaviours. Those reporting stronger religious beliefs also reported making fewer risk-taking behaviours. Similarly, McIlroy et al. [139], found stronger beliefs in the influence of God over one's life to be related to safer attitudes and pedestrian behaviours in Kenya and Bangladesh. Note that the opposite pattern was found in China, Thailand, the UK, and Vietnam. Once again, research conducted in the setting of interest is crucial for successful intervention design and implementation.

Road transport is a highly complex sociotechnical system, with influence from a wide variety of individuals and organisations at varying levels of system abstraction [140]. Although it is the end user that carries the weight of the road traffic injury and fatality burden, it is generally not the end user that makes safety intervention decisions [141]. Yet it is this group that generally shoulders the blame [142]. This is not unique to low- and middle-income countries; however, change is beginning to be seen in some high-income countries with acceptance of the 'safe system' or 'vision zero' philosophy. The approach's primary central tenet is the idea that the end user is fallible, and that the system should be designed in a way to reduce the likelihood of crashes and reduce the consequences of crashes that happen. Crucially, blame is not placed solely on the end user, rather it is shared among system actors [143]. Although this way of thinking is gaining traction in the Europe, North America, and Australia, there is as yet a complete lack of systems-level research in low-income settings [144]. To reiterate the sociotechnical perspective, road safety is not the domain of one actor or group of actors alone, but it is the concern of many entities, at many levels of the system, from the end user to the policy maker [140].

### 3.10. Current Capacity for Research and Practice

Capacity for research and practice in the field of road safety is one of the key issues impeding progress in this area in the context of LICs. This is a recurrent theme in all global and regional reports published to date on road safety [1,8,91,92]. Capacity is needed for research as well as planning and implementation of appropriate safety improvement programmes, and any work being done in such settings ought to embed capacity development into it [145,146], as well as consider how to overcome institutional barriers [147].

Initiatives such as the Road Traffic Injuries Research Network [148] and the UNECE programmes in developing countries, in partnership with ECLAC (Latin America and Caribbean region) and ESCAP (Asia and the Pacific region) [149] have aimed to reduce the burden of road crashes in developing countries by identifying and promoting effective, evidenced-based interventions and supporting research capacity building in road safety research in LMICs. Building on the experience gained from such initiatives, there is a need to develop formal training programs that are readily accessible by individuals residing in LMICs. Based on the literature review and through liaising with experts and stakeholders, we have identified two main areas where improving capacity is most needed.

I Limited trained human resources: Many LICs around the world lack adequate human resources in the various areas necessary for effective action on road safety—research, program planning and implementation, as well as monitoring and evaluation [92,149]. One main reason for this is the lack of formal training programs in these settings that are specifically targeted towards the skills necessary for effective and appropriate action on an issue as complex and multi-sectoral as road safety [150]. This limits the amount and type of research conducted locally. For example, road safety studies that focus on data derived from LICs are extremely under-represented in peer-reviewed publications [145], and when they exist, they are often far away from advanced methodological techniques adopted in studies conducted in the developed world. To this end, it is important to train practitioners and researchers residing in LICs in the different areas necessary to effect programmatic change, such as road safety management, research on risk factors for road safety, evaluation of interventions, etc. Incorporating some of these capacity building activities in ongoing engagements and initiatives in LMICs will also have the added benefit of enhancing collaboration between HIC and LMIC researchers and practitioners.

II Lack of data: As set out in Sections 3.1, 3.3, and 3.4, another gap identified in LICs is the limited amount of disaggregated data available for understanding the safety condition and monitoring safety improvement programs when implemented. Capacity development efforts for road safety in LMICs ought to focus on improving data systems such that there is valid, reliable, and timely data available to not only assess the safety condition, but also serve as a basis for assessing the effectiveness of programs or interventions implemented.

## 4. Expert Consultation Results

As described in Section 2.4, six road safety experts, with extensive experience in road safety in LICs, participated in our survey. They provided insight into those topics that were most important to consider for future research in LICs. Two main topics emerged as the most important issues: (i) data collection and management techniques and (ii) governance and legislation. The range of sub-topics and proposed solutions are reported in Table 5. The outcomes from the survey and other discussions also provided context for topics for which literature is unlikely to be available, but which helps provide the background to the future research agenda set out in Section 5.

**Table 5.** Expert elicitation results showing important topics for future road safety.

| Topic | Potential Solutions |
|---|---|
| **Data**: | |
| Inconsistent data collection and management techniques (including under-reporting) | - Improved training and resources for police and hospital staff<br>- Dedicated specialists in crash data reporting<br>- Use of modern technology to ensure accuracy of reporting and efficiency of data storage<br>- Use of multiple data sources<br>- Public awareness raising of importance of accurate data |
| Under-use of road crash cost estimates | - Establish specialised team to promote road safety, and ensure cost implications are explicitly considered in road safety interventions<br>- Disseminate cost evaluation results to reveal shortcomings in current approaches |
| **Governance**: | |
| Lack of integrated approach to road safety issues | - Develop and apply an integrated multi-sectoral, multi-governance framework, possibly using a systems-based approach, including feedback mechanisms and knowledge exchange<br>- Consider how to integrate research into vulnerable road users, policy, and education |
| Lack of accountability of leadership | - Review and re-evaluate current governance mechanisms to improve transparency and avoid redundancy and overlapping of responsibilities<br>- Vision Zero approach<br>- Implement Lead Road Safety Agency to coordinate and champion road safety activities at national, municipal and local levels |
| Poor legislation | - Apply best practice techniques from the developed world |
| Poorly resourced sector | - Increased levels of trained personnel, equipment and knowledge base<br>- Consider use of outsourcing to private sector for certain aspects of road traffic management |

Respondents agreed that the complexities of road safety implementation meant that there are unlikely to be any 'quick wins', although investigating how governance could be improved and held more accountable could initiate action to improve the impact of road safety interventions in the short term. Further responses suggested that the effective use of increased resources and awareness of global best practice could help provide rapid insight into the priority issues, particularly with respect to the safety of vulnerable road users and the expected growth of traffic in developing countries. Additionally, reviewing current practices could help inform government and other stakeholders of the key issues in their specific location, and highlight where current governance mechanisms can be improved, with the aim of achieving greater political and funding commitments from decision-makers, resulting in better legislation and enforcement.

## 5. Discussion: Implications and Future Directions

This section provides a discussion of our findings and their implications grouped in seven major areas. A number of topics emerged from our analysis of the State of Knowledge review above combined with discussions with experts and stakeholders, which can help form a future research agenda. These topics relate to both empirical and methodological frontiers; therefore, they will lead to noteworthy improvements in the way road safety research will be conducted in the context of LICs. Note that topics discussed below are inter-related; nevertheless, their theoretical and empirical weights vary from one topic to another.

### 5.1. *Under-Reporting of Crashes*

This is a major issue in LICs; research is needed not only to understand causes of under-reporting but also to develop methodological approaches that can better address the issue. The use of capture-recapture techniques has been shown to be a useful tool to estimate the levels of under-reporting in low-income countries, and as such could be used when authorities wish to understand the true nature of road traffic crash rates in their country. To better address the under-reporting problem, alternative statistical methods that overcome the limitations of the capture-recapture approach (discussed in Section 3.1) should be examined and/or developed. It is also advisable to carry out a review of police and hospital data availability in LICs and LMICs, in order to understand better the nature of the types of data generally available. Following such a review, a general 'toolkit' could be developed offering guidance and methodologies for the analysis of under-reporting in LICs and LMICs. Such guidance could incorporate minimum data requirements, software tools and reporting templates in order to standardise such reporting in all LICs and LMICs in the future.

### 5.2. *Traffic Injuries Sustained in the Crash*

Understanding the causes, severities, long-term implications, and costs to society of disabilities resulting from road crashes could help refine road safety policies in LICs and improve impact analysis of road safety interventions. There are inconsistencies in the metrics and methods used to assess injury severities (as discussed in Section 3.4), and further work could seek to build on any review of road safety and healthcare data recording systems suggested in Section 5.1 (under-reporting), and aim to assess the most appropriate metrics for LICs to use in analyses of disabilities. Further research could help to develop methodologies for road safety and healthcare practitioners in LICs to understand not only the trends associated with disabilities resulting from road crashes but also their impact on society and the economy. It may be appropriate to focus such future research on vulnerable groups. Also, a surprising gap in previous road safety research conducted in LICs relates to the lack of studies that aim at analysing and understanding differing injury-severity levels properly. That is, how different factors increase or decrease the likelihood of injury-severity sustained by road users. Such analyses are common in the developed world and help draw a complete picture, allowing decision makers to design cost-effective countermeasures to reduce injuries once a crash occurs. In this regard, further research is thus needed in LICs.

### 5.3. Road Crash Costing

There are limitations in the methods used to apply cost estimates to road crashes in LMICs [151] and in LICs in particular. Wesson et al. [152] carried out a review of economic evaluations in LMICs, finding only three studies that assessed the costs of road safety interventions, and six studies that were cost-effectiveness analyses. Only one of these papers referred specifically to LICs [153]. Similarly, Banstola and Mytton [103] only found five studies assessing cost-effectiveness in LMICs, with only two interventions showing moderate evidence of being cost-effective. More recently, Mukama et al. [154] assessed the costs of unintentional injuries to children in a Ugandan slum, noting that costs associated with road traffic crashes are higher than those for incidents occurring at a school, due to the severe nature of most road traffic injuries requiring specialised care and hence higher treatment costs. This lack of LIC-based research indicates that further research is needed to identify relevant methods of road crash cost estimations or cost-benefit analyses of road safety interventions applicable in those settings.

### 5.4. Characterisations of the Vehicle Fleet

With respect to the vehicle safety features and risk compensation concerns discussed in Section 3.6, further research is needed to better understand risk compensation issues in LICs to be able to take advantage of emerging safety innovations more fully. In addition, it is important to explore issues surrounding the second hand vehicle market and how countries that are heavily reliant on imported second-hand vehicles can regulate more appropriately, as well as vehicle technologies. In this regard, scientific research is needed to quantify road safety implications of dumping old vehicles in LICs.

### 5.5. Challenges of Data Collection and Analysis

As set out in Section 3, and identified by the expert survey, many of the issues pertinent to this study relate to the challenges of data collection, management and analysis. While it may be possible to draw on experiences of global good practice, it may also be relevant to develop specific methods and analysis techniques that apply in the LIC-context. The following sections discuss these issues in greater detail.

#### 5.5.1. Expected Increase in Traffic Volume and Its Implications

Previous research indicates that as a country's economy grows and traffic volume (and consequently exposure) increases, road safety deteriorates. However, the relationship is not linear and varies from one jurisdiction to another [155,156]. It is important to quantify the rate of deterioration in road safety as traffic exposure increases; and consequently, investigate how we can reduce this rate. For example, using advanced statistical methods, a study conducted by Heydari et al. [157] shows how with the same set of variables available in the data one can understand variation in crash frequency as traffic exposure increases. This is important since an increase in traffic volume in LICs seems inevitable in the near future. Further research is thus needed to understand the relationship between road safety and traffic exposure in LICs. To this end, a series of safety performance functions should be developed for different road infrastructures in LICs, similar to those developed in the Highway Safety Manual 2010 [158]. Note that safety performance functions are the main ingredient for the six-step safety management process described in the Highway Safety Manual [158]. They are used to quantify road safety, to understand factors affecting safety, and to identify hazardous locations that should then be prioritised for safety improvement programs.

#### 5.5.2. Accounting for Data-Related Limitations and Unobserved Heterogeneity

As previously discussed, in LICs, crash data tends to be lacking or when it exists, it is often limited, in terms of sample size limitations and/or the lack of risk factors available in the data. Therefore, issues discussed in Section 3.8 are often more prevalent and frequent in LICs compared to the developed world. The number of road safety studies conducted in LICs is limited, and rarely do those studies employ rigorous road safety and statistical techniques to address the aforementioned issues properly. One aspect of future research should be to continue work in this area by developing statistical methods, especially those applicable to LIC-contexts, and help provide robust analyses of road safety data.

#### 5.5.3. Feasibility of Proactive Approaches to Road Safety in LICs

Although we recognise challenges for implementing proactive road safety approaches in LICs, it is important to investigate how this could help mitigate crash risk propensities in LICs. Building on the successful experience of developed countries in implementing such approaches, while considering resource and cost constraints applicable to LIC-contexts, it would be possible to adapt proactive methods in LICs to optimise their benefits with a minimum cost. Initially, this could be treated as a range of feasibility studies to extend the knowledge base to LICs. Perhaps, it would be possible to focus on monitoring a limited number of road infrastructures and based on that provide valuable insights, which in turn allow for designing tailored safety improvements transferable to other similar locations within each jurisdiction. To summarise, proactive safety approaches should be considered in the future if a major improvement in road safety in LICs is expected to be achieved.

#### 5.5.4. Alternative Approaches to Obtain Data

With respect to data issues and limitations encountered by LICs, which also relates to limited capacity in these countries, a valuable and cost-effective approach could be based on using street imagery, for example, available in Google street view or shared on social networks. Previous research has successfully taken advantage of street imagery, for example, to identify travel patterns [159]. Estimating travel patterns and/or traffic exposure by different modes of transport and/or road user types could be promising when such information is not available (e.g., when data are not collected through traditional data collection techniques). Also, automatic data collection, for example, using traffic analyser sensors has been shown to provide valuable data that allows the analyst to conduct road safety studies [160,161]. This is particularly important given the expected increase in traffic exposure in LICs since traffic exposure is known as a major determinant of road safety. Therefore, it would be interesting to investigate how this could be employed to collect relevant data and improve safety in LICs.

### *5.6. Social and Behavioural Approaches to Road Safety*

As described in Section 3.2, the majority of road safety research is performed in high-income countries; hence, the majority of methods are biased towards these settings. As such, there is a strong requirement for research methods developed in low-income settings. This can also be said for road safety interventions. Given differences in cultures, attitudes, and behaviour constraints, it may not be the case that what has worked in Europe or America will also work in LICs. As such, interventions should be designed preferably based on research conducted in the country in which the intervention is to be implemented. Therefore, there is a clear need for more in-depth research on the social and behavioural factors that influence road safety in low-income settings. There is also a strong need to embed sociotechnical systems thinking in crash analyses, and in road transport policy, planning, and construction. This is beginning to happen in high-income countries; however, LICs are being left behind in this respect. As such, a concerted effort to apply contemporary sociotechnical systems methods to the analysis of road transport in LICs is needed. The design of road safety interventions should be based on a good understanding of the context of application and a consideration of all the factors that influence outcomes. Neither of these points are typically addressed in LICs; as such, they represent important topics for research.

### *5.7. How to Build Sustainable Capacity Effectively and in a Timely Manner?*

Two of the future road safety challenges identified by the WHO [10] are 'Building Capacity' and 'Strengthening Data Collection', and effective training programmes could help improve both of these aspects of road safety. As discussed in Section 3.10, limited trained human resources and a lack of data are two areas where capacity building is most warranted. Although some previous studies have investigated related issues in LICs, there is a need to continue work in this area in order to systematically identify and/or define the most cost-effective and sustainable strategies that can be in place in a timely fashion.

## 6. Summary

The general objective of our research was to help improve road safety in LICs, identifying and targeting the most important problems encountered by these countries and defining critical future research directions that help enhance safety effectively given the expected increase in traffic exposure in LICs. Following recent developments in the field, including the UN Decade of Action for Road Safety (2010–2020), this article contributes to the literature by reviewing the state of knowledge and recommending a future research agenda for further improvements. Our study reveals where some of the major knowledge gaps exist for those topics that have been part of the research arena for some time. To this end, not only have we conducted a literature review, but we have also liaised with international organisations, local authorities, experts and professionals, particularly where the body of literature does not exist. Thus, we have identified some of the most critical road safety concerns in LICs; consequently, we have suggested some areas of future research that could be considered to inform an agenda for future action.

Specifically, we identified ten focus areas: (i) under-reporting of road crashes in LICs; (ii) lessons learned from global best practices; (iii) vulnerable groups and gender disaggregation; (iv) disabilities due to road crashes; (v) economic burden of road crashes; (vi) vehicle safety standards and the dumping of old vehicles; (vii) proactive approaches to road safety; (viii) limited data conditions, omitted variables and unobserved heterogeneity; (ix) reaching out to other sectors, considering social and behavioural aspects; and (x) capacity limitations of road safety research. Based on our analyses, we conclude that road safety is inherently a multi-sectoral issue, where interventions will need to involve multiple strategies and stakeholders. The most successful programmes globally have been those that have integrated systems of legislation, regulation and enforcement, combining robust data collection and management systems, economic evaluation systems to inform investment decisions, significant technical and enforcement capacity, and a substantial knowledge base of the social, medical and behavioural implications of road safety interventions.

Seven major research directions were identified based on the above-mentioned focus areas. In summary, in terms of policy recommendations, given the poor resources and lack of capacity for data management available in LICs, improving the quality of data in these countries would be one of the initial steps to make any improvements, either through investing in capacity for analysis and research, or through the development of modern techniques of data collection. Raising public awareness of the importance of accurate data and the reporting of road crashes is another central issue that should be considered by governments. We also suggest investigations into data analysis techniques considering both the statistical foundations upon which such analyses are built, and the use of proactive measures to prioritise investments, all carried out with a focus on existing cost and resource limitations in LICs, and using evidence-based techniques to promote effective changes. Combining expert insights and experiences with research from LICs, LMICs, and developed countries should provide the basis for a robust approach and a future research agenda that will help improve road safety in LICs.

**Author Contributions:** S.H., A.H., J.T., and A.M.B. were involved in drawing the conceptual plan of the paper. S.H. and A.H. wrote the first draft of the manuscript. R.M. wrote the sections relating to social and behavioural aspects. A.M.B. and J.T. provided valuable inputs on the content of the article.

**Funding:** This research was funded by the UK Department for International Development.

**Acknowledgments:** This research was funded by UK AID through the UK Department for International Development under the High Volume Transport Applied Research Programme, managed by IMC Worldwide. The authors are grateful to the panel of experts, who participated in the survey, and anonymous reviewers for their valuable comments.

**Conflicts of Interest:** The authors declare no conflict of interest.

## References

1. World Health Organization. *Global Status Report on Road Safety 2018*; World Health Organization: Geneva, Switzerland, 2018.
2. Mannering, F.L.; Shankar, V.; Bhat, C.R. Unobserved heterogeneity and the statistical analysis of highway accident data. *Anal. Methods Accid. Res.* **2016**, *11*, 1–16. [CrossRef]
3. World Health Organization. *Global Plan. for the Decade of Action for Road Safety 2011–2020*; World Health Organization: Geneva, Switzerland, 2011.
4. Global Road Safety Partnership. *Road Map: Strategic Plan 2015–2020*; Global Road Safety Partnership: Geneva, Switzerland, 2016.
5. World Bank. *Global Road Safety Facility Annual Report 2018*; World Bank: Washington, DC, USA, 2018.
6. Bloomberg Philanthropies Road Safety. Available online: https://www.bloomberg.org/program/public-health/road-safety/ (accessed on 24 April 2019).
7. Hyder, A.A.; Allen, K.A.; Peters, D.H.; Chandran, A.; Bishai, D. Large-scale road safety programmes in low-and middle-income countries: An opportunity to generate evidence. *Glob. Public Health* **2013**, *8*, 504–518. [CrossRef] [PubMed]
8. Hyder, A.A.; Paichadze, N.; Toroyan, T.; Peden, M.M. Monitoring the Decade of Action for Global Road Safety 2011–2020: An update. *Glob. Public Health* **2017**, *12*, 1492–1505. [CrossRef] [PubMed]
9. World Road Association Safe System Principles. Road Safety Manual. Available online: https://roadsafety.piarc.org/en/road-safety-management-safe-system-approach/safe-system-principles (accessed on 28 October 2019).
10. World Health Organization. Decade of Action for Road Safety 2011–2020. Available online: https://www.who.int/roadsafety/decade_of_action/en/ (accessed on 28 October 2019).
11. Moher, D.; Shamseer, L.; Clarke, M.; Ghersi, D.; Liberati, A.; Petticrew, M.; Shekelle, P.; Stewart, L.A.; Group, P.-P. Preferred reporting items for systematic review and meta-analysis protocols (PRISMA-P) 2015 statement. *Syst. Rev.* **2015**, *4*, 1. [CrossRef] [PubMed]
12. IRTAD. *Reporting on Serious Road Traffic Casualties*; International Traffic Safety Data and Analysis Group: Paris, France, 2011.
13. Magoola, J.; Kobusingye, O.; Bachani, A.M.; Tumwesigye, N.M.; Kimuli, D.; Paichadze, N. Estimating road traffic injuries in Jinja district, Uganda, using the capture-recapture method. *Int. J. Inj. Contr. Saf. Promot.* **2018**, *25*, 341–346. [CrossRef] [PubMed]
14. Abegaz, T.; Berhane, Y.; Worku, A.; Assrat, A.; Assefa, A. Road traffic deaths and injuries are under-reported in Ethiopia: A capture-recapture method. *PLoS ONE* **2014**, *9*, e103001. [CrossRef] [PubMed]
15. Dandona, R.; Kumar, G.A.; Ameer, M.A.; Reddy, G.B.; Dandona, L. Under-reporting of road traffic injuries to the police: Results from two data sources in urban India. *Inj. Prev.* **2008**, *14*, 360–365. [CrossRef]
16. Samuel, J.C.; Sankhulani, E.; Qureshi, J.S.; Baloyi, P.; Thupi, C.; Lee, C.N.; Miller, W.C.; Cairns, B.A.; Charles, A.G. Under-reporting of road traffic mortality in developing countries: Application of a capture-recapture statistical model to refine mortality estimates. *PLoS ONE* **2012**, *7*, e31091. [CrossRef]
17. Bhatti, J.A.; Razzak, J.A.; Lagarde, E.; Salmi, L.R. Differences in police, ambulance, and emergency department reporting of traffic injuries on Karachi-Hala road, Pakistan. *BMC Res. Notes* **2011**, *4*, 75. [CrossRef]
18. Bonnet, E.; Nikiéma, A.; Traoré, Z.; Sidbega, S.; Ridde, V. Technological solutions for an effective health surveillance system for road traffic crashes in Burkina Faso. *Glob. Health Action* **2017**, *10*, 1295698. [CrossRef]
19. Patel, A.; Krebs, E.; Andrade, L.; Rulisa, S.; Vissoci, J.R.N.; Staton, C.A. The epidemiology of road traffic injury hotspots in Kigali, Rwanda from police data. *BMC Public Health* **2016**, *16*, 697. [CrossRef] [PubMed]
20. Amoros, E.; Martin, J.-L.; Laumon, B. Estimating non-fatal road casualties in a large French county, using the capture–recapture method. *Accid. Anal. Prev.* **2007**, *39*, 483–490. [CrossRef] [PubMed]

21. Amoros, E.; Martin, J.-L.; Laumon, B. Under-reporting of road crash casualties in France. *Accid. Anal. Prev.* **2006**, *38*, 627–635. [CrossRef] [PubMed]
22. Amoros, E.; Martin, J.-L.; Lafont, S.; Laumon, B. Actual incidences of road casualties, and their injury severity, modelled from police and hospital data, France. *Eur. J. Public Health* **2008**, *18*, 360–365. [CrossRef] [PubMed]
23. Amoros, E.; Soler, G.; Pascal, L.; Ndaiye, A.; Gadegbeku, B.; Martin, J.-L. Estimation of the number of seriously injured road users in France, 2006–2015. *Rev. Epidemiol. Sante Publique* **2018**, *66*, S334. [CrossRef]
24. Broughton, J.; Keigan, M.; Yannis, G.; Evgenikos, P.; Chaziris, A.; Papadimitriou, E.; Bos, N.M.; Hoeglinger, S.; Pérez, K.; Amoros, E.; et al. Estimation of the real number of road casualties in Europe. *Saf. Sci.* **2010**, *48*, 365–371. [CrossRef]
25. Bauer, R.; Steiner, M.; Khnelt-Leddihn, A.; Rogmans, W.; Kisser, R. Under-reporting of vulnerable road users in official EU road accident statistics—Implications for road safety and added value of EU IDB hospital data. *Inj. Prev.* **2018**, *24*, A266.
26. Chini, F.; Farchi, S.; Camilloni, L.; Borgia, P.; Antoniozzi, T.; Ciaramella, I.; Valenti, M.; Rossi, P. Road traffic injuries in one local health unit in the Lazio region: Results of a surveillance system integrating police and health data. *Int. J. Health Geogr.* **2009**, *8*, 21. [CrossRef]
27. Ferrando, J.; Plasencia, A.; Ricart, I.; Oros, M.; Borrell, C. The contribution of independent health and police sources of information to the ascertainment of traffic deaths in a European urban setting. In Proceedings of the 43rd Annual Conference of the Association for the Advancement of Automotive Medicine, Barcelona, Spain, 20–21 September 1999.
28. Reurings, M.C.B.; Stipdonk, H.L. Estimating the number of serious road injuries in The Netherlands. *Ann. Epidemiol.* **2011**, *21*, 648–653. [CrossRef]
29. Ward, H.; Lyons, R.; Thoreau, R. *Road Safety Research Report No. 69. Under-Reporting of Road Casualties—Phase 1*; Department for Transport: London, UK, 2006.
30. Yannis, G.; Papadimitriou, E.; Chaziris, A.; Broughton, J. Modeling road accident injury under-reporting in Europe. *Eur. Transp. Res. Rev.* **2014**, *6*, 425–438. [CrossRef]
31. Conderino, S.; Fung, L.; Sedlar, S.; Norton, J.M. Linkage of traffic crash and hospitalization records with limited identifiers for enhanced public health surveillance. *Accid. Anal. Prev.* **2017**, *101*, 117–123. [CrossRef] [PubMed]
32. Sciortino, S.; Vassar, M.; Radetsky, M.; Knudson, M.M. San Francisco pedestrian injury surveillance: Mapping, under-reporting, and injury severity in police and hospital records. *Accid. Anal. Prev.* **2005**, *37*, 1102–1113. [CrossRef] [PubMed]
33. Alsop, J.; Langley, J. Under-reporting of motor vehicle traffic crash victims in New Zealand. *Accid. Anal. Prev.* **2001**, *33*, 353–359. [CrossRef]
34. Meuleners, L.B.; Lee, A.H.; Cercarelli, L.R.; Legge, M. Estimating crashes involving heavy vehicles in Western Australia, 1999–2000: A capture–recapture method. *Accid. Anal. Prev.* **2006**, *38*, 170–174. [CrossRef] [PubMed]
35. Cercarelli, L.R.; Rosman, D.L.; Ryan, G.A. Comparison of accident and emergency with police road injury data. *J. Trauma* **1996**, *40*, 805–809. [CrossRef]
36. Lujic, S.; Finch, C.; Boufous, S.; Hayen, A.; Dunsmuir, W. How comparable are road traffic crash cases in hospital admissions data and police records? An examination of data linkage rates. *Aust. N. Z. J. Public Health* **2008**, *32*, 28–33. [CrossRef]
37. Rosman, D.L. The Western Australian Road Injury Database (1987–1996): Ten years of linked police, hospital and death records of road crashes and injuries. *Accid. Anal. Prev.* **2001**, *33*, 81–88. [CrossRef]
38. Watson, A.; Watson, B.; Vallmuur, K. Estimating under-reporting of road crash injuries to police using multiple linked data collections. *Accid. Anal. Prev.* **2015**, *83*, 18–25. [CrossRef]
39. Ma, S.; Li, Q.; Zhou, M.; Duan, L.; Bishai, D. Road traffic injury in China: A review of national data sources. *Traffic Inj. Prev.* **2012**, *13*, 57–63. [CrossRef]
40. Nakahara, S.; Wakai, S. Underreporting of traffic injuries involving children in Japan. *Inj. Prev.* **2001**, *7*, 242–244. [CrossRef]
41. Ahmed, A.; Sadullah, A.F.M.; Yahya, A.S. Errors in accident data, its types, causes and methods of rectification-analysis of the literature. *Accid. Anal. Prev.* **2017**, *130*, 3–21. [CrossRef] [PubMed]
42. Morrison, A.; Stone, D.H. Capture-recapture: A useful methodological tool for counting traffic related injuries? *Inj. Prev.* **2000**, *6*, 299–304. [CrossRef] [PubMed]

43. Van Hest, R.; Grant, A.; Abubakar, I. Quality assessment of capture–recapture studies in resource-limited countries. *Trop. Med. Int. Health* **2011**, *16*, 1019–1041. [CrossRef] [PubMed]
44. Carter, K.L.; Williams, G.; Tallo, V.; Sanvictores, D.; Madera, H.; Riley, I. Capture-recapture analysis of all-cause mortality data in Bohol, Philippines. *Popul. Health Metr.* **2011**, *9*, 9. [CrossRef]
45. Yamamoto, T.; Hashiji, J.; Shankar, V.N. Underreporting in traffic accident data, bias in parameters and the structure of injury severity models. *Accid. Anal. Prev.* **2008**, *40*, 1320–1329. [CrossRef]
46. Ye, F.; Lord, D. Investigation of effects of underreporting crash data on three commonly used traffic crash severity models. *Transp. Res. Rec. J. Transp. Res. Board* **2011**, *2241*, 51–58. [CrossRef]
47. Patil, S.; Geedipally, S.R.; Lord, D. Analysis of crash severities using nested logit model—Accounting for the underreporting of crashes. *Accid. Anal. Prev.* **2012**, *45*, 646–653. [CrossRef]
48. Sango, H.A.; Testa, J.; Meda, N.; Contrand, B.; Traore, M.S.; Staccini, P.; Lagarde, E. Mortality and morbidity of urban road traffic crashes in Africa: Capture-recapture estimates in Bamako, Mali, 2012. *PLoS ONE* **2016**, *11*, e0149070. [CrossRef]
49. Wegman, F. The future of road safety: A worldwide perspective. *IATSS Res.* **2017**, *40*, 66–71. [CrossRef]
50. Traffic General Directorate. *Spanish Road Safety Strategy 2011–2020*; Traffic General Directorate: Madrid, Spain, 2010.
51. Wang, C.; Quddus, M.A.; Ison, S.G. The effect of traffic and road characteristics on road safety: A review and future research direction. *Saf. Sci.* **2013**, *57*, 264–275. [CrossRef]
52. Safer Africa SaferAfrica Project. Available online: http://www.saferafrica.eu/ (accessed on 7 June 2019).
53. Bonnet, E.; Lechat, L.; Ridde, V. What interventions are required to reduce road traffic injuries in Africa? A scoping review of the literature. *PLoS ONE* **2018**, *13*, e0208195. [CrossRef] [PubMed]
54. Gupta, M.; Menon, G.R.; Devkar, G.; Thomson, H. *Regulatory and Road Engineering Interventions for Preventing Road Traffic Injuries and Fatalities among Vulnerable (NON-Motorised and Motorised Two-Wheel) Road Users in Low-and Middle-Income Countries*; EPPI-Centre, Social Science Research Unit, Institute of Education, University of London: London, UK, 2016.
55. Staton, C.; Vissoci, J.; Gong, E.Y.; Toomey, N.; Wafula, R.; Abdelgadir, J.; Zhou, Y.; Liu, C.; Pei, F.D.; Zick, B.; et al. Road traffic injury prevention initiatives: A systematic review and metasummary of effectiveness in low and middle income countries. *PLoS ONE* **2016**, *11*, e0144971.
56. Esperato, A.; Bishai, D.; Hyder, A.A. Projecting the health and economic impact of road safety initiatives: A case study of a multi-country project. *Traffic Inj. Prev.* **2012**, *13*, 82–89. [CrossRef] [PubMed]
57. Abegaz, T.; Berhane, Y.; Worku, A.; Assrat, A. Effectiveness of an improved road safety policy in Ethiopia: An interrupted time series study. *BMC Public Health* **2014**, *14*, 539. [CrossRef] [PubMed]
58. Ahmad, H.; Naeem, R.; Feroze, A.; Zia, N.; Shakoor, A.; Khan, U.R.; Mian, A.I. Teaching children road safety through storybooks: An approach to child health literacy in Pakistan. *BMC Pediatr.* **2018**, *18*, 1–8. [CrossRef] [PubMed]
59. Li, Q.; Alonge, O.; Hyder, A.A. Children and road traffic injuries: Can't the world do better? *Arch. Dis. Child.* **2016**, *101*, 1063–1070. [CrossRef]
60. Zimmerman, K.; Jinadasa, D.; Maegga, B.; Guerrero, A. Road traffic injury on rural roads in Tanzania: Measuring the effectiveness of a road safety program. *Traffic Inj. Prev.* **2015**, *16*, 456–460. [CrossRef]
61. Salmon, R.; Eckersley, W. Where there's no green man: Child road-safety education in Ethiopia. *Dev. Pract.* **2010**, *20*, 726–733. [CrossRef]
62. Bradbury, A.; Quimby, A. Community road safety education: An international perspective. *Munic. Eng.* **2008**, *161*, 137–143. [CrossRef]
63. Blantari, J.; Asiamah, G.; Appiah, N.; Mock, C. An evaluation of the effectiveness of televised road safety messages in Ghana. *Int. J. Inj. Control Saf. Promot.* **2005**, *12*, 23–29. [CrossRef]
64. Bachani, A.M.; Mogere, S.; Hyder, A.A.; Hung, Y.W.; Nyamari, J.; Akunga, D. Helmet wearing in Kenya: Prevalence, knowledge, attitude, practice and implications. *Public Health* **2017**, *144*, S23–S31. [CrossRef] [PubMed]
65. Bachani, A.M.; Risko, C.B.; Gnim, C.; Coelho, S.; Hyder, A.A. Knowledge, attitudes, and practices around drinking and driving in Cambodia: 2010–2012. *Public Health* **2017**, *144*, S32–S38. [CrossRef] [PubMed]
66. Stewart, K.; Bivans, B.; Nguyen, L.H.; Onigbogi, L. Drink driving prevention initiatives in low-and middle-income countries: Challenges and progress in Nigeria and Vietnam. In Proceedings of the 20th International Conference on Alcohol, Drugs and Traffic Safety, Brisbane, Australia, 25–28 August 2013.

67. Bachani, A.M.; Hung, Y.W.; Mogere, S.; Akungah, D.; Nyamari, J.; Stevens, K.A.; Hyder, A.A. Prevalence, knowledge, attitude and practice of speeding in two districts in Kenya: Thika and Naivasha. *Injury* **2013**, *44*, S24–S30. [CrossRef]
68. Kraemer, J.D.; Honermann, B.J.; Roffenbender, J.S. Cyclists' helmet usage and characteristics in central and southern Malawi: A cross-sectional study. *Int. J. Inj. Control Saf. Promot.* **2012**, *19*, 373–377. [CrossRef] [PubMed]
69. Johnson, M. Assessing a country's drink driving situation: An overview of the method used in 6 low-and middle-income countries. *Traffic Inj. Prev.* **2012**, *13*, 96–100. [CrossRef] [PubMed]
70. Tulu, G.S.; Haque, M.M.; Washington, S.; King, M.J. Investigating pedestrian injury crashes on modern roundabouts in Addis Ababa, Ethiopia. *Transp. Res. Rec. J. Transp. Res. Board* **2015**, *2512*, 1–10. [CrossRef]
71. John Hopkins International Injury Research Unit. *Risky Road Behaviors in Addis Ababa*; John Hopkins International Injury Research Unit: Baltimore, MD, USA, 2019.
72. UNRSC—United Nations Road Safety Collaboration Publications. Available online: https://www.who.int/roadsafety/publications/en/ (accessed on 6 June 2019).
73. Wismans, J.; Skogsmo, I.; Nilsson-Ehle, A.; Lie, A.; Thynell, M.; Lindberg, G. Commentary: Status of road safety in Asia. *Traffic Inj. Prev.* **2016**, *17*, 217–225. [CrossRef]
74. World Health Organization. *Global Status Report on Road Safety 2013*; World Health Organization: Geneva, Switzerland, 2013.
75. Ojo, T.; Adetona, C.O.; Agyemang, W.; Afukaar, F.K. Pedestrian risky behavior and safety at zebra crossings in a Ghanaian metropolitan area. *Traffic Inj. Prev.* **2019**, *20*, 216–219. [CrossRef]
76. Nantulya, V.M.; Reich, M.R. The neglected epidemic: Road traffic injuries in developing countries. *BMJ* **2002**, *324*, 1139–1141. [CrossRef]
77. Nantulya, V.M.; Reich, M.R. Equity dimensions of road traffic injuries in low-and middle-income countries. *Inj. Control Saf. Promot.* **2003**, *10*, 13–20. [CrossRef]
78. Lagarde, E. Road traffic injury is an escalating burden in Africa and deserves proportionate research efforts. *PLoS Med.* **2007**, *4*, 170. [CrossRef] [PubMed]
79. Towner, E.; Towner, J. Child injury in a changing world. *Glob. Public Health* **2009**, *4*, 402–413. [CrossRef] [PubMed]
80. Odero, W.; Garner, P.; Zwi, A. Road traffic injuries in developing countries: A comprehensive review of epidemiological studies. *Trop. Med. Int. Health* **1997**, *2*, 445–460. [CrossRef] [PubMed]
81. Peden, M.; Kobusingye, O.; Monono, M.E. Africa's roads—The deadliest in the world. *SAMJ S. Afr. Med. J.* **2013**, *103*, 228–229. [CrossRef] [PubMed]
82. Turner, J.; Fletcher, J. *TI-UP Enquiry: Gender and Road Safety*; TI-UP Resource Centre: London, UK, 2007.
83. UN DESA. *The World's Women 2015: Trends and Statistics*; UN Department of Economic and Social Affairs, Ed.; UN DESA: New York, NY, USA, 2015.
84. Damsere-Derry, J.; Palk, G.; King, M. Road accident fatality risks for "vulnerable" versus "protected" road users in northern Ghana. *Traffic Inj. Prev.* **2017**, *18*, 736–743. [CrossRef] [PubMed]
85. Uteng, T.P. *Gender and Mobility in the Developing World*; World Bank: Washington, DC, USA, 2012.
86. Remacle, E.; Sanon, C.; Aketch, S. *Safer Africa: Thematic Fact Sheet on Gender*; European Commission: Brussels, Belgium, 2018.
87. Venter, C.; Vokolkova, V.; Michalek, J. Gender, residential location, and household travel: Empirical findings from low-income urban settlements in Durban, South Africa. *Transp. Rev.* **2007**, *27*, 653–677. [CrossRef]
88. Peters, D. *Gender and Sustainable Urban Mobility*; UN Habitat: Kenya, Nairobi, 2013.
89. Hyder, A.A.; Razzak, J.A. The challenges of injuries and trauma in Pakistan: An opportunity for concerted action. *Public Health* **2013**, *127*, 699–703. [CrossRef]
90. Üzümcüoğlu, Y.; Özkan, T.; Lajunen, T.; Morandi, A.; Orsi, C.; Papadakaki, M.; Chliaoutakis, J. Life quality and rehabilitation after a road traffic crash: A literature review. *Transp. Res. Part. F Traffic Psychol. Behav.* **2016**, *40*, 1–13. [CrossRef]
91. Peden, M. *World Report on Road Traffic Injury Prevention*; World Health Organization: Geneva, Switzerland, 2004.
92. Bachani, A.M.; Peden, M.; Gururaj, G.; Norton, R.; Hyder, A.A. Road traffic injuries. In *Injury Prevention and Environmental Health. Disease Control Priorities Volume 7*; Mock, C.N., Nugent, R., Kobusingye, O., Smith, K.R., Eds.; World Bank Publications: Washington, DC, USA, 2017.

93. Zafar, S.N.; Canner, J.K.; Nagarajan, N.; Kushner, A.L.; Gupta, S.; Canner, J.K.; Tran, T.M.; Nagarajan, N.; Stewart, B.T.; Kamara, T.B.; et al. Road traffic injuries: Cross-sectional cluster randomized countrywide population data from 4 low-income countries. *Int. J. Surg.* **2018**, *52*, 237–242. [CrossRef]
94. Polinder, S.; Haagsma, J.; Bos, N.; Panneman, M.; Wolt, K.K.; Brugmans, M.; Weijermars, W.; van Beeck, E. Burden of road traffic injuries: Disability-adjusted life years in relation to hospitalization and the maximum abbreviated injury scale. *Accid. Anal. Prev.* **2015**, *80*, 193–200. [CrossRef]
95. Petrucelli, E.; States, J.D.; Hames, L.N. The abbreviated injury scale: Evolution, usage and future adaptability. *Accid. Anal. Prev.* **1981**, *13*, 29–35. [CrossRef]
96. Alemany, R.; Ayuso, M.; Guillén, M. Impact of road traffic injuries on disability rates and long-term care costs in Spain. *Accid. Anal. Prev.* **2013**, *60*, 95–102. [CrossRef] [PubMed]
97. Rissanen, R.; Berg, H.-Y.; Hasselberg, M. Quality of life following road traffic injury: A systematic literature review. *Accid. Anal. Prev.* **2017**, *108*, 308–320. [CrossRef] [PubMed]
98. Bachani, A.M.; Galiwango, E.; Kadobera, D.; Bentley, J.A.; Bishai, D.; Wegener, S.; Hyder, A.A. A new screening instrument for disability in low-income and middle-income settings: Application at the Iganga-Mayuge Demographic Surveillance System (IM-DSS), Uganda. *BMJ Open* **2014**, *4*, e005795. [CrossRef] [PubMed]
99. Bachani, A.M.; Galiwango, E.; Kadobera, D.; Bentley, J.A.; Bishai, D.; Wegener, S.; Zia, N.; Hyder, A.A. Characterizing disability at the Iganga-Mayuge Demographic Surveillance System (IM-DSS), Uganda. *Disabil. Rehabil.* **2016**, *38*, 1291–1299. [CrossRef] [PubMed]
100. Palmera-Suárez, R.; López-Cuadrado, T.; Brockhaus, S.; Fernández-Cuenca, R.; Alcalde-Cabero, E.; Galán, I. Severity of disability related to road traffic crashes in the Spanish adult population. *Accid. Anal. Prev.* **2016**, *91*, 36–42. [CrossRef] [PubMed]
101. Palmera-Suárez, R.; López-Cuadrado, T.; Fernández-Cuenca, R.; Alcalde-Cabero, E.; Galán, I. Inequalities in the risk of disability due to traffic injuries in the Spanish adult population, 2009–2010. *Injury* **2018**, *49*, 549–555. [CrossRef]
102. Lund, J.; Bjerkedal, T. Permanent impairments, disabilities and disability pensions related to accidents in Norway. *Accid. Anal. Prev.* **2001**, *33*, 19–30. [CrossRef]
103. Banstola, A.; Mytton, J. Cost-effectiveness of interventions to prevent road traffic injuries in low- and middle-income countries: A literature review. *Traffic Inj. Prev.* **2017**, *18*, 357–362. [CrossRef]
104. Martin, A.; Lagarde, E.; Salmi, L.R. Burden of road traffic injuries related to delays in implementing safety belt laws in low-and lower-middle-income countries. *Traffic Inj. Prev.* **2018**, *19*, S1–S6. [CrossRef]
105. Jacobs, G.; Aeron-Thomas, A.; Astrop, A. *Estimating Global Road Fatalities*; Transport Research Laboratory: Crowthorne, UK, 2000.
106. The World Bank. *The High Toll of Traffic Injuries: Unacceptable and Preventable*; The World Bank: Washington, DC, USA, 2017.
107. Ameratunga, S.; Hijar, M.; Norton, R. Road-traffic injuries: Confronting disparities to address a global-health problem. *Lancet* **2006**, *367*, 1533–1540. [CrossRef]
108. Aeron-Thomas, A.; Jacobs, G.D.; Sexton, B.F.; Gururaj, G.; Rahman, F. *The Involvement and Impact of Road Crashes on the Poor: Bangladesh and India Case Studies*; TRL Limited: Wokingham, UK, 2004.
109. Bishai, D.; Bachani, A.M. Injury costing frameworks. In *Injury Research: Theories, Methods and Approaches*; Li, G., Baker, S.P., Eds.; Springer: New York, NY, USA, 2012.
110. Winston, C.; Maheshri, V.; Mannering, F. An exploration of the offset hypothesis using disaggregate data: The case of airbags and antilock brakes. *J. Risk Uncertain.* **2006**, *32*, 83–99. [CrossRef]
111. Lv, J.; Lord, D.; Zhang, Y.; Chen, Z. Investigating Peltzman effects in adopting mandatory seat belt laws in the US: Evidence from non-occupant fatalities. *Transp. Policy* **2015**, *44*, 58–64. [CrossRef]
112. Nicita, A.; Benedettini, S. The costs of avoiding accidents: Selective compliance and the "Peltzman Effect" in Italy. *Int. Rev. Law Econom.* **2012**, *32*, 256–270.
113. Garbacz, C. Impact of the New Zealand seat belt law. *Econ. Inq.* **1991**, *29*, 310–316. [CrossRef]
114. Health Effects Institute. *State of Global Air 2018. A Special Report on Global Exposure to Air Pollution and Its Disease Burden*; Health Effects Institute: Boston, MA, USA, 2018.
115. Sayed, T.; Saunier, N.; Lovegrove, G.; de Leur, P. Advances in proactive road safety planning. In Proceedings of the 20th Canadian Multidisciplinary Road Safety Conference, Niagara Falls, ON, Canada, 6–9 June 2010.
116. Uzondu, C.; Jamson, S.; Lai, F. Investigating unsafe behaviours in traffic conflict situations: An observational study in Nigeria. *J. Traffic Transp. Eng.* **2019**, *6*, 482–492. [CrossRef]

117. Uzondu, C.; Jamson, S.; Lai, F. Exploratory study involving observation of traffic behaviour and conflicts in Nigeria using the Traffic Conflict Technique. *Saf. Sci.* **2018**, *110*, 273–284. [CrossRef]
118. Chang, A.; Saunier, N.; Laureshyn, A. *Proactive Methods for Road Safety Analysis*; SAE Technical Paper WP-0005; SAE International: Troy, MI, USA, 2017.
119. Dhibi, M. Road safety determinants in low and middle income countries. *Int. J. Inj. Control Saf. Promot.* **2019**, *26*, 99–107. [CrossRef] [PubMed]
120. St-Aubin, P.; Saunier, N.; Miranda-Moreno, L. Large-scale automated proactive road safety analysis using video data. *Transp. Res. Part. C Emerg. Technol.* **2015**, *58*, 363–379. [CrossRef]
121. De Leur, P.; Sayed, T. A framework to proactively consider road safety within the road planning process. *Can. J. Civ. Eng.* **2003**, *30*, 711–719. [CrossRef]
122. Mitra, S.; Washington, S. On the significance of omitted variables in intersection crash modeling. *Accid. Anal. Prev.* **2012**, *49*, 439–448. [CrossRef] [PubMed]
123. Mannering, F.L.; Bhat, C.R. Analytic methods in accident research: Methodological frontier and future directions. *Anal. Methods Accid. Res.* **2014**, *1*, 1–22. [CrossRef]
124. Heydari, S.; Miranda-Moreno, L.F.; Lord, D.; Fu, L. Bayesian methodology to estimate and update safety performance functions under limited data conditions: A sensitivity analysis. *Accid. Anal. Prev.* **2014**, *64*, 41–51. [CrossRef] [PubMed]
125. Heydari, S. A flexible discrete density random parameters model for count data: Embracing unobserved heterogeneity in highway safety analysis. *Anal. Methods Accid. Res.* **2018**, *20*, 68–80. [CrossRef]
126. Nordfjærn, T.; Şimşekoğlu, Ö.; Rundmo, T. Culture related to road traffic safety: A comparison of eight countries using two conceptualizations of culture. *Accid. Anal. Prev.* **2014**, *62*, 319–328. [CrossRef] [PubMed]
127. Nordfjærn, T.; Jrgensen, S.; Rundmo, T. A cross-cultural comparison of road traffic risk perceptions, attitudes towards traffic safety and driver behaviour. *J. Risk Res.* **2011**, *14*, 657–684. [CrossRef]
128. McIlroy, R.C.; Nam, V.H.; Bunyasi, B.; Jikyong, U.; Kokwaro, G.O.; Wu, J.; Hoque, M.S.; Plant, K.A.; Preston, J.M.; Stanton, N.A. Vulnerable road users in low-, middle-, and high-income countries: Examining the relationships between pedestrian behaviours and attitudes towards traffic safety. *Accid. Anal. Prev.* **2019**, *131*, 80–94. [CrossRef] [PubMed]
129. Kouabenan, D.R. Beliefs and the perception of risks and accidents. *Risk Anal.* **1998**, *18*, 243–252. [CrossRef]
130. Dixey, R.A. "Fatalism", accident causation and prevention: Issues for health promotion from an exploratory study in a Yoruba town, Nigeria. *Health Educ. Res.* **1999**, *14*, 197–208. [CrossRef]
131. Peltzer, K. Seatbelt use and belief in destiny in a sample of South African black and white drivers. *Psychol. Rep.* **2003**, *93*, 732–734. [CrossRef]
132. Omari, K.; Baron-Epel, O. Low rates of child restraint system use in cars may be due to fatalistic beliefs and other factors. *Transp. Res. Part. F Traffic Psychol. Behav.* **2013**, *16*, 53–59. [CrossRef]
133. Maghsoudi, A.; Boostani, D.; Rafeiee, M. Investigation of the reasons for not using helmet among motorcyclists in Kerman, Iran. *Int. J. Inj. Control Saf. Promot.* **2018**, *25*, 58–64. [CrossRef] [PubMed]
134. Peltzer, K.; Renner, W. Superstition, risk-taking and risk perception of accidents among South African taxi drivers. *Accid. Anal. Prev.* **2003**, *35*, 619–623. [CrossRef]
135. Nordfjærn, T.; Jørgensen, S.; Rundmo, T. Cultural and socio-demographic predictors of car accident involvement in Norway, Ghana, Tanzania and Uganda. *Saf. Sci.* **2012**, *50*, 1862–1872. [CrossRef]
136. Şimşekoglu, Ö.; Nordfjærn, T.; Zavareh, M.F.; Hezaveh, A.M.; Mamdoohi, A.R.; Rundmo, T. Risk perceptions, fatalism and driver behaviors in Turkey and Iran. *Saf. Sci.* **2013**, *59*, 187–192. [CrossRef]
137. Ngueutsa, R.; Kouabenan, D.R. Fatalistic beliefs, risk perception and traffic safe behaviors. *Rev. Eur. Psychol. Appliquée* **2017**, *67*, 307–316. [CrossRef]
138. Yildirim, Z. *Religiousness, Conservatism and Their Relationship with Traffic Behaviour*; Middle East Technical University: Ankara, Turkey, 2007.
139. McIlroy, R.C.; Kokwaro, G.O.; Nam, V.H.; Jikyong, U.; Wu, J.; Hoque, M.S.; Preston, J.P.; Plant, K.A.; Stanton, N.A. How do fatalistic beliefs affect the attitudes and behaviours of vulnerable road users in different countries? A cross-cultural pedestrian study. Unpublished work.
140. McIlroy, R.C.; Plant, K.A.; Hoque, M.S.; Wu, J.; Kokwaro, G.O.; Vũ, N.H.; Stanton, N.A.; Jianping, W.; Kokwaro, G.O.; Vũ, N.H.; et al. Who is responsible for global road safety? A cross-cultural comparison of Actor Maps. *Accid. Anal. Prev.* **2019**, *122*, 8–18. [CrossRef]

141. Salmon, P.M.; Read, G.J.M.; Stevens, N.J. Who is in control of road safety? A STAMP control structure analysis of the road transport system in Queensland, Australia. *Accid. Anal. Prev.* **2016**, *96*, 140–151. [CrossRef]
142. Fahlquist, J.N. Responsibility ascriptions and Vision Zero. *Accid. Anal. Prev.* **2006**, *38*, 1113–1118. [CrossRef]
143. Kristianssen, A.-C.; Andersson, R.; Belin, M.-Å.; Nilsen, P. Swedish Vision Zero policies for safety—A comparative policy content analysis. *Saf. Sci.* **2018**, *103*, 260–269. [CrossRef]
144. Jadaan, K.; Al-Braizat, E.; Al-Rafayah, S.; Gammoh, H.; Abukahlil, Y. Traffic safety in developed and developing countries: A comparative analysis. *J. Traffic Logist. Eng.* **2018**, *6*, 1–5. [CrossRef]
145. Perel, P.; Ker, K.; Ivers, R.; Blackhall, K. Road safety in low-and middle-income countries: A neglected research area. *Inj. Prev.* **2007**, *13*, 227. [CrossRef] [PubMed]
146. Bhalla, K.; Shotten, M. Building road safety institutions in low-and middle-income countries: The case of Argentina. *Health Syst. Reform* **2019**, *1–13*, 121–133. [CrossRef] [PubMed]
147. Khademi, N.; Choupani, A.-A. Investigating the road safety management capacity: Toward a lead agency reform. *IATSS Res.* **2018**, *42*, 105–120. [CrossRef]
148. Hyder, A.A.; Norton, R.; Pérez-Núñez, R.; Mojarro-Iñiguez, F.R.; Peden, M.; Kobusingye, O.; Road Traffic Injuries Research Network's Group. The Road Traffic Injuries Research Network: A decade of research capacity strengthening in low-and middle-income countries. *Health Res. Policy Syst.* **2016**, *14*, 14. [CrossRef]
149. UNECE. *Strengthening the National Road Safety Management Capacities of Selected Developing Countries and Countries with Economies in Transition*; UNECE: Geneva, Switzerland, 2014.
150. World Health Organization. *Capacity Building for Preventing Injuries and Violence: Strategic Plan 2009–2013*; World Health Organization: Geneva, Switzerland, 2008.
151. Wijnen, W.; Stipdonk, H. Social costs of road crashes: An international analysis. *Accid. Anal. Prev.* **2016**, *94*, 97–106. [CrossRef]
152. Wesson, H.K.H.; Boikhutso, N.; Bachani, A.M.; Hofman, K.J.; Hyder, A.A. The cost of injury and trauma care in low-and middle-income countries: A review of economic evidence. *Health Policy Plan.* **2013**, *29*, 795–808. [CrossRef]
153. Bishai, D.; Hyder, A.A.; Ghaffar, A.; Morrow, R.H.; Kobusingye, O. Rates of public investment for road safety in developing countries: Case studies of Uganda and Pakistan. *Health Policy Plan.* **2003**, *18*, 232–235. [CrossRef]
154. Mukama, T.; Ssemugabo, C.; Ali Halage, A.; Gibson, D.G.; Paichadze, N.; Ndejjo, R.; Sempebwa, J.; Kobusingye, O. Costs of unintentional injuries among children in an urban slum community in Kampala city, Uganda. *Int. J. Inj. Control Saf. Promot.* **2019**, *26*, 129–136. [CrossRef]
155. Hauer, E. *Observational before/after Studies in Road Safety. Estimating the Effect of Highway and Traffic Engineering Measures on Road Safety*; Emerald Group Publishing: Bingley, UK, 1997.
156. Wiebe, D.J.; Ray, S.; Maswabi, T.; Kgathi, C.; Branas, C.C. Economic development and road traffic fatalities in two neighbouring African nations. *Afr. J. Emerg. Med.* **2016**, *6*, 80–86. [CrossRef]
157. Heydari, S.; Fu, L.; Thakali, L.; Joseph, L. Benchmarking regions using a heteroskedastic grouped random parameters model with heterogeneity in mean and variance: Applications to grade crossing safety analysis. *Anal. Methods Accid. Res.* **2018**, *19*, 33–48. [CrossRef]
158. AASHTO. *Highway Safety Manual*; American Association of State Highway and Transportation Officials: Washington, DC, USA, 2010.
159. Goel, R.; Garcia, L.M.T.; Goodman, A.; Johnson, R.; Aldred, R.; Murugesan, M.; Brage, S.; Bhalla, K.; Woodcock, J. Estimating city-level travel patterns using street imagery: A case study of using Google Street View in Britain. *PLoS ONE* **2018**, *13*, e0196521. [CrossRef] [PubMed]
160. Eluru, N.; Chakour, V.; Chamberlain, M.; Miranda-Moreno, L.F. Modeling vehicle operating speed on urban roads in Montreal: A panel mixed ordered probit fractional split model. *Accid. Anal. Prev.* **2013**, *59*, 125–134. [CrossRef] [PubMed]
161. Heydari, S.; Miranda-Moreno, L.; Fu, L. Is speeding more likely during weekend night hours? Evidence from sensor-collected data in Montreal. *Can. J. Civ. Eng.* **2019**. [CrossRef]

MDPI
St. Alban-Anlage 66
4052 Basel
Switzerland
Tel. +41 61 683 77 34
Fax +41 61 302 89 18
www.mdpi.com

*Sustainability* Editorial Office
E-mail: sustainability@mdpi.com
www.mdpi.com/journal/sustainability

www.ingramcontent.com/pod-product-compliance
Lightning Source LLC
LaVergne TN
LVHW070058170726
843515LV00007B/1562
*9783039430888*